T0318552

Glucose Monitoring Devices

Glucose Monitoring Devices

Measuring Blood Glucose to Manage and Control Diabetes

Edited by

Chiara Fabris, PhD
Assistant Professor
Center for Diabetes Technology
Department of Psychiatry and Neurobehavioral Sciences
University of Virginia
Charlottesville, Virginia
United States

Boris Kovatchev, PhD
Professor and Director
Center for Diabetes Technology
Department of Psychiatry and Neurobehavioral Sciences
University of Virginia
Charlottesville, Virginia
United States

Academic Press is an imprint of Elsevier
125 London Wall, London EC2Y 5AS, United Kingdom
525 B Street, Suite 1650, San Diego, CA 92101, United States
50 Hampshire Street, 5th Floor, Cambridge, MA 02139, United States
The Boulevard, Langford Lane, Kidlington, Oxford OX5 1GB, United Kingdom

Library of Congress Cataloging-in-Publication Data
A catalog record for this book is available from the Library of Congress

British Library Cataloguing-in-Publication Data
A catalogue record for this book is available from the British Library

ISBN: 978-0-12-816714-4

For information on all Academic Press publications visit our website at https://www.elsevier.com/books-and-journals

Publisher: Mara Conner
Acquisitions Editor: Fiona Geraghty
Editorial Project Manager: Isabella C. Silva
Production Project Manager: Kiruthika Govindaraju
Cover Designer: Miles Hitchen

Typeset by TNQ Technologies

Contents

Contributors

Giada Acciaroli, PhD
Department of Information Engineering, University of Padova, Padova, Italy

David Ahn, MD
Program Director, Mary & Dick Allen Diabetes Center, Hoag Memorial Hospital Presbyterian, Newport Beach, CA, United States

Tadej Battelino, MD, PhD
Department of Endocrinology, Diabetes and Metabolism, University children's hospital University Medical Centre Ljubljana, Ljubljana, Slovenia; Faculty of Medicine, University of Ljubljana, Ljubljana, Slovenia

Rachel Brandt, BSc
Illinois Institute of Technology, Biomedical Engineering, Chicago, IL, United States

Marc D. Breton, PhD
Assistant Professor, Center for Diabetes Technology, Department of Psychiatry and Neurobehavioral Sciences, University of Virginia, Charlottesville, VA, United States

Enrique Campos-Náñez, PhD
Principal Algorithm Engineer, Research & Development, Dexcom Inc, Charlottesville, VA, United States

Ali Cinar, PhD
Professor, Chemical and Biological Engineering Department, Illinois Institute of Technology, Chicago, IL, United States

William L. Clarke, MD
Profesor, Emeritus of Pediatric Endocrinology, Department of Pediatrics, University of Virginia, Charlottesville, VA, United States

Claudio Cobelli, PhD
Department of Information Engineering, University of Padova, Padova, Italy

Andrew DeHennis, PhD
Sr. Director of Engineering, R&D, Product Development Senseonics Incorporated, Germantown, MD, United States

Simone Del Favero, PhD
Assistant Professor, Department of Information Engineering, Padova, Italy

Laya Ekhlaspour, MD
Instructor, Pediatric Endocrinology, Stanford University, Palo Alto, CA, United States

Chiara Fabris, PhD
Assistant Professor, Center for Diabetes Technology, Department of Psychiatry and Neurobehavioral Sciences, University of Virginia, Charlottesville, VA, United States

Andrea Facchinetti, PhD
Department of Information Engineering, University of Padova, Padova, Italy

Gregory P. Forlenza, MD
Assistant Professor, Barbara Davis Center, University of Colorado Denver, Aurora, CO, United States

Kurt Fortwaengler, PMP
Disease Modeling, Global Market Access, Roche Diabetes Care, Mannheim, Germany

Satish Garg, MD
Professor of Pediatrics and Medicine, Barbara Davis Center for Diabetes Adult Clinic, University of Colorado Anschutz Medical Center, Aurora, CO, United States

Iman Hajizadeh, MSc
Research Assistant and PhD Student, Chemical and Biological Engineering, Illinois Institute of Technology, Chicago, IL, United States

Nicole Hobbs, BSc
Graduate Research Assistant, Department of Biomedical Engineering, Illinois Institute of Technology, Chicago, IL, United States

David Klonoff, MD, FACP, FRCP (Edin), Fellow AIMBE
Medical Director, Diabetes Research Institute, Mills-Peninsula Medical Center, San Mateo, CA, United States

Boris Kovatchev, PhD
Professor and Director, Center for Diabetes Technology, Department of Psychiatry and Neurobehavioral Sciences, University of Virginia, Charlottesville, Virginia, United States

Mark Mortellaro, PhD
Director of Chemistry, Senseonics Incorporated, Germantown, MD, United States

Laura M. Nally, MD
Associate Professor, Pediatric Endocrinology, Yale Children's Diabetes Program, Yale University School of Medicine, New Haven, CT, United States

Nick Oliver, FRCP
Wynn Professor of Human Metabolism, Consultant in Endocrinology, Diabetes and Internal Medicine, Imperial College London, St. Mary's Hospital Medical School Building, London, United Kingdom

Mudassir Rashid, PhD, BEng
Senior Research Associate, Department of Chemical and Biological Engineering, Illinois Institute of Technology, Chicago, IL, United States

Monika Reddy, MBChB, MRCP (UK), PhD
Honorary Senior Clinical Lecturer, Consultant in Endocrinology, Diabetes and Internal Medicine, Imperial College London, St. Mary's Hospital Medical School Building, London, United Kingdom

Amanda Rewers, MD
Research Assistant, Barbara Davis Center for Diabetes Adult Clinic, Aurora, CO, United States

Sediqeh Samadi, MSc
Illinois Institute of Technology, Chemical and Biological Engineering, Chicago, IL, United States

Mert Sevil, MSc
Research Assistant and PhD Student, Biomedical Engineering, Illinois Institute of Technology, Chicago, IL, United States

Viral N. Shah, MD
Assistant Professor of Pediatrics and Medicine, Barbara Davis Center for Diabetes Adult Clinic, University of Colorado Anschutz Medical Center, Aurora, CO, United States

Jennifer L. Sherr, MD, PhD
Instructor, Pediatric Endocrinology, Yale Children's Diabetes Program, Yale University School of Medicine, New Haven, CT, United States

Darja Smigoc Schweiger, MD, PhD
Department of Endocrinology, Diabetes and Metabolism, University children's hospital University Medical Centre Ljubljana, Ljubljana, Slovenia; Faculty of Medicine, University of Ljubljana, Ljubljana, Slovenia

Giovanni Sparacino, PhD
Department of Information Engineering, University of Padova, Padova, Italy

Chukwuma Uduku, MBBS, BSc, MRCP
Clinical Research Fellow and Specialist Registrar in Endocrinology, Diabetes and Internal Medicine, Imperial College London, St. Mary's Hospital Medical School Building, London, United Kingdom

Martina Vettoretti, PhD
Department of Information Engineering, University of Padova, Padova, Italy

About the Authors

Dr. Chiara Fabris is Assistant Professor at the University of Virginia School of Medicine and member of the faculty at the Center for Diabetes Technology. She holds a Master's and a Doctoral Degree in Bioengineering from the University of Padova (Padova, Italy) and has significant experience in mathematical modeling and simulation—especially regarding the glucose/insulin regulation system—and algorithm development. Over the past 4 years, Dr. Fabris has been awarded an Advanced Postdoctoral Fellowship and a Career Development Award by the Juvenile Diabetes Research Foundation, which supported the development and clinical testing of decision support systems to help people with diabetes manage their disease. Dr. Fabris is involved in several projects focused on optimization of treatments for diabetes and diabetes data science.

Dr. Boris Kovatchev is Professor at the University of Virginia School of Medicine and School of Engineering and founding director of the Center for Diabetes Technology. He has a 30-year track record in mathematical modeling, biosimulation, and algorithm development. Currently, he is Principal Investigator of several projects dedicated to Diabetes Data Science and the development of artificial pancreas and decision support systems, including the large-scale NIH International Diabetes Closed-Loop Trial and the UVA Strategic Investment Fund project "Precision Individualized Medicine for Diabetes". Dr. Kovatchev is author of over 200 peer-reviewed publications and holds 85 patents.

SECTION 1

Self-monitoring of blood glucose (SMBG) devices

CHAPTER

Introduction to SMBG

1

Darja Smigoc Schweiger, MD, PhD, Tadej Battelino, MD, PhD

Department of Endocrinology, Diabetes and Metabolism, University children's hospital University Medical Centre Ljubljana, Ljubljana, Slovenia; Faculty of Medicine, University of Ljubljana, Ljubljana, Slovenia

Historical perspective and principles of blood glucose control

As Benedict developed a copper reagent for urine glucose, urine glucose testing has been the main method for diabetes monitoring for 50 years [1]. Later, a more convenient and specific "dip-and-read" urine glucose oxidase-based reagent strip (Clinistix) was introduced [2]. However, urine tests had several well-recognized limitations. High glucose levels were detected only when the renal threshold for glucose was exceeded over a period of several hours and the results were affected by fluid intake and urine concentration. Moreover, the test did not detect low glucose levels [3]. In the 1960s, first blood glucose (BG) test strips (Dextrostix) were developed. The exposure to blood resulted in a colorimetric reaction proportional to blood glucose concentration. The color change that occurred was compared to a color chart providing a semiquantitative assessment of blood glucose levels [4]. The first blood glucose meter, the Ames Reflectance Meter, was introduced in 1970. The meter exhibited quantitative blood glucose results based on Dextrostix test strips and reflectance photometry, thus eliminating visual reading errors. The results were displayed by a moving pointer on three analog scales [5]. The device was only available for testing in a doctor's office and hospital emergency departments [6]. Although the meter was heavy, expensive, and cumbersome to use, it ushered the development in an era of blood glucose monitoring systems. In 1972, more convenient Eyetone glucometer using Dextrostix test strips was developed, which was more precise, lighter, and easier to operate [7]. In 1974, Boehringer Mannheim launched Reflomat, a reflectance meter with modified reagent test strips (Reflotest), equipped to accept smaller volumes of blood, which was removed more easily and thus more suitable for at-home self-monitoring of blood glucose (SMBG) [6]. Dextrometer and Glucochek launched, in 1980, were the first glucometers with digital display [8]. Technological advances during 1980s made glucometers smaller and easier to use with built-in software to store and retrieve results [6]. The One Touch meter introduced in 1987 was regarded as the first second-generation blood glucose meter because it utilized an improved sampling procedure that eliminated

Glucose Monitoring Devices. https://doi.org/10.1016/B978-0-12-816714-4.00001-6

blood removal step and the need for time reactions [9]. Toward the end of the 1980s, test strips changed dramatically when electrochemical principles to measure blood glucose were introduced. Furthermore, the introduction of electrochemical technology led to the development of the third generation of glucose monitoring systems [10]. The landmark in glucose self-monitoring was the release of the first electrochemical blood glucose monitor, ExacTech by Medisense, in 1987. The device used an enzyme electrode strip containing glucose oxidase and ferrocene as an electron transfer mediator. A current generated at the electrode was detected by an amperometric sensor [11].

Today, most glucometers are electrochemical, using commercial screen-printed strips based on the same principle. They require a smaller blood sample and provide results in a few seconds. Glucose oxidase and glucose dehydrogenase are two types of enzymes that have been used for commercial electrochemical blood glucose test strips. Test strips using glucose oxidase technology are susceptible to dissolved oxygen concentrations and can only be used with capillary blood in a normal range of oxygen levels. Glucose dehydrogenase-based test strips are not sensitive to oxygen [12]. However, coenzyme pyrroloquinoline quinone and glucose dehydrogenase containing test strips lack specificity as they cross-react with maltose, galactose, and xylose. Therefore they must not be used by patients on peritoneal dialysis [13]. The most common electrochemical detection methods for glucose measurement are amperometry and coulometry [12]. Coulometric strips have demonstrated to operate over the wider ranges of hematocrit values and with the minimized effect of temperature, high concentrations of paracetamol, uric acid, and vitamin C [14]. The performance of glucometers has further improved with simplified sampling and testing procedures to minimize user interaction errors. Meters using no-coding technology are precalibrated to report whole blood or plasma equivalent results [15]. Most current meters are plasma calibrated and automatically convert results into plasma equivalent results [16]. Modern electrochemical blood glucose test strips use the capillary gap to automatically draw blood into the test surface, which requires only a small volume of blood (just about 0.3 μL) and has automatic fill detection ensuring that sufficient volume of blood is provided to the strip. The average test time has been reduced to just less than 5 s [17]. In addition, lower blood volume requirements allow alternative sites for blood glucose testing such as arm or thigh that are likely to be less painful and provide similar results to the fingertip [18]. However, when blood glucose is changing rapidly, significant differences in blood glucose results can be anticipated due to the time lag of up to 20 min at alternative sites [19]. Therefore testing at alternative sites is not recommended within the early postmeal period, immediately after exercise or when blood glucose is suspected to be low [20]. Some fully automated devices have integral lacing device and extract blood by drawing a vacuum over a lanced site [21]. Newer meters offer data-storage software that can be downloaded and used by diabetes management systems for the graphical display of trends, statistics, and sharing of reports [22]. Downloading information from blood glucose meters enables the analysis of large amounts of data that reveal glycemic patterns and support persons with diabetes and healthcare professionals to make appropriate management strategies [23]. Data retrieval has further improved with wireless connectivity to smartphone apps [24]. The analytical quality of personal blood glucose meters used for at-home monitoring is important as appropriate therapeutic decisions rely on accurate glucose readings. Standardized quality among manufacturers is required by the regulatory recommendations and analytical performance criteria. In 2003, the International Organization for Standardization (ISO) criteria for glucose meters were introduced. The ISO 15197: 2003 standard recommended an allowable error of ±15 mg/dL for blood glucose levels <75 mg/dL and ±20% for blood glucose levels ≥75 mg/dL [25]. These criteria were updated in ISO 15197:2013 standard, which required an allowable error of ±15 mg/dL for BG concentrations <100 mg/dL and ±15% for BG concentrations ≥100 mg/dL [26]. In the United States, the Food and Drug Administration (FDA) standard finalized in 2016 recommended that at least 95% of measurement results shall fall within ±15% of the reference value at blood glucose concentrations <100 mg/dL and ±15% at ≥100 mg/dL, thus requiring greater hypoglycemia accuracy than the ISO 15197:2013 [27].

The evidence base for SMBG in type 1 diabetes

Richard Bernstein was the first reported person with type 1 diabetes (T1D) to adopt a glucometer for personal use. With frequent glucose monitoring, he was able to refine insulin doses and diet regimen to maintain essentially normal blood glucose levels and prevent hypoglycemia. However, he failed to publish his personal experience using SMBG until he earned a medical degree in the early 1980s [28]. In the mid-1970s, people with diabetes for the first time started using reflectance glucometers Eyetone and Reflomat at home for SMBG. In 1978, first experiences in teaching people with insulin-dependent diabetes to measure their own blood glucose concentrations were published [29–34]. Direct measurement of blood glucose by people with diabetes at home provided sufficiently accurate results for easier and more predictable adjustment of insulin doses over the urine-glucose analysis [35]. Frequent SMBG as a guide to multiple injections of insulin has considerably improved metabolic control and could guard against undue hypoglycemia [36]; it was well accepted by persons with diabetes and improved their understanding of diabetes and motivation to become more involved in their own care [37]. Due to the growing evidence in the late 1970s that chronic complications of diabetes can be minimized with glycemic control, daily SMBG gained wider acceptance [38]. In addition, improved glycemic control could objectively be assessed by the measurement of glycated hemoglobin levels [39]. Over the next decade, SMBG proved to be one of the major technological advances in addition to multiple daily insulin injections and the newly developed insulin pumps that established intensive insulin therapy, a therapeutic strategy that has become increasingly used in an attempt to achieve near-normal glycemia [40,41]. In the 1980s, smaller, more portable, easier to use, and cheaper devices made SMBG more applicable, and their use steadily increased [6]. In view of this widespread use of SMBG, the American Diabetes Association (ADA) convened the first consensus statement on SMBG in 1987 [42]. The landmark Diabetes Control and Complications Trial (DCCT) was the first long-term randomized prospective study to ascertain whether intensive therapy aimed at near-normal glycemic control could reduce microvascular complications as compared to standard diabetes care among people with T1D. Near-normal glycemic control included preprandial blood glucose concentrations between 70 and 120 mg/dL, postprandial concentrations of less than 180 mg/dL, a weekly 3 a.m. measurement greater than 65 mg/dL, and hemoglobin A1c (HbA1c), measured monthly, within the normal range (less than 6.05%). Intensive glycemic control was guided by frequent SMBG (≥4 times daily) as a tool for insulin dose titration to achieve normal blood glucose levels, whether in standard therapy once-daily SMBG generally did not guide insulin supplementation. In 1993, the DCCT confirmed that intensive diabetes management dramatically reduced the risk of microvascular complications in T1D. Thus the study resolved the controversy about the effect of glycemic control on microvascular complications of diabetes [43]. Following the DCCT, intensive therapy became the standard of care in the management of T1D and the value of SMBG as an integral part of intensive therapy was generally accepted [44]. Eleven years after the conclusion of the DCCT, the follow-up observational Epidemiology of Diabetes and its Complications (EDIC) study of the DCCT cohort demonstrated the long-lasting favorable effect of intensive therapy on the risk of macrovascular complications despite the minor differences in mean HbA1c between the groups over the follow-up period [45]. The long-lasting beneficial effects of intensive therapy on the incidence of cardiovascular disease—termed "metabolic memory"—continues after over 30 years of follow-up [46].

Due to the higher glucose variability in persons with T1D, greater SMBG frequency generally correlated with lower HbA1c. In addition, reanalyzed DCCT data demonstrated that within-day blood glucose standard deviationas a measure of glycemic variability predicted hypoglycemia independently of HbA1c [47]. Following the DCCT, several studies have confirmed a strong association between increased frequency of SMBG and lower HbA1c levels [48–50]. Moreover, one additional SMBG per day resulted in an HbA1c reduction of 0.26% corrected for age, gender, diabetes duration, insulin therapy, and center difference [51]. Data analysis of more than 20,000 children and adults from the T1D Exchange Registry showed a strong association between a higher number of SMBG measurements per day and lower HbA1c across a wide age range. The association was present in both continuous subcutaneous insulin infusion (CSII) and multiple daily injections (MDI) users. The difference between measuring 3–4 times per day and measuring ≥10 times per day has been shown to affect HbA1c of about 1%. The association between SMBG and HbA1c appeared to level-off at approximately 10 SMBG measurements per day [52]. Similarly, adults with T1D under excellent control (HbA1c $< 6.5\%$) performed SMBG more frequently, including more frequent SMBG measurements before giving a bolus compared to individuals under poor control (HbA1c $\geq 8.5\%$) [53].

Although the DCCT did not enroll children of 13 years old and younger, it demonstrated higher HbA1c values both in the conventionally and intensively treated adolescent cohort compared with adults, as well as more acute complications, such as ketoacidosis and severe hypoglycemia [43]. Several studies suggested that frequent SMBG is associated with improved glycemic control and less acute complications in youth with T1D. A prospective, 1-year study, which involved 300 subjects of 7–16 years old demonstrated that glycemic control improved significantly as the frequency of SMBG increased. The decrease from an HbA1c of 9.1%–8.0% has been shown between those measuring at most once per day and those measuring 5 or more times per day. In addition, the incidence of hypoglycemia and hospitalization rate was higher in those with the poorest glycemic control [54]. In the same way, the association between frequency of SMBG and glucose control has been reported for adolescents [55], children visiting a diabetes camp [56] and 1 year following diagnosis of T1D [57]. Furthermore, analysis of the German/Austrian Diabetes Patienten Verlaufsdokumentation (DPV) database of 26,723 children and adolescents with T1D, aged 0–18 years, showed—after adjustment for multiple confounders—that more frequent SMBG was significantly associated with better metabolic control, with a drop of HbA1c of 0.2% for one additional SMBG per day and decreased rate of diabetes ketoacidosis. However, increasing the SMBG frequency above five per day was associated with a decrease in average HbA1c only in the group on CSII [58]. Age-dependent analysis from the DPV database across two decades demonstrated an increase in the frequency of SMBG in all-age groups, both in intensified conventional therapy and insulin pump users [59].

The evidence base for SMBG in type 2 diabetes

Similarly, SMBG was used in major clinical studies of people with type 2 diabetes (T2D) for adaptation of treatment in intensive glycemic management. However, the role of SMBG in optimal glycemic control and clinical outcomes is less clear in T2D. In the UK Prospective Diabetes Study (UKPDS), improved blood glucose control significantly decreased rates of microvascular complications and decreased the progression of diabetic microvascular diseases in participants newly diagnosed with T2D followed for 10 years [60]. In the prospective 6-year Kumamoto study, intensive insulin therapy targeting both fasting and postprandial glucose effectively delayed the onset and progression of diabetic microvascular complications with almost comparable results to those in the DCCT [61]. Extended follow-up of the UKPDS trial revealed the enduring effects of intensive glycemic control on microvascular complications and long-term reductions in myocardial infarction and all-cause mortality [62]. Conversely, results from randomized controlled trials Action to Control Cardiovascular Risk in Diabetes (ACCORD), Action in Diabetes and Vascular Disease: Preterax and Diamicron MR Controlled Evaluation, and Veterans Affairs Diabetes Trial suggested the lack of significant reduction in cardiovascular disease events with intensive glycemic control in T2D participants followed for 3.5–5.6 years [63–65] and ACCORD was halted due to the increased rate of mortality in the intensive glycemic control group [66]. Thus ADA's Standards of Medical Care in Diabetes emphasize individualization of blood glucose and glycemic targets, suggesting that less stringent goals may be appropriate for some individuals with T2D [67]. Two observational studies investigated the association of SMBG with clinical outcomes. Data from the Fremantle Diabetes Study showed no independent cross-sectional relationship between HbA1c and SMBG frequency regardless of treatment [68]. In addition, assessment of longitudinal data over a 5-year period revealed that SMBG was not independently associated with improved survival and, after adjustment, cardiac mortality was even higher in SMBG users not treated with insulin [69]. On the other hand, the Self-monitoring of Blood Glucose and Outcome in Patients with Type 2 Diabetes (ROSSO) study, which followed participant from diagnosis of T2D with a mean follow-up period of 6.5 years, reported a lower total rate of nonfatal (micro- and macrovascular) as well as fatal events in the SMBG group in comparison with the non-SMBG group [70]. In the large observational Kaiser Permanente study, SMBG performed at least daily was associated with lower HbA1c levels among individuals with pharmacologically treated T2D compared to less frequent monitoring. In nonpharmacologically treated participants, SMBG at any frequency was associated with lower HbA1c compared to no SMBG [49]. A longitudinal study with a 4-year follow-up found evidence for improvements in HbA1c with more frequent monitoring in new SMBG users regardless of diabetes therapy and among pharmacologically treated prevalent users [71]. In an observational retrospective study of 657 individuals with T2D, targeted HbA1c values of <7% were associated with greater use of SMBG test strips in the noninsulin-treated group. Of interest, there were no significant differences in the insulin-treated group [72]. Data from the DPV database showed that more frequent SMBG was associated with HbA1c reduction of 0.16% for one additional SMBG per day in individuals with T2D treated with insulin, while no benefit on metabolic control was observed in those not treated with insulin [51]. In a cross-sectional study of 1480 participants with T2D, increased frequency of SMBG was related to increased HbA1c and a higher proportion of insulin users. However, within each treatment category, there was no relationship between the frequency of SMBG and HbA1c for those treated with insulin, oral agents, or diet alone [73].

Although SMBG has been found to be effective in the management of T1D and insulin-treated T2D, the clinical benefits of SMBG have been debated for nearly 75% of people living with T2D, who are not using insulin and manage their disease with lifestyle modification and oral medications. Several randomized trials and meta-analyses have been conducted to evaluate the clinical benefit and cost-effectiveness of routine SMBG in noninsulin-treated people with T2D. The effect of SMBG in noninsulin-treated T2D has not been consistent in randomized control trials, and many studies have found no clinically relevant effect of SMBG on glycemic control. The Diabetes Glycemic Education and Monitoring (DiGEM) randomized controlled trial [74] assessed the effectiveness of two strategies of SMBG in improving glycemic control in noninsulin-treated individuals with T2D versus usual care alone. In the study, 453 participants with mean baseline HbA1c levels of 7.5% and median duration of diabetes of 3 years were randomized to one of three interventions: no SMBG, SMBG standardized with advice to contact their doctor for interpretation of results, and SMBG that involved additional training of participants in interpretation and application of the results into self-care. The differences in HbA1c levels between the three groups were not significant at 12 months. Investigators concluded that SMBG has little effect on glycemic control in people with stable, near-target metabolic control. In an economic evaluation of the DiGEM study [75], SMBG was significantly more expensive than standardized usual care for noninsulin-treated T2D. As there were no significant differences in HbA1c, the analysis implied that SMBG is unlikely to be cost-effective if added to standardized usual care in insulin-independent T2D. The efficacy of SMBG in patients with newly diagnosed T2D was assessed in the ESMON study [76]; the prospective randomized controlled trial assessed the effect of SMBG on glycemic control and psychological indices in 184 individuals with newly diagnosed T2D over 12 months. Subjects were recruited soon after the diagnosis of T2D and randomized to SMBG or non-SMBG (control) group. Intensive education and treatment resulted in a decrease of mean HbA1c levels after 12 months in both groups; however, there were no significant differences in HbA1c between groups at any time point. Moreover, SMBG was associated with a 6% higher score on the depression subscale of the well-being questionnaire. In the Monitor Trial [77]—a pragmatic randomized controlled trial conducted in 15 primary care practices—450 participants with noninsulin-treated T2D and HbA1c between 6.5% and 9.5% were randomized to one of three interventions: no SMBG, once-daily SMBG, or once-daily SMBG with enhanced patient feedback that featured automatic tailored messages delivered via the meter. At baseline, >85% of study participants had been receiving care for diabetes for >1 year and the mean HbA1c level was about 7.5%. After a year of follow-up, no significant differences in HbA1c levels among the three groups were reported. In addition, there were no significant differences between the study groups in terms of health-related quality of life and adverse events such as hypoglycemia frequency, nor was there any difference in insulin initiation. The authors concluded that routine SMBG does not significantly improve HbA1c levels or quality of life for most individuals with noninsulin-treated T2D. However, the trial evaluated once-daily SMBG, which may not provide sufficient information about daily glucose excursions.

Many trials looking at the clinical effectiveness of SMBG in noninsulin-treated people with T2D did not include structured SMBG regimens. Structured SMBG is a systematic approach in which SMBG is performed periodically, according to a defined regimen, such as before and after meals or exercise. Blood glucose values provide feedback to make appropriate treatment decisions and lifestyle adjustments [78]. Randomized controlled trials that have utilized structured SMBG as an intervention reported greater HbA1c reduction compared with programs without structured SMBG. The Structured Testing Program study [79] was a randomized prospective trial that evaluated the efficacy of two strategies of SMBG in persons with noninsulin-treated T2D in a primary care setting. In the study, 483 poorly controlled (mean HbA1c 8.9%) participants with noninsulin-treated T2D were assigned to a structured testing group or an active control group. Both groups received enhanced usual care. In addition, the structured testing group was instructed to perform a seven-point SMBG profile on three consecutive days before each scheduled study visit using the ACCU-CHEK 360 degrees View tool. Structured SMBG data were at least quarterly interpreted and used for treatment modifications. At 1 year, the intervention SMBG group showed a significantly greater mean reduction in HbA1c. Furthermore, participants actively adherent to the structured SMBG protocol experienced significantly greater improvements in reported diabetes self-confidence and increases in general well-being with respect to patients receiving enhanced usual care [80]. In the Role of Self-Monitoring of Blood Glucose and Intensive Education in Patients with Type 2 Diabetes Not Receiving Insulin (ROSES) trial [81], 62 participants were randomly assigned to either SMBG with intensive education or no monitoring with usual care. The participants in the intervention group received education on how to adjust nutrition and physical activity according to SMBG readings. Participants received counseling during additional monthly telephone contact. After 6 months, HbA1c reduction was significantly greater in the intervention group compared with the control group with a significant mean difference of 0.5%. Additionally, significantly greater reductions were observed in weight loss. In the prospective randomized trial, St. Carlos study [82], 161 newly diagnosed T2D participants were assigned to either an SMBG-based intervention or an HbA1c-based control group. The intervention group used SMBG as an educational and therapeutic tool to promote lifestyle changes and adjust pharmacological treatment. The control group received standard treatment based on HbA1c values without SMBG. After 1 year of follow-up, the SMBG intervention group showed a significant reduction in median HbA1c level and body mass index (BMI). There was no change in median HbA1c or BMI in the control group. The 12-month Prospective Randomized Trial on Intensive SMBG Management Added Value in Noninsulin-Treated T2DM Patients study enrolled 1024 participants with noninsulin-treated T2D with median baseline HbA1c of 7.3% [83]. The intervention group performed structured monitoring with four-point SMBG profiles 3 days per week. The active control group performed four-point SMBG profiles at baseline and at 6 and 12 months. At 1 year, the intervention SMBG group had a greater HbA1c reduction compared to the control group with a between-group difference of −0.12%. In the per-protocol population, consisting of all randomized patients who completed the study without major protocol violations and were compliant with the SMBG regimen, the between-group difference was −0.21%. This study demonstrated that structured SMBG improved glycemic control in individuals with relatively well-controlled noninsulin-treated T2D. Furthermore, psychosocial data analysis demonstrated that structured SMBG was not associated with a deterioration of quality of life [84]. In a randomized controlled trial of 446 participants with established T2D not on insulin therapy and suboptimal glycemic control (HbA1c $\geq$ 7.5%), the use of structured SMBG alone or with additional monthly telecare support was compared to a control group receiving usual diabetes care. In both of the structured SMBG groups, glycemic management was based on SMBG results alone. At 12 months, the use of structured SMBG provided a significant reduction in HbA1c of 0.8% compared to the control group, whereas no additional benefit in glycemic control over the use of structured SMBG was observed with the addition of once-monthly TeleCare support [85].

Combining the results of individual studies and pooling large amounts of data gives us insights into the overall measure of the effect of SMBG for noninsulin-treated T2D. Recent meta-analyses have generally shown a small, short-term reduction in HbA1 in those individuals performing SMBG compared to those who did not. The first meta-analysis based on individual participant data from six randomized controlled trials compared SMBG with no SMBG in individuals with noninsulin-treated T2D [86]. SMBG reduced HbA1c levels at 3, 6, and 12 months compared with no self-monitoring by 0.18%, 0.25%, and 0.23%, respectively. The effect of SMBG on HbA1c levels was consistent across predefined subgroups of participants according to age, baseline HbA1c level, sex, and duration of diabetes. No clinically significant reductions occurred in clinical indices such as blood pressure and total cholesterol. The authors concluded that clinical management of noninsulin-treated diabetes using SMBG compared with no SMBG resulted in a very modest reduction in HbA1c levels, which probably has no clinical significance and therefore does not provide convincing evidence to support the routine use of SMBG for people with noninsulin-treated T2D. A Cochrane review [87] included 12 randomized controlled trials and examined the utility of SMBG in individuals with T2D who did not require insulin therapy. Pooled analysis showed that SMBG led to a statistically significant decrease in HbA1c of 0.3% after 6 months in participants who have had diabetes for more than 1 year. Two trials that extended follow-up to 12 months revealed a nonsignificant reduction of HbA1c (0.1%). In participants with newly diagnosed T2D, a significant reduction of HbA1c (0.5%) was observed at 12 months in favor of SMBG. It was concluded that SMBG is beneficial in lowering HbA1c in individuals with newly diagnosed T2D who are not using insulin. However, for those with established diabetes for more than a year, the glycemic effect of SMBG was small at 6 months and disappeared after 12 months of monitoring. There was also no evidence that SMBG affects patient-oriented outcomes such as general health-related quality of life, general well-being or patient satisfaction. In two trials [71–75,88] that analyzed the cost of SMBG, total estimated costs in the first year of SMBG were 12 times higher if compared with usual care or self-monitoring of urine glucose. Following the Cochrane review, a meta-analysis including 15 newer randomized controlled trials and 3383 participants with noninsulin-treated T2D [89] found that SMBG intervention improved HbA1c with a 0.33% mean difference compared to controls in the overall effect. In contrast to the Cochrane review, SMBG improved HbA1c in the short and long term, as well as regardless of diabetes duration. In addition to HbA1c, significant reductions in BMI and total cholesterol were observed. The study did not track the intensity of education, lifestyle or dietary interventions, use of medications, or frequency of SMBG testing. As the differences in SMBG regimens and the use of SMBG data to adjust blood-glucose-lowering medications may affect glucose control, the aim of the meta-analysis by Mannucci et al. [90] was to assess the effect of SMBG on HbA1c in noninsulin-treated T2D considering these potential confounders. SMBG in comparison with no SMBG led to a small reduction (0.17%) in HbA1c. However, when SMBG data were used to adjust medical treatment, a greater reduction in HbA1c levels was observed (0.3%). In the randomized controlled trials comparing structured and unstructured SMBG, in which structured SMBG was also coupled with adjustment of medications, the difference in HbA1c reduction between groups was 0.27% in favor of structured SMBG. Another review [91] took a look at SMBG in T2D without intensive insulin treatment to establish whether SMBG improves glycemic control. The review included 24 randomized controlled trials and 5454 people with T2D not using intensive insulin regimens. Studies, where people were using basal-bolus insulin injections, were excluded, although four studies included people on less intensive insulin treatment. This meta-analysis showed a small benefit of SMBG for HbA1c reduction in a short term (0.31% lower HbA1c at 12 weeks and 0.34% lower at 24 weeks), with the greatest benefit seen in those with poor glycemic control. However, the benefit of SMBG did not last beyond 6 months; at a 1-year follow-up, there was no difference between the groups.

Guidelines for SMBG

For people living with diabetes, medical guidelines recommend SMBG at varying frequencies, depending on the type of diabetes, antihyperglycemic therapy and adequacy of glycemic control. In addition, the frequency and timing of SMBG should be individualized to a person's specific needs and goals. Daily SMBG is essential in insulin-treated individuals, providing the means to assess the progress of treatment and avoid hypoglycemia. For individuals on intensive insulin regimens, the ADA recommends performing SMBG before meals and snacks, at bedtime, occasionally postprandially, before exercise, when they suspect low blood glucose, after treating low blood glucose until they are normoglycemic and before critical tasks such as driving. Consequently, testing may be required 6–10 times per day to optimize intensive control [92]. Similarly, to optimize intensive diabetes management in children, adolescents, and young adults aged <25 years, the International Society for Pediatric and Adolescent Diabetes recommends self-monitoring of glucose at least 6–10 times a day: before meals and snacks, 2–3 h after food intake, during the action profiles of insulin, with vigorous exercise, at bedtime, during the night and on awakening, during intercurrent illness, to confirm hypoglycemia and monitor recovery, before driving a car or operating hazardous machinery. In addition, frequent and regular reviewing of results and pattern recognition is necessary to make appropriate treatment adjustments [93].

As pregnancy complicated by either gestational diabetes mellitus or preexisting T1D or T2D is associated with risks to maternal and fetal complications, maternal glycemic targets are stricter to maintain glucose control throughout pregnancy as close to normal. All women with diabetes need to monitor fasting and postprandial blood glucose levels daily. The frequency of monitoring will increase up to 10 times per day for women with T1D and women with T2D on an intensive insulin regimen, who should also test blood glucose preprandially [94,95].

Due to the insufficient evidence, the ADA guidelines do not recommend when to prescribe SMBG and at what frequency the testing is needed in persons with T2D using basal insulin with or without oral agents; thus the optimal regimen is not clear [92]. Nevertheless, in persons with T2D on basal insulin therapy, assessing fasting plasma glucose with SMBG enables insulin dose titration to reach glycemic targets in the absence of hypoglycemia [96,97]. In the Reduction with an Initial Glargine Intervention (ORIGIN) trial, fasting capillary glucose levels were recorded daily until target values were achieved, and then at least twice per week [98].

SMBG in noninsulin-treated T2D helps to educate people about their condition, allows them to monitor the impact of food, physical activity, lifestyle choices, and medications on blood glucose levels, empowers self-management, and guides healthcare professionals to adjust therapeutic regimens. However, the clinical efficacy of SMBG in noninsulin-treated T2D has varied between studies. This emphasizes the challenge to understand the true benefit of SMBG in noninsulin-treated T2D and warrants further well-designed randomized controlled trials and longitudinal observational studies. ADA guidelines recommend SMBG use combined with education and support as a guide to successful therapy for some people with T2D not using insulin, and leave options open for individualized care [92,99]. Diabetes UK, similarly, recommends a targeted approach to SMBG use in noninsulin-treated individuals with T2D, based on the individual assessment of need [100]. The National Institute for Health and Care Excellence, however, advises against daily glucose testing in adults with T2D, unless the person is on insulin, is pregnant or is at risk of hypoglycemic episodes [101]. Some guidelines recommend SMBG in persons whose regimens include sulfonylurea due to the higher risk of hypoglycemia [96,102,103]. The International Diabetes Federation (IDF) recommends using structured SMBG, where patients and healthcare providers have the knowledge, skills, and willingness to incorporate SMBG and therapy adjustments into diabetes care plans [104]. Considering postmeal plasma glucose as a key predictor of cardiovascular events and all-cause mortality, the IDF recommends to evaluate postprandial glucose levels, as a key component to improve glycemic control regardless of insulin treatment prescription [105]. As the relative contribution of postprandial hyperglycemia to HbA1c levels is greater at HbA1c levels that are closer to 7% [106], postprandial testing should be considered for achieving HbA1c under 7% [101]. The American Association of Clinical Endocrinologists and the American College of Endocrinology (AACE/ACE) states that people with diabetes should be given advice regarding when and how frequently to monitor their blood glucose. Healthcare providers should review recorded SMBG data in a logbook or downloaded to the personal computer to make appropriate therapeutic adjustments if glycemic control is not at goal [96]. To optimize treatment, structured SMBG schedules, which enable to identify daily glycemic patterns have been proposed [104].

Structured SMBG has been shown to effectively reduce HbA1c levels in noninsulin-treated individuals with suboptimal initial glycemic control [79]. To enhance standardized approaches to SMBG in noninsulin-treated T2D, a European expert panel has recommended on length and frequency of SMBG performance depending on clinical circumstances and quality of glycemic control by using two SMBG schemes (a less intensive and an intensive). In an intensive seven-point SMBG regimen, SMBG is performed before and after each meal and at bedtime. In a less intensive SMBG regimen testing blood glucose before and after meals is performed for alternating meals over the course of a week. After obtaining sufficient glucose profiles, the frequency of SMBG should be reevaluated and timing and frequency of monitoring further individualized because in stable and good glucose control infrequent monitoring may be needed [107]. Current guidelines and consensus statements on the SMBG in persons with T2D are outlined in Table 1.1.

Table 1.1 Current clinical guidelines and consensus statements for self-monitoring of blood glucose in type 2 diabetes.

Guideline	Recommendations
American Diabetes Association[a]	• Most patients using intensive insulin regimens (multiple daily injections or insulin pump therapy) should assess glucose levels using self-monitoring of blood glucose (or continuous glucose monitoring) before meals and snacks, at bedtime, occasionally postprandially, before exercise, when they suspect low blood glucose, after treating low blood glucose until they are normoglycemic, and before critical tasks such as driving. • When prescribed as part of a broad educational program, self-monitoring of blood glucose may help to guide treatment decisions and/or self-management for patients taking less frequent insulin injections. • When prescribing self-monitoring of blood glucose, ensure that patients receive ongoing instruction and regular evaluation of technique, results, and their ability to use data from self-monitoring of blood glucose to adjust therapy. Similarly, continuous glucose monitoring use requires robust and ongoing diabetes education, training, and support. • The evidence is insufficient regarding when to prescribe SMBG and how often testing is needed for insulin-treated patients who do not use intensive insulin regimens, such as those with type 2 diabetes using basal insulin with or without oral agents.

Table 1.1 Current clinical guidelines and consensus statements for self-monitoring of blood glucose in type 2 diabetes.—*cont'd*

Guideline	Recommendations
	• In people with type 2 diabetes not using insulin, routine glucose monitoring may be of limited additional clinical benefit. For some individuals, glucose monitoring can provide insight into the impact of diet, physical activity, and medication management on glucose levels. Glucose monitoring may also be useful in assessing hypoglycemia, glucose levels during intercurrent illness, or discrepancies between measured A1C and glucose levels when there is concern an A1C result may not be reliable in specific individuals. A key consideration is that performing SMBG alone does not lower blood glucose levels. To be useful, the information must be integrated into clinical and self-management plans.
ADA-EASD Consensus Report[b]	• Regular SMBG may help with self-management and medication adjustment, particularly in individuals taking insulin. • SMBG plans should be individualized. People with diabetes and the healthcare team should use the data in an effective and timely manner. • In people with type 2 diabetes not using insulin, routine glucose monitoring is of limited additional clinical benefit while adding burden and cost. However, for some individuals, glucose monitoring can provide insight into the impact of lifestyle and medication management on blood glucose and symptoms, particularly when combined with education and support.
International Diabetes Federation Global Guideline for Type 2 Diabetes[c]	• SMBG should only be made available to people with diabetes when they have the knowledge, skills, and willingness to use the information obtained through testing to actively adjust treatment, enhance understanding of diabetes, and assess the effectiveness of the management plan on glycemic control. • The purpose(s) of performing SMBG and using SMBG data should be agreed between the person with diabetes and the healthcare provider. • SMBG on an ongoing basis should be available to those people with diabetes using insulin. • SMBG should be considered for people using oral glucose-lowering medications as an

Continued

Table 1.1 Current clinical guidelines and consensus statements for self-monitoring of blood glucose in type 2 diabetes.—*cont'd*

Guideline	Recommendations
	optional component of self-management, and in association with HbA1c testing: **o** To provide information on, and help avoid, hypoglycemia. **o** To assess changes in blood glucose control due to medications and lifestyle changes. **o** To monitor the effects of foods on postprandial glycemia. **o** To monitor changes in blood glucose levels during intercurrent illness. • Regular use of SMBG should not be considered part of routine care where diabetes is well controlled by nutrition therapy or oral medications alone. • SMBG protocols (intensity and frequency) should be individualized to address each individual's specific educational/behavioral/clinical requirements, and provider requirements for data on glycemic patterns to monitor therapeutic decision-making. • Structured assessment of self-monitoring skills, the quality and use made of the results obtained, and of the equipment used, should be made annually.
International Diabetes Federation Self-monitoring of blood glucose in noninsulin treated type 2 diabetes[d]	• SMBG should be used only when individuals with diabetes (and/or their care-givers) and/or their healthcare providers have the knowledge, skills, and willingness to incorporate SMBG monitoring and therapy adjustment into their diabetes care plan to attain the agreed treatment goals. • SMBG should be considered at the time of diagnosis to enhance the understanding of diabetes as part of individuals' education and to facilitate timely treatment initiation and titration optimization. • SMBG should also be considered as part of ongoing diabetes self-management education to assist people with diabetes to better understand their disease and provide a means to actively and effectively participate in its control and treatment, modifying behavioral and pharmacological interventions as needed, in consultation with their healthcare provider. • SMBG protocols (intensity and frequency) should be individualized to address each individual's specific educational/behavioral/

Table 1.1 Current clinical guidelines and consensus statements for self-monitoring of blood glucose in type 2 diabetes.—*cont'd*

Guideline	Recommendations
	clinical requirements (to identify/prevent/manage acute hyper- and hypoglycemia) and provider requirements for data on glycaemic patterns and to monitor the impact of therapeutic decision-making. • The purpose(s) of performing SMBG and using SMBG data should be agreed between the person with diabetes and the healthcare provider. These agreed-upon purposes/goals and actual review of SMBG data should be documented. • SMBG use requires an easy procedure for patients to regularly monitor the performance and accuracy of their glucose meter
National Institute for Health and Care Excellence[e]	• Take the Driver and Vehicle Licensing Agency (DVLA) At a glance guide to the current medical standards of fitness to drive into account when offering self-monitoring of blood glucose levels for adults with type 2 diabetes. • Do not routinely offer self-monitoring of blood glucose levels for adults with type 2 diabetes unless: o the person is on insulin or o there is evidence of hypoglycemic episodes or o the person is on oral medication that may increase their risk of hypoglycemia while driving or operating machinery or o the person is pregnant, or is planning to become pregnant. • Consider short-term self-monitoring of blood glucose levels in adults with type 2 diabetes (and review treatment as necessary): o when starting treatment with oral or intravenous corticosteroids or o to confirm suspected hypoglycemia. • Be aware that adults with type 2 diabetes who have acute intercurrent illness are at risk of worsening hyperglycemia. Review treatment as necessary. • If adults with type 2 diabetes are self-monitoring their blood glucose levels, conduct a structured assessment at least annually. The assessment should include the following: o the person's self-monitoring skills o the quality and frequency of testing

Continued

Table 1.1 Current clinical guidelines and consensus statements for self-monitoring of blood glucose in type 2 diabetes.—*cont'd*

Guideline	Recommendations
	o checking that the person knows how to interpret the blood glucose results and what action to take **o** the impact on the person's quality of life **o** the continued benefit to the person **o** the equipment used.
Diabetes UK[f]	• SMBG in people with Ttype 2 diabetes on insulin and medication that carries a risk of hypoglycemia should be regarded as an integral part of treatment and should not be restricted. • For people not in these treatment groups, SMBG should be available based on an individual assessment of need. Arbitrary withdrawal of SMBG in those who clearly benefit from doing so should not occur. People have the right to ask for a review and challenge these decisions if necessary. • DVLA guidance requires some treatment groups to conduct testing for licensing and safety reasons. • Use and frequency of testing, choice of meter, and target blood glucose should be agreed between the person with diabetes and their healthcare team. Local medicine management policies should allow sufficient choice and flexibility per individual circumstances to be taken into account. • Structured assessment of self-monitoring skills, the equipment used, and the quality and use made of the results obtained should be performed at least annually, or more frequently according to need. • SMBG should be integrated with a care package, accompanied by education, and should enable the individual to either interpret and adjust treatment accordingly or inform their healthcare team.
AACE/ACE Outpatient Glucose Monitoring Consensus Statement[g]	Type 2—Receiving insulin/sulfonylureas, glinides • Structured BGM is recommended. • BGM in patients on intensive insulin: fasting, premeal, bedtime, and periodically in the middle of the night. • BGM in patients on insulin ± other diabetes medication: at a minimum, when fasting and at bedtime.

Table 1.1 Current clinical guidelines and consensus statements for self-monitoring of blood glucose in type 2 diabetes.—*cont'd*

Guideline	Recommendations
	• BGM in patients on basal insulin +1 daily prandial or premixed insulin injection: at minimum when fasting and before the prandial or premixed insulin, and periodically at other times (i.e., premeal, bedtime, 3 a.m.). • Additional testing before exercise or critical tasks (e.g., driving) as needed.
	Type 2—Low risk of hypoglycemia • Daily BGM not recommended. • Initial periodic structured BGM (e.g., at meals and bedtime) may be useful in helping patients understand the effectiveness of MNT/lifestyle therapy. • Once at the A1C goal, less frequent monitoring is acceptable.

A1C, *glycated hemoglobin;* BGM, *blood glucose monitoring;* CMG, *continuous glucose monitoring;* DVLA, *Driver and Vehicle Licensing Agency;* HbA1c, *glycated hemoglobin;* MDI, *multiple daily injections;* MNT, *medical nutrition therapy;* SMBG, *self-monitoring of blood glucose.*
[a] *American Diabetes Association. 7. Diabetes Technology: Standards of Medical Care in Diabetes (2019).*
[b] *Management of hyperglycemia in type 2 diabetes: ADA-EASD Consensus Report (2018).*
[c] *International Diabetes Federation. Global guideline for type 2 diabetes (2012).*
[d] *International Diabetes Federation. Self-monitoring of blood glucose in noninsulin-treated type 2 diabetes (2009).*
[e] *National Institute for Health and Care Excellence. Type 2 diabetes in adults: management (2015).*
[f] *Diabetes UK, Position statement: Self-monitoring of blood glucose levels for adults with Type 2 diabetes (2017).*
[g] *American Association of Clinical Endocrinologists and American College of Endocrinology. Outpatient Glucose Monitoring Consensus Statement (2016).*

The shortcomings of SMBG and future perspective

Studies indicate that a substantial proportion of individuals with both T1D and T2D does not perform SMBG at recommended frequencies. In a large cohort study of 24,312 adults with diabetes, SMBG adherence rates with ADA recommendations were reported to be as low as 34% for T1D, 54% for insulin-treated T2D and 20% for noninsulin-treated T2D [49]. Similarly, a youth analysis of uploaded data from insulin pumps demonstrated only a 31% adherence rate to the recommendation to enter four or more SMBG readings per day [108]. Data form the T1D Exchange Registry showed that 34% of participants performed SMBG less than four times per day [109]. Moreover, recently updated data demonstrated that 41% of 13,344 participants not using a continuous glucose monitor performed SMBG zero to three times per day [110]. In Sweden, where glucose meters and strips are being generally available at no cost, less than 50% of adults with T1D perform SMBG four times per day or more [111]. Similarly, the analysis of SMBG data from over 13,000 people with T2D found that SMBG is underutilized both in insulin-treated and noninsulin-treated individuals. In addition, postprandial glucose values were seldom checked suggesting nonadherence to the structured SMBG schemes [112].

There are several factors that may influence SMBG adherence. In a study by Vincze et al., environmental barriers, such as lifestyle interference, inconvenience, painfulness, and cost, were significantly associated with adherence to SMBG [113]. On the other hand, in a study by Fisher et al., SMBG information, motivation, and behavioral skills deficits were significantly correlated with SMBG frequency among individuals both with T1D and T2D, and accounted for 25% of the variability in SMBG frequency among individuals with T1D and for 9% of the variance in SMBG frequency among individuals with T2D. Moreover, a substantial proportion of individuals was unconvinced of SMBG usefulness [114]. Similarly, in a Swedish survey, 30% of adults with T1D were not aware that four or more SMBG measurements were recommended, implying a need for appropriate education addressing current guidelines [111]. Self-management interventions, such as education, problem-solving, contingency management, goal setting, cognitive-behavioral therapy, and motivational interviewing, demonstrated at least short-term improvements in adherence to recommended SMBG frequency [115]. Although the pain associated with finger pricking has been reduced with modern lancing devices [112], approximately 34% of individuals with T1D and 35% with T2D viewed SMBG as painful [116]. On the contrary, in a self-reported Swedish survey, only 14% of adults with T1D stated SMBG-associated pain as the main reason for not performing SMBG according to recommendations [111]. The questionnaire-based survey from 517 individuals with T1D and 1648 with T2D showed that individuals experiencing SMBG-associated pain had more mental distress, lower health-related quality of life, higher HbA1c, and appreciated the importance of SMBG less [117].

SMBG may potentially have adverse psychological effects in some individuals [76,118,119]. In noninsulin-treated T2D, SMBG frequency of one or more times per day was associated with higher levels of distress, worries, and depressive symptoms [118]. A systematic review of SMBG in T2D revealed a lack of education on how to interpret SMBG results together with the omission of appropriate lifestyle and treatment adjustments [119]. Hence, structured SMBG with sufficient education on how to interpret and respond to SMBG results was not associated with a deterioration of quality of life in noninsulin-treated individuals with T2D [84].

Underutilization of glucose data from both SMBG and continuous glucose monitoring (CGM), lack of easy and standardized glucose data collection, analysis, visualization, and guided clinical decision-making were found to be key contributors to poor glycemic control among individuals with T1D [120]. Recent data from T1D Exchange showed that 71% of non-CGM users and 60% of CGM users never downloaded the blood glucose meter outside of the doctor's office despite the increased use of the devices. Similarly, the use of a mobile medical application was as low as 16%, suggesting uncommon use of data reporting and analysis to assist diabetes self-management [110]. Easier to use seamless connectivity of blood glucose meter with smartphone application and cloud-based storage of data may increase the use of downloaded data in diabetes self-management [24].

Accuracy of SMBG is the basis of proper diabetes management. The requirements regarding the accuracy of blood glucose monitoring systems (BGMSs) are defined in the ISO standards [26] and FDA guidance [27]. Although the approval by regulatory agencies is based on their performance before market introduction, postmarketing assessment of 18 commercially available BGMSs revealed that only 6 of the 18 BGMSs fulfilled the latest ISO standard (ISO 15197-2013) [121]. In addition, user errors have been identified as a common cause of measurement inaccuracy [122,123]. Nevertheless, BGMSs have to achieve the same accuracy levels when used by trained personnel and intended users (laypersons). In the study evaluating measurement accuracy of four different BGMSs in the hands of lay users and trained personnel, BGMSs accuracy varied markedly depending on the operator. Common lay user errors included not checking the test strip codes, incorrect application of blood, and not using the blood drop immediately [124]. The authors concluded that BGMSs insensitive to operator errors would be a useful improvement. Inaccurate blood glucose levels can lead to incorrect treatment decisions and adverse clinical outcomes [125]. In silico study showed a 10-fold increase in missed hypoglycemic episodes when SMBG errors ranged from 10% to 20% [126]. Similarly, a 30-day in silico study in T1D individuals on insulin pump therapy showed that large error rate increased episodes of severe hypoglycemia, but had little effect on HbA1c. On the other hand, glucose meter's systematic bias affected HbA1c as well as a number of severe hypoglycemia events. Both bias and error exhibited a significant effect on total daily insulin and the number of necessary glucose measurements per day [127].

SMBG a few times a day can reveal only rough patterns of daily blood glucose variation. Consequently, high blood glucose excursions may be overlooked and low blood glucose values may be undetected, especially in individuals with impaired awareness of hypoglycemia [128]. CGM is a step further to optimal glycemic control [129]. In addition to current glucose, it provides information on direction and velocity of glucose change [130]. CGM can improve metabolic control, reduces the risk of hypoglycemia, increases time in normoglycemic range [131], and improves quality of life [132]. A clear clinical benefit has been demonstrated in outcome trials for people with T1D and T2D, using either MDI or CSII therapy [133]. In addition, real-world use of CGM demonstrated reduction of costs related to diabetes, which may outweigh CGM-related costs [134]. As CGM can now be used for insulin treatment decision-making [135] and future devices might be factory calibrated, it may be anticipated to replace SMBG in routine use for a wide range of people with diabetes.

References

[1] Benedict S. A reagent for the detection of reducing sugars. The Journal of Biological Chemistry 1908;5:485–7.

[2] Free A, Adams E, Kercher M, Free H, Cook M. Simple specific test for urine glucose. Clinical Chemistry 1957;3:163–8.

[3] Goldstein D, Little R, Lorenz R, Malone J, Nathan D, Peterson C, et al. Tests of glycemia in diabetes. Diabetes Care 2004;27:1761–73. https://doi.org/10.2337/diacare.27.7.1761.

[4] Free A, Free H. Self testing, an emerging component of clinical chemistry. Clinical Chemistry 1984;30:829–38.

[5] Cheah J, Wong A. A rapid and simple blood sugar determination using the Ames reflectance meter and Dextrostix system: a preliminary report. Singapore Medical Journal 1974;15:51–2.

[6] Clarke S, Foster J. A history of blood glucose meters and their role in self-monitoring of diabetes mellitus. British Journal of Biomedical Science 2012;69:83–93. https://doi.org/10.1080/09674845.2012.12002443.

[7] Yamada S. Historical achievements of self-monitoring of blood glucose technology development in Japan. Journal of Diabetes Science and Technology 2011;5:1300–6. https://doi.org/10.1177/193229681100500541.

[8] Moodley N, Ngxamngxa U, Turzyniecka M, Pillay T. Historical perspectives in clinical pathology: a history of glucose measurement. Journal of Clinical Pathology 2015;68:258–64. https://doi.org/10.1136/jclinpath-2014-202672.

[9] Leroux M, Desjardins P. Ward level evaluation of the "One Touch" glucose meter. Clinical Chemistry 1988;34:1928.

[10] Maloberti F, Davies A. A short history of circuits and systems. Gistrup: River Publishers; 2016.

[11] Matthews D, Bown E, Watson A, Holman R, Steemson J, Hughes S, et al. Pen-sized digital 30-second blood glucose meter. The Lancet 1987;329:778–9. https://doi.org/10.1016/s0140-6736(87)92802-9.

[12] Vadgama P, Peteu S. Detection challenges in clinical diagnostics. Cambridge: The Royal Society of Chemistry; 2013.

[13] Wens R, Taminne M, Devriendt J, Collart F, Broeders N, Mestrez F, et al. A previously undescribed side effect of icodextrin: overestimation of glycemia by glucose analyzer. Peritoneal Dialysis International 1998;18:603–9.

[14] Feldman B, McGarraugh G, Heller A, Bohannon N, Skyler J, DeLeeuw E, et al. FreeStyle™: a small-volume electrochemical glucose sensor for home blood glucose testing. Diabetes Technology and Therapeutics 2000;2:221–9. https://doi.org/10.1089/15209150050025177.

[15] Segil J. Handbook of biomechatronics. 1st ed. London: Academic Press; 2018.

[16] The Diabetes Research In Children Network (Direcent) Study Group. A multicenter study of the accuracy of the one Touch® Ultra® home glucose meter in children with type 1 diabetes. Diabetes Technology and Therapeutics 2003;5:933–41. https://doi.org/10.1089/152091503322640971.

[17] Schlesinger M. Applications of electrochemistry in medicine. New York: Springer; 2013.

[18] Suzuki Y, Atsumi Y, Matusoka K. Alternative site testing increases compliance of SMBG (preliminary study of 3 years cohort trials). Diabetes Research and Clinical Practice 2003;59:233–4. https://doi.org/10.1016/s0168-8227(02)00193-6.

[19] Öberg P, Togawa T, Spelman F. Sensors applications. Weinheim: Wiley-VCH Verlag; 2004.

[20] Bina D, Anderson R, Johnson M, Bergenstal R, Kendall D. Clinical impact of prandial state, exercise, and site preparation on the equivalence of alternative-site blood glucose testing. Diabetes Care 2003;26:981–5. https://doi.org/10.2337/diacare.26.4.981.

[21] Fineberg S, Bergenstal R, Bernstein R, Laffel L, Schwartz S. Use of an automated device for alternative site blood glucose monitoring. Diabetes Care 2001;24: 1217–20. https://doi.org/10.2337/diacare.24.7.1217.

[22] Rodbard D. Evaluating quality of glycemic control. Journal of Diabetes Science and Technology 2014;9:56–62. https://doi.org/10.1177/1932296814551046.

[23] Hirsch I, Bode B, Childs B, Close K, Fisher W, Gavin J, et al. Self-monitoring of blood glucose (SMBG) in insulin- and non–insulin-using adults with diabetes: consensus recommendations for improving SMBG accuracy, utilization, and research. Diabetes Technology and Therapeutics 2008;10:419–39. https://doi.org/10.1089/dia.2008.0104.

[24] Bailey T, Wallace J, Pardo S, Warchal-Windham M, Harrison B, Morin R, et al. Accuracy and user performance evaluation of a new, wireless-enabled blood glucose monitoring system that links to a smart mobile device. Journal of Diabetes Science and Technology 2017;11:736–43. https://doi.org/10.1177/1932296816680829.

[25] Freckmann G, Schmid C, Baumstark A, Rutschmann M, Haug C, Heinemann L. Analytical performance requirements for systems for self-monitoring of blood glucose with focus on system accuracy. Journal of Diabetes Science and Technology 2015;9: 885–94. https://doi.org/10.1177/1932296815580160.

[26] International Organization for Standardization: in vitro diagnostic test systems—requirements for blood-glucose monitoring systems for self-testing in managing diabetes mellitus. ISO 15197:2013, Iso.Org. (n.d.). Available from: https://www.iso.org/obp/ui/#iso:std:54976:en [Accessed 27 May 2019].

[27] Self-monitoring blood glucose test systems for over-the-counter use. U.S. Food And Drug Administration; 2016. Available from: https://www.fda.gov/downloads/MedicalDevices/DeviceRegulationandGuidance/GuidanceDocuments/UCM380327.pdf. [Accessed 27 May 2019].

[28] Bernstein R. Virtually continuous euglycemia for 5 yr in a labile juvenile-onset diabetic patient under noninvasive closed-loop control. Diabetes Care 1980;3:140–3. https://doi.org/10.2337/diacare.3.1.140.

[29] Danowski T, Sunder J. Jet injection of insulin during self-monitoring of blood glucose. Diabetes Care 1978;1:27–33. https://doi.org/10.2337/diacare.1.1.27.

[30] Sönksen P, Judd S, Lowy C. Home monitoring of blood-glucose. The Lancet 1978;311: 729–32. https://doi.org/10.1016/s0140-6736(78)90854-1.

[31] Walford S, Gale E, Allison S, Tattersall R. Self-monitoring of blood-glucose. The Lancet 1978;311:732–5. https://doi.org/10.1016/s0140-6736(78)90855-3.

[32] Skyler J, Lasky I, Skyler D, Robertson E, Mintz D. Home blood glucose monitoring as an aid in diabetes management. Diabetes Care 1978;1:150–7. https://doi.org/10.2337/diacare.1.3.150.

[33] Ikeda Y, Tajima N, Minami N, Ide Y, Yokoyama J, Abe M. Pilot study of self-measurement of blood glucose using the dextrostix-eyetone system for juvenile-onset diabetes. Diabetologia 1978;15:91–3. https://doi.org/10.1007/bf00422251.
[34] Howe-Davies S, Holman R, Phillips M, Turner R. Home blood sampling for plasma glucose assay in control of diabetes. British Medical Journal 1978;2:596–8. https://doi.org/10.1136/bmj.2.6137.596.
[35] Tattersall R. Home blood glucose monitoring. Diabetologia 1979;16:71–4. https://doi.org/10.1007/bf01225453.
[36] Danowski T, Ohlsen P, Fisher E, Sunder J. Parameters of good control in diabetes mellitus. Diabetes Care 1980;3:88–93. https://doi.org/10.2337/diacare.3.1.88.
[37] Peterson C, Jones R, Dupuis A, Levine B, Bernstein R, O'Shea M. Feasibility of improved blood glucose control in patients with insulin-dependent diabetes mellitus. Diabetes Care 1979;2:329–35. https://doi.org/10.2337/diacare.2.4.329.
[38] Cahill G, Etzwiler D, Freinkel N. Control and diabetes. New England Journal of Medicine 1976;294:1004–5. https://doi.org/10.1056/nejm197604292941811.
[39] Franklin Bunn H. Evaluation of glycosylated hemoglobin in diabetic patients. Diabetes 1981;30:613–7. https://doi.org/10.2337/diab.30.7.613.
[40] Skyler J, Skyler D, Seigler D, O'Sullivan M. Algorithms for adjustment of insulin dosage by patients who monitor blood glucose. Diabetes Care 1981;4:311–8. https://doi.org/10.2337/diacare.4.2.311.
[41] Hirsch I, Farkas-Hirsch R, Skyler J. Intensive insulin therapy for treatment of type I diabetes. Diabetes Care 1990;13:1265–83. https://doi.org/10.2337/diacare.13.12.1265.
[42] Consensus statement on self-monitor ing of blood glucose. Diabetes Care 1987;10:95–9. https://doi.org/10.2337/diacare.10.1.95.
[43] The effect of intensive treatment of diabetes on the development and progression of long-term complications in insulin-dependent diabetes mellitus. New England Journal of Medicine 1993;329:977–86. https://doi.org/10.1056/nejm199309303291401.
[44] Self-monitoring of blood glucose. Diabetes Care 1996;19:S62–6. https://doi.org/10.2337/diacare.19.1.s62.
[45] Intensive diabetes treatment and cardiovascular disease in patients with type 1 diabetes. New England Journal of Medicine 2005;353:2643–53. https://doi.org/10.1056/nejmoa052187.
[46] Intensive diabetes treatment and cardiovascular outcomes in type 1 diabetes: the DCCT/EDIC study 30-year follow-up. Diabetes Care 2016;39:686–93. https://doi.org/10.2337/dc15-1990.
[47] Kilpatrick E, Rigby A, Goode K, Atkin S. Relating mean blood glucose and glucose variability to the risk of multiple episodes of hypoglycaemia in type 1 diabetes. Diabetologia 2007;50:2553–61. https://doi.org/10.1007/s00125-007-0820-z.
[48] Evans J, Newton R, Ruta D, MacDonald T, Stevenson R, Morris A. Frequency of blood glucose monitoring in relation to glycaemic control: observational study with diabetes database. British Medical Journal 1999;319:83–6. https://doi.org/10.1136/bmj.319.7202.83.
[49] Karter A, Ackerson L, Darbinian J, D'Agostino R, Ferrara A, Liu J, et al. Self-monitoring of blood glucose levels and glycemic control: the Northern California Kaiser Permanente Diabetes registry. The American Journal of Medicine 2001;111:1–9. https://doi.org/10.1016/s0002-9343(01)00742-2.

[50] Hansen M, Pedersen-Bjergaard U, Heller S, Wallace T, Rasmussen Å, Jørgensen H, et al. Frequency and motives of blood glucose self-monitoring in type 1 diabetes. Diabetes Research and Clinical Practice 2009;85:183–8. https://doi.org/10.1016/j.diabres.2009.04.022.
[51] Schütt M, Kern W, Krause U, Busch P, Dapp A, Grziwotz R, et al. Is the frequency of self-monitoring of blood glucose related to long-term metabolic control? Multicenter analysis including 24 500 patients from 191 centers in Germany and Austria. Experimental and Clinical Endocrinology and Diabetes 2006;114:384–8. https://doi.org/10.1055/s-2006-924152.
[52] Miller K, Beck R, Bergenstal R, Goland R, Haller M, McGill J, et al. Evidence of a strong association between frequency of self-monitoring of blood glucose and hemoglobin A1c levels in T1D Exchange clinic registry participants. Diabetes Care 2013;36:2009–14. https://doi.org/10.2337/dc12-1770.
[53] Simmons J, Chen V, Miller K, McGill J, Bergenstal R, Goland R, et al. Differences in the management of type 1 diabetes among adults under excellent control compared with those under poor control in the T1D Exchange clinic registry. Diabetes Care 2013;36:3573–7. https://doi.org/10.2337/dc12-2643.
[54] Levine B, Anderson B, Butler D, Antisdel J, Brackett J, Laffel L. Predictors of glycemic control and short-term adverse outcomes in youth with type 1 diabetes. The Journal of Pediatrics 2001;139:197–203. https://doi.org/10.1067/mpd.2001.116283.
[55] Moreland E, Tovar A, Zuehlke J, Butler D, Milaszewski K, Laffel L. The impact of physiological, therapeutic and psychosocial variables on glycemic control in youth with type 1 diabetes mellitus. Journal of Pediatric Endocrinology and Metabolism 2004;17. https://doi.org/10.1515/jpem.2004.17.11.1533.
[56] Haller M, Stalvey M, Silverstein J. Predictors of control of diabetes: monitoring may be the key. The Journal of Pediatrics 2004;144:660–1. https://doi.org/10.1016/j.jpeds.2003.12.042.
[57] Redondo M, Connor C, Ruedy K, Beck R, Kollman C, Wood J, et al. Pediatric diabetes consortium type 1 diabetes new onset (NeOn) study: factors associated with HbA1c levels one year after diagnosis. Pediatric Diabetes 2013;15:294–302. https://doi.org/10.1111/pedi.12061.
[58] Ziegler R, Heidtmann B, Hilgard D, Hofer S, Rosenbauer J, Holl R. Frequency of SMBG correlates with HbA1c and acute complications in children and adolescents with type 1 diabetes. Pediatric Diabetes 2011;12:11–7. https://doi.org/10.1111/j.1399-5448.2010.00650.x.
[59] Bohn B, Karges B, Vogel C, Otto K, Marg W, Hofer S, et al. 20 Years of pediatric benchmarking in Germany and Austria: age-dependent analysis of longitudinal follow-up in 63,967 children and adolescents with type 1 diabetes. PLoS One 2016;11:e0160971. https://doi.org/10.1371/journal.pone.0160971.
[60] Intensive blood-glucose control with sulphonylureas or insulin compared with conventional treatment and risk of complications in patients with type 2 diabetes (UKPDS 33). The Lancet 1998;352:837–53. https://doi.org/10.1016/s0140-6736(98)07019-6.
[61] Ohkubo Y, Kishikawa H, Araki E, Miyata T, Isami S, Motoyoshi S, et al. Intensive insulin therapy prevents the progression of diabetic microvascular complications in Japanese patients with non-insulin-dependent diabetes mellitus: a randomized prospective 6-year study. Diabetes Research and Clinical Practice 1995;28:103–17. https://doi.org/10.1016/0168-8227(95)01064-k.

[62] Holman R, Paul S, Bethel M, Matthews D, Neil H. 10-Year follow-up of intensive glucose control in type 2 diabetes. New England Journal of Medicine 2008;359: 1577–89. https://doi.org/10.1056/nejmoa0806470.
[63] Duckworth W, Abraira C, Moritz T, Reda D, Emanuele N, Reaven P, et al. Glucose control and vascular complications in Veterans with type 2 diabetes. New England Journal of Medicine 2009;360:129–39. https://doi.org/10.1056/nejmoa0808431.
[64] Intensive blood glucose control and vascular outcomes in patients with type 2 diabetes. New England Journal of Medicine 2008;358:2560–72. https://doi.org/10.1056/nejmoa0802987.
[65] Effects of intensive glucose lowering in type 2 diabetes. New England Journal of Medicine 2008;358:2545–59. https://doi.org/10.1056/nejmoa0802743.
[66] Ismail-Beigi F, Craven T, Banerji M, Basile J, Calles J, Cohen R, et al. Effect of intensive treatment of hyperglycaemia on microvascular outcomes in type 2 diabetes: an analysis of the ACCORD randomised trial. The Lancet 2010;376:419–30. https://doi.org/10.1016/s0140-6736(10)60576-4.
[67] Skyler J, Bergenstal R, Bonow R, Buse J, Deedwania P, Gale E, et al. Intensive glycemic control and the prevention of cardiovascular events: implications of the ACCORD, ADVANCE, and VA diabetes trials. Journal of the American College of Cardiology 2009;53:298–304. https://doi.org/10.1016/j.jacc.2008.10.008.
[68] Davis W, Bruce D, Davis T. Is self-monitoring of blood glucose appropriate for all type 2 diabetic patients?: the Fremantle diabetes study. Diabetes Care 2006;29:1764–70. https://doi.org/10.2337/dc06-0268.
[69] Davis W, Bruce D, Davis T. Does self-monitoring of blood glucose improve outcome in type 2 diabetes? The Fremantle Diabetes Study. Diabetologia 2007;50:510–5. https://doi.org/10.1007/s00125-006-0581-0.
[70] Martin S, Schneider B, Heinemann L, Lodwig V, Kurth H, Kolb H, et al. Self-monitoring of blood glucose in type 2 diabetes and long-term outcome: an epidemiological cohort study. Diabetologia 2005;49:271–8. https://doi.org/10.1007/s00125-005-0083-5.
[71] Karter A, Parker M, Moffet H, Spence M, Chan J, Ettner S, et al. Longitudinal study of new and prevalent use of self-monitoring of blood glucose. Diabetes Care 2006;29: 1757–63. https://doi.org/10.2337/dc06-2073.
[72] Elgart J, González L, Prestes M, Rucci E, Gagliardino J. Frequency of self-monitoring blood glucose and attainment of HbA1c target values. Acta Diabetologica 2015;53: 57–62. https://doi.org/10.1007/s00592-015-0745-9.
[73] Harris M. Frequency of blood glucose monitoring in relation to glycemic control in patients with type 2 diabetes. Diabetes Care 2001;24:979–82. https://doi.org/10.2337/diacare.24.6.979.
[74] Farmer A, Wade A, Goyder E, Yudkin P, French D, Craven A, et al. Impact of self monitoring of blood glucose in the management of patients with non-insulin treated diabetes: open parallel group randomised trial. British Medical Journal 2007;335: 132. https://doi.org/10.1136/bmj.39247.447431.be.
[75] Simon J, Gray A, Clarke P, Wade A, Neil A, Farmer A. Cost effectiveness of self monitoring of blood glucose in patients with non-insulin treated type 2 diabetes: economic evaluation of data from the DiGEM trial. British Medical Journal 2008;336:1177–80. https://doi.org/10.1136/bmj.39526.674873.be.
[76] O'Kane M, Bunting B, Copeland M, Coates V. Efficacy of self monitoring of blood glucose in patients with newly diagnosed type 2 diabetes (ESMON study): randomised controlled trial. British Medical Journal 2008;336:1174–7. https://doi.org/10.1136/bmj.39534.571644.be.

[77] Young L, Buse J, Weaver M, Vu M, Mitchell C, Blakeney T, et al. Glucose self-monitoring in non−insulin-treated patients with type 2 diabetes in primary care settings. JAMA Internal Medicine 2017;177:920. https://doi.org/10.1001/jamainternmed.2017.1233.

[78] Parkin C, Buskirk A, Hinnen D, Axel-Schweitzer M. Results that matter: structured vs. unstructured self-monitoring of blood glucose in type 2 diabetes. Diabetes Research and Clinical Practice 2012;97:6−15. https://doi.org/10.1016/j.diabres.2012.03.002.

[79] Polonsky W, Fisher L, Schikman C, Hinnen D, Parkin C, Jelsovsky Z, et al. Structured self-monitoring of blood glucose significantly reduces A1C levels in poorly controlled, noninsulin-treated type 2 diabetes: results from the structured testing program study. Diabetes Care 2011;34:262−7. https://doi.org/10.2337/dc10-1732.

[80] Fisher L, Polonsky W, Parkin C, Jelsovsky Z, Petersen B, Wagner R. The impact of structured blood glucose testing on attitudes toward self-management among poorly controlled, insulin-naïve patients with type 2 diabetes. Diabetes Research and Clinical Practice 2012;96:149−55. https://doi.org/10.1016/j.diabres.2011.12.016.

[81] Franciosi M, Lucisano G, Pellegrini F, Cantarello A, Consoli A, Cucco L, et al. ROSES: role of self-monitoring of blood glucose and intensive education in patients with Type 2 diabetes not receiving insulin. A pilot randomized clinical trial. Diabetic Medicine 2011;28:789−96. https://doi.org/10.1111/j.1464-5491.2011.03268.x.

[82] Durán A, Martín P, Runkle I, Pérez N, Abad R, Fernández M, et al. Benefits of self-monitoring blood glucose in the management of new-onset Type 2 diabetes mellitus: the St Carlos Study, a prospective randomized clinic-based interventional study with parallel groups. Journal of Diabetes 2010;2:203−11. https://doi.org/10.1111/j.1753-0407.2010.00081.x.

[83] Bosi E, Scavini M, Ceriello A, Cucinotta D, Tiengo A, Marino R, et al. Intensive structured self-monitoring of blood glucose and glycemic control in noninsulin-treated type 2 diabetes: the PRISMA randomized trial. Diabetes Care 2013;36:2887−94. https://doi.org/10.2337/dc13-0092.

[84] Russo G, Scavini M, Acmet E, Bonizzoni E, Bosi E, Giorgino F, et al. The burden of structured self-monitoring of blood glucose on diabetes-specific quality of life and locus of control in patients with noninsulin-treated type 2 diabetes: the PRISMA study. Diabetes Technology and Therapeutics 2016;18:421−8. https://doi.org/10.1089/dia.2015.0358.

[85] Parsons S, Luzio S, Harvey J, Bain S, Cheung W, Watkins A, et al. Effect of structured self-monitoring of blood glucose, with and without additional TeleCare support, on overall glycaemic control in non-insulin treated Type 2 diabetes: the SMBG Study, a 12-month randomized controlled trial. Diabetic Medicine 2019;36:578−90. https://doi.org/10.1111/dme.13899.

[86] Farmer A, Perera R, Ward A, Heneghan C, Oke J, Barnett A, et al. Meta-analysis of individual patient data in randomised trials of self monitoring of blood glucose in people with non-insulin treated type 2 diabetes. British Medical Journal 2012;344:e486. https://doi.org/10.1136/bmj.e486. e486.

[87] Malanda U, Welschen L, Riphagen I, Dekker J, Nijpels G, Bot S. Self-monitoring of blood glucose in patients with type 2 diabetes mellitus who are not using insulin. Cochrane Database of Systematic Reviews 2012. https://doi.org/10.1002/14651858.cd005060.pub3.

[88] Allen B, DeLong E, Feussner J. Impact of glucose self-monitoring on non-insulin-treated patients with type II diabetes mellitus: randomized controlled trial comparing blood and urine testing. Diabetes Care 1990;13:1044–50. https://doi.org/10.2337/diacare.13.10.1044.

[89] Zhu H, Zhu Y, Leung S. Is self-monitoring of blood glucose effective in improving glycaemic control in type 2 diabetes without insulin treatment: a meta-analysis of randomised controlled trials. British Medical Journal Open 2016;6:e010524. https://doi.org/10.1136/bmjopen-2015-010524.

[90] Mannucci E, Antenore A, Giorgino F, Scavini M. Effects of structured versus unstructured self-monitoring of blood glucose on glucose control in patients with non-insulin-treated type 2 diabetes: a meta-analysis of randomized controlled trials. Journal of Diabetes Science and Technology 2017;12:183–9. https://doi.org/10.1177/1932296817719290.

[91] Machry R, Rados D, Gregório G, Rodrigues T. Self-monitoring blood glucose improves glycemic control in type 2 diabetes without intensive treatment: a systematic review and meta-analysis. Diabetes Research and Clinical Practice 2018;142:173–87. https://doi.org/10.1016/j.diabres.2018.05.037.

[92] 7. Diabetes technology: standards of medical care in diabetes—2019. Diabetes Care 2018;42:S71–80. https://doi.org/10.2337/dc19-s007.

[93] DiMeglio L, Acerini C, Codner E, Craig M, Hofer S, Pillay K, et al. ISPAD Clinical Practice Consensus Guidelines 2018: glycemic control targets and glucose monitoring for children, adolescents, and young adults with diabetes. Pediatric Diabetes 2018;19:105–14. https://doi.org/10.1111/pedi.12737.

[94] 14. Management of diabetes in pregnancy: standards of medical care in diabetes—2019. Diabetes Care 2018;42:S165–72. https://doi.org/10.2337/dc19-s014.

[95] Recommendations | Diabetes in pregnancy: management from preconception to the postnatal period | Guidance. NICE, Nice.Org.Uk; 2015. Available from: https://www.nice.org.uk/guidance/ng3/chapter/1-Recommendations. [Accessed 1 June 2019].

[96] Bailey T, Grunberger G, Bode B, Handelsman Y, Hirsch I, Jovanovič L, et al. American association of clinical endocrinologists and American college of endocrinology 2016 outpatient glucose monitoring consensus statement. Endocrine Practice 2016;22:231–61. https://doi.org/10.4158/ep151124.cs.

[97] Rosenstock J, Davies M, Home P, Larsen J, Koenen C, Schernthaner G. A randomised, 52-week, treat-to-target trial comparing insulin detemir with insulin glargine when administered as add-on to glucose-lowering drugs in insulin-naive people with type 2 diabetes. Diabetologia 2008;51:408–16. https://doi.org/10.1007/s00125-007-0911-x.

[98] Rationale, design, and baseline characteristics for a large international trial of cardiovascular disease prevention in people with dysglycemia: the ORIGIN Trial (outcome reduction with an initial glargine intervention). American Heart Journal 2008;155:26.e1–26.e13. https://doi.org/10.1016/j.ahj.2007.09.009.

[99] Davies M, D'Alessio D, Fradkin J, Kernan W, Mathieu C, Mingrone G, et al. Management of hyperglycemia in type 2 diabetes, 2018. A consensus report by the American Diabetes Association (ADA) and the European Association for the Study of Diabetes (EASD). Diabetes Care 2018;41:2669–701. https://doi.org/10.2337/dci18-0033.

[100] Self-monitoring of blood glucose levels for adults with Type 2 diabetes (March 2017). Diabetes UK; 2017. Available from: https://www.diabetes.org.uk/professionals/position-statements-reports/diagnosis-ongoing-management-monitoring/self-monitoring-of-blood-glucose-levels-for-adults-with-type-2-diabetes. [Accessed 31 May 2019].

[101] Recommendations | Type 2 diabetes in adults: management | Guidance | NICE. Nice.Org.Uk; 2015. Available from: https://www.nice.org.uk/guidance/ng28/chapter/1-recommendations#individualised-care. [Accessed 1 June 2019].

[102] IDF clinical practice recommendations for managing Type 2 diabetes in primary care. Idf.Org.; 2017. Available from: https://www.idf.org/e-library/guidelines/128-idf-clinical-practice-recommendations-for-managing-type-2-diabetes-in-primary-care.html. [Accessed 1 June 2019].

[103] Berard L, Siemens R, Woo V. Monitoring glycemic control. Canadian Journal of Diabetes 2018;42:S47–53. https://doi.org/10.1016/j.jcjd.2017.10.007.

[104] Self-monitoring of blood glucose in non-insulin treated type 2 diabetes. Idf.Org; 2009. Available from: https://www.idf.org/e-library/guidelines/85-self-monitoring-of-blood-glucose-in-non-insulin-treated-type-2-diabetes.html. [Accessed 1 June 2019].

[105] Guideline for management of post meal glucose in diabetes. Idf.Org; 2011. Available from: https://www.idf.org/e-library/guidelines/82-management-of- postmeal-glucose.html. [Accessed 1 June 2019].

[106] Woerle H, Neumann C, Zschau S, Tenner S, Irsigler A, Schirra J, et al. Impact of fasting and postprandial glycemia on overall glycemic control in type 2 diabetes. Diabetes Research and Clinical Practice 2007;77:280–5. https://doi.org/10.1016/j.diabres.2006.11.011.

[107] Schnell O, Alawi H, Battelino T, Ceriello A, Diem P, Felton A, et al. Addressing schemes of self-monitoring of blood glucose in type 2 diabetes: a European perspective and expert recommendation. Diabetes Technology and Therapeutics 2011;13: 959–65. https://doi.org/10.1089/dia.2011.0028.

[108] O'Connell M, Donath S, Cameron F. Poor adherence to integral daily tasks limits the efficacy of CSII in youth. Pediatric Diabetes 2011. https://doi.org/10.1111/j.1399-5448.2010.00740.x.

[109] Miller K, Foster N, Beck R, Bergenstal R, DuBose S, DiMeglio L, et al. Current state of type 1 diabetes treatment in the U.S.: updated data from the T1D Exchange clinic registry. Diabetes Care 2015;38:971–8. https://doi.org/10.2337/dc15-0078.

[110] Foster N, Beck R, Miller K, Clements M, Rickels M, DiMeglio L, et al. State of type 1 diabetes management and outcomes from the T1D Exchange in 2016–2018. Diabetes Technology and Therapeutics 2019;21:66–72. https://doi.org/10.1089/dia.2018.0384.

[111] Moström P, Ahlén E, Imberg H, Hansson P, Lind M. Adherence of self-monitoring of blood glucose in persons with type 1 diabetes in Sweden. British Medical Journal Open Diabetes Research and Care 2017;5:e000342. https://doi.org/10.1136/bmjdrc-2016-000342.

[112] Rossi M, Lucisano G, Ceriello A, Mazzucchelli C, Musacchio N, Ozzello A, et al. Real-world use of self-monitoring of blood glucose in people with type 2 diabetes: an urgent need for improvement. Acta Diabetologica 2018;55:1059–66. https://doi.org/10.1007/s00592-018-1186-z.

[113] Vincze G, Barner J, Lopez D. Factors associated with adherence to self-monitoring of blood glucose among persons with diabetes. The Diabetes Educator 2004;30:112–25. https://doi.org/10.1177/014572170403000119.

[114] Fisher W, Kohut T, Schachner H, Stenger P. Understanding self-monitoring of blood glucose among individuals with type 1 and type 2 diabetes. The Diabetes Educator 2011;37:85–94. https://doi.org/10.1177/0145721710391479.

[115] Patton S. Adherence to glycemic monitoring in diabetes. Journal of Diabetes Science and Technology 2015;9:668–75. https://doi.org/10.1177/1932296814567709.

[116] Heinemann L. Finger pricking and pain: a never ending story. Journal of Diabetes Science and Technology 2008;2:919–21. https://doi.org/10.1177/193229680800200526.
[117] Tanaka N, Yabe D, Murotani K, Ueno S, Kuwata H, Hamamoto Y, et al. Mental distress and health-related quality of life among type 1 and type 2 diabetes patients using self-monitoring of blood glucose: a cross-sectional questionnaire study in Japan. Journal of Diabetes Investigation 2018;9:1203–11. https://doi.org/10.1111/jdi.12827.
[118] Franciosi M, Pellegrini F, De Berardis G, Belfiglio M, Cavaliere D, Di Nardo B, et al. The impact of blood glucose self-monitoring on metabolic control and quality of life in type 2 diabetic patients: an urgent need for better educational strategies. Diabetes Care 2001;24:1870–7. https://doi.org/10.2337/diacare.24.11.1870.
[119] Clar C, Barnard K, Cummins E, Royle P, Waugh N. Self-monitoring of blood glucose in type 2 diabetes: systematic review. Health Technology Assessment 2010;14. https://doi.org/10.3310/hta14120.
[120] Bergenstal R, Ahmann A, Bailey T, Beck R, Bissen J, Buckingham B, et al. Recommendations for standardizing glucose reporting and analysis to optimize clinical decision making in diabetes: the ambulatory glucose profile. Journal of Diabetes Science and Technology 2013;7:562–78. https://doi.org/10.1177/193229681300700234.
[121] Klonoff D, Parkes J, Kovatchev B, Kerr D, Bevier W, Brazg R, et al. Investigation of the accuracy of 18 marketed blood glucose monitors. Diabetes Care 2018;41:1681–8. https://doi.org/10.2337/dc17-1960.
[122] Klonoff D, Blonde L, Cembrowski G, Chacra A, Charpentier G, Colagiuri S, et al. Consensus report: the current role of self-monitoring of blood glucose in non-insulin-treated type 2 diabetes. Journal of Diabetes Science and Technology 2011;5:1529–48. https://doi.org/10.1177/193229681100500630.
[123] Erbach M, Freckmann G, Hinzmann R, Kulzer B, Ziegler R, Heinemann L, et al. Interferences and limitations in blood glucose self-testing. Journal of Diabetes Science and Technology 2016;10:1161–8. https://doi.org/10.1177/1932296816641433.
[124] Freckmann G, Baumstark A, Jendrike N, Rittmeyer D, Pleus S, Haug C. Accuracy evaluation of four blood glucose monitoring systems in the hands of intended users and trained personnel based on ISO 15197 requirements. Diabetes Technology and Therapeutics 2017;19:246–54. https://doi.org/10.1089/dia.2016.0341.
[125] Boettcher C, Dost A, Wudy S, Flechtner-Mors M, Borkenstein M, Schiel R, et al. Accuracy of blood glucose meters for self-monitoring affects glucose control and hypoglycemia rate in children and adolescents with type 1 diabetes. Diabetes Technology and Therapeutics 2015;17:275–82. https://doi.org/10.1089/dia.2014.0262.
[126] Breton M, Kovatchev B. Impact of blood glucose self-monitoring errors on glucose variability, risk for hypoglycemia, and average glucose control in type 1 diabetes: an in silico study. Journal of Diabetes Science and Technology 2010;4:562–70. https://doi.org/10.1177/193229681000400309.
[127] Campos-Náñez E, Fortwaengler K, Breton M. Clinical impact of blood glucose monitoring accuracy: an in-silico study. Journal of Diabetes Science and Technology 2017; 11:1187–95. https://doi.org/10.1177/1932296817710474.
[128] Kovatchev B, Cobelli C. Glucose variability: timing, risk analysis, and relationship to hypoglycemia in diabetes. Diabetes Care 2016;39:502–10. https://doi.org/10.2337/dc15-2035.
[129] Bergenstal R. Continuous glucose monitoring: transforming diabetes management step by step. The Lancet 2018;391:1334–6. https://doi.org/10.1016/s0140-6736(18)30290-3.

[130] Price D, Walker T. The rationale for continuous glucose monitoring-based diabetes treatment decisions and non-adjunctive continuous glucose monitoring use. European Endocrinology 2016;12:24. https://doi.org/10.17925/ee.2016.12.01.24.

[131] Dovč K, Bratina N, Battelino T. A new horizon for glucose monitoring. Hormone Research in Paediatrics 2015;83:149–56. https://doi.org/10.1159/000368924.

[132] Charleer S, Mathieu C, Nobels F, De Block C, Radermecker R, Hermans M, et al. Effect of continuous glucose monitoring on glycemic control, acute admissions, and quality of life: a real-world study. Journal of Clinical Endocrinology and Metabolism 2018;103:1224–32. https://doi.org/10.1210/jc.2017-02498.

[133] Rodbard D. Continuous glucose monitoring: a review of recent studies demonstrating improved glycemic outcomes. Diabetes Technology and Therapeutics 2017;19. https://doi.org/10.1089/dia.2017.0035. S-25-S-37.

[134] Battelino T. Continuous glucose monitoring efficacy in routine use. Journal of Clinical Endocrinology and Metabolism 2018;103:2414–6. https://doi.org/10.1210/jc.2018-00275.

[135] Aleppo G, Laffel L, Ahmann A, Hirsch I, Kruger D, Peters A, et al. A practical approach to using trend arrows on the dexcom G5 CGM system for the management of adults with diabetes. Journal of the Endocrine Society 2017;1:1445–60. https://doi.org/10.1210/js.2017-00388.

CHAPTER

Analytical performance of SMBG systems

2

David Ahn, MD [1], David Klonoff, MD, FACP, FRCP (Edin), Fellow AIMBE [2]

[1]*Program Director, Mary & Dick Allen Diabetes Center, Hoag Memorial Hospital Presbyterian, Newport Beach, CA, United States;* [2]*Medical Director, Diabetes Research Institute, Mills-Peninsula Medical Center, San Mateo, CA, United States*

Introduction

The foundation for the current optimal management of diabetes mellitus centers heavily on self-monitoring of blood glucose (SMBG) using blood glucose meters (BGMs) [1]. The overwhelming majority of people living with type 1 diabetes and type 2 diabetes still utilize SMBG, and even alternative methods of glucose monitoring such as continuous glucose monitoring still benefit from calibration with traditional BGMs. Therefore the analytical performance and accuracy of SMBG systems are of the utmost importance because substandard performance would pose a risk to patients that rely on this information for (1) making treatment decisions about insulin dosing and taking other actions based on the blood glucose level; (2) detecting critical events such as severe hypoglycemia and hyperglycemia; (3) calibrating continuous glucose monitors that can inform treatment decisions; and (4) controlling closed-loop insulin delivery systems that depend on accurate calibration of continuous glucose monitors [2].

Because of the important clinical role that SMBG plays, standards have been created and utilized by governing bodies such as the United States' Food and Drug Administration (FDA) [3] and the International Organization for Standardization (ISO) [4] to regulate and clear BGMs intended for over-the-counter commercial release in their respective territories. Despite these performance standards being consistently used to clear certain meters before market release, multiple studies have demonstrated that many of these meters, after they have been cleared, perform below the same international or FDA standards [5–7].

As a result of these concerning findings, a panel of experts in clinical chemistry, clinical diabetes, and regulatory science created a consensus protocol for the Diabetes Technology Society (DTS)-BGM Surveillance Program [8] to provide an independent assessment of BGM accuracy after receiving clearance by the FDA. The study was the largest accuracy study exclusively testing FDA-cleared BGM systems ever reported. Their findings were published in August 2018 [9].

Glucose Monitoring Devices. https://doi.org/10.1016/B978-0-12-816714-4.00002-8

The process for premarket approval of SMBG devices

In Europe, medical devices such as SMBG meters require a Conformite Europeenne mark before they can be cleared for market release. As part of the application process, the BGM manufacturer asserts that their product fulfills established standards with respect to health protection, safety, and environmental protection. The ISO created the standard ISO 15197 in 2003 [10] for SMBG systems intended to be used by people with diabetes for therapy adjustments. This standard included a requirement for analytical performance evaluation. In 2013 [4], this ISO 15197 standard was revised to require more stringent accuracy criteria. This revision went into effect in Europe in 2016.

Similarly, in the United States, the FDA oversees the premarket certification process for medical device clearance, in addition to the postmarketing surveillance after product clearance. In the premarket certification process for blood glucose monitoring systems, under section 510(k) of the FD&C Act, a manufacturer must submit a 510(k) application to FDA at least 90 days before commercially releasing a new product. During this period, the FDA will determine whether the BGM meets the criteria for Class II market clearance based on whether the device is substantially equivalent to a legally marketed (predicate) device. Once determined to be Class II, the meter is subject to the same requirements as predicate devices. The manufacturer cannot commercialize the device until this clearance takes place.

As Class II devices, new glucose meters must demonstrate that they comply with the standards set forth by the FDA for SMBG over-the-counter use. Before 2014, the FDA also utilized the ISO 15197 2003 standard before introducing their own draft guidance in 2014 for personal SMBG (over-the-counter) BGMs, which was replaced by final guidance in 2016. The FDA's guidance employs more rigorous performance standards than the ISO 15197 2013 standard being used in Europe. Of note, on November 30, 2018, the FDA released a new draft of a guidance document for BGM accuracy in point of care (POC) and SMBG (over-the-counter) devices, with an open review period for comments until February 28, 2019 [11]. This new draft did not change the major accuracy thresholds or study expectations. For the purpose of this chapter, we will reference the 2016 guidelines as the current FDA guidelines.

Analytical performance according to ISO 15197

When evaluating the analytical performance of an SMBG system for personal use, a device is assumed to being properly handled by well-trained personnel. The original 2003 version of the ISO 15197 standard evaluated measurement precision and system accuracy. In 2013, the newer update to the ISO 15197 standard required that analytical performance be evaluated for no less than three test strip lots, rather than the previous minimum of one test strip lot. The 2013 update also added an evaluation of SMBG performance in the presence of potentially interfering endogenous (e.g., hematocrit variability) or exogenous (e.g., acetaminophen) substances and conditions (e.g., ketoacidosis) and an evaluation of the stability of reagents and materials, but these changes only minimally affect the analytical performance of BGMs.

Precision

This term represents how closely and consistently repeated measurements performed by SMBG on a given sample align with each other. According to the latest ISO 15197 2013 guidelines, testing must demonstrate repeatability on at least five venous blood samples with defined glucose concentrations in the hypoglycemic, euglycemic, and hyperglycemic ranges. These guidelines also assess *intermediate measurement* precision by requiring daily measurements over 10 days.

Interference evaluation

The 2013 update to ISO 15197 introduced the evaluation of performance in the presence of various hematocrit ranges and other possible interfering substances that might be found in the blood such as medications or physiologically occurring substances. For hematocrit, the BGM must be tested in five different hematocrit ranges at each of three defined glucose concentrations. Other interfering substances must be tested for at least two defined glucose concentrations. In each of these different testing environments, the ISO 15197:2013 standard allows for up to: 1) 10 mg/dL difference between the test sample and control sample for glucose concentrations $\leq$ 100 mg/dL and 2) 10% difference for glucose concentrations $>$100 mg/dL.

Accuracy

Accuracy refers to how closely the BGM system's measurements align with reference values obtained by a gold standard technique. ISO 15197:2013 requires that at least 95% of measurements fall within $\pm$15 mg/dL of the reference value for blood glucose (BG) concentrations $<$100 mg/dL and within 15% at BG concentrations $\geq$ 100 mg/dL. The previous 2003 version was less rigorous, requiring $\pm$15 mg/dL of reference for BG concentrations $<$75 mg/dL and $\pm$20% for concentrations $\geq$ 75 mg/dL.

In addition, the 2013 update introduced the requirement that at least 99% of measurement results fall within the consensus error grid (CEG) zones A and B [12]. The CEG has five zones, characterized by different clinical risks to the patient. By emphasizing that 99% of results fall in zones A (no effect on clinical action) and B (little or no effect on the clinical outcome), this addition reduces the maximum amount of *clinically* unacceptable results to 1%, along with the maximum amount of *analytically* unacceptable results to 5%.

Furthermore, the 2013 update increased the minimum number of test strips lots being evaluated for accuracy from one to three, allowing for a more comprehensive assessment to compensate for lot-to-lot variability in performance, which has been shown to significantly affect the accuracy of a system [13].

Analytical performance according to FDA

In the United States, the FDA oversees and regulates SMBG systems and blood glucose monitoring test systems for prescription POC use. FDA regulates these for all phases of market release: premarket notification for market approval, manufacturing and performance standards, and postmarketing surveillance. Before 2014, the FDA utilized the ISO 15197:2003 standard for approving glucose meters in the United States. The FDA introduced draft guidance in 2014 and later finalized this guidance in 2016, implementing its own standard for BGMs for self-monitoring that was similar to the ISO 15197:2013. Similar to its modern ISO counterpart, the 2016 FDA guidelines require 95% of measurements to fall within ±15% of the reference BG concentration for BG values ≥ 100 mg/dL. However, the FDA also requires this same level of performance (±15%) for BG values across the entire range, as opposed to the more forgiving ISO 2013 criteria of an absolute amount of ±15 mg/dL for BG values < 100 mg/dL. It is important to note that the FDA criteria are especially stringent for the severe hypoglycemic range. For example, 15% of 54 mg/dL only allows for a variance of 8 mg/dL. The accuracy requirements are unchanged in the latest 2018 draft guidance. A summary of the various analytical accuracy requirements are found in Table 2.1.

Similar to ISO 15197:2013, the latest 2016 and 2018 FDA guidelines require daily precision testing over 10 days. The 2016 guidance requires that a minimum of three test strip lots be scrutinized, but the 2018 draft guidance proposes a reduction to only two test strip lots.

Table 2.1 International and food and drug administration analytical accuracy requirements for blood glucose monitors for self-monitoring.

Accuracy standard	Minimum accuracy requirements
ISO 15197 2013	95% of glucose values must be within −15 mg/dL of reference for values <75 mg/dL or within 20% of reference for values ±75 mg/dL
ISO 15197 2016	95% of glucose values must be within −15 mg/dL of reference for values <100 mg/dL or within 15% of reference for values ±100 mg/dL
FDA 2016 for over-the-counter BGMs (FDA 2016)	95% of glucose values must be within 15% of reference, and 99% of glucose values must be within 20% of reference
FDA 2018 Draft Guidance for over-the-counter BGMs (FDA 2018)	Same as FDA 2016

Postmarket analytical performance

Despite the rigorous process that manufacturers must endure to obtain initial marketing clearance for their SMBG devices by regulatory agencies, the postmarketing surveillance of such devices has not been mandated because the 510(k) process has very little room for product recall unless there is proof that the manufacturer falsified data when obtaining clearance or there is proof that the product ingredients are being adulterated following clearance. In the absence of surveillance programs, independent studies have repeatedly demonstrated that many FDA-cleared systems do not meet the latest ISO or FDA guidelines, and some do not even meet the less strict ISO 15197:2003 guideline.

A 2018 review by King and colleagues analyzed 58 studies published between January 1, 2010 and July 12, 2017 that assessed the accuracy of 143 BGMs, 59 of which had been cleared by the FDA [14]. The number of BGMs evaluated per study was 4.86. In addition to assessing whether the studies demonstrated that the evaluated BGMs met ISO 15197:2003 or ISO 15197:2013 guidelines, the article also stratified the results by FDA approval status, year of market clearance, and whether the study was supported by device manufacturers (as opposed to independently supported).

The review article discovered that newer meters outperformed older meters and that FDA-approved meters outperformed non-FDA approved meters. However, out of the 59 FDA approved BGMs, only 75.4% managed to pass the older ISO 15197 2003 standard. When evaluated according to the newer ISO 15197 2013 standard, only 46.4% of FDA-cleared BGMs passed the test. The authors concluded that even fewer would pass the most recent 2016 FDA guidelines because they are even more rigorous than the ISO 15197 2013 standard.

The analysis also demonstrated that FDA-cleared BGMs assessed in manufacturer-supported studies were statistically significantly more likely to pass the requirements set forth in ISO 15197 2003 (OR = 22.4, CI 8.73–21.6, $P < .001$) and ISO 15197 2013 (OR = 23.08, CI 10.2–60.0, $P < .001$), when compared with independent studies. Furthermore, manufacturer-supported studies were discovered to be statistically significantly more likely to meet ISO 15197 2003 (OR = 37.2, CI 4.62–299, $P < .01$) and ISO 15197 2013 (OR = 56.1, CI 11.9–263, $P < .01$) criteria when testing their own product versus when testing a competitor's product.

Prompted by these concerning studies, the Diabetes Technology Society gathered a panel of experts in clinical chemistry, clinical diabetes management, and regulatory affairs to create a consensus protocol for the DTS-BGM Surveillance Program that would independently assess the BGM accuracy of systems postmarket release through a surveillance program. Findings from this study were published in 2018 [9] that evaluated 18 BGM systems that had been cleared by the FDA. These 18 systems represented approximately 90% of all commercially available systems used from 2013 to 2015. This study included 1035 subjects across three clinical sites and only used testing supplies that were obtained from consumer outlets (as opposed to being supplied by the manufacturer) to best assess real-world performance. Because the 2016 FDA guideline had not yet been formally adopted at the time of study design, this study evaluated meter analytical performance against the then-latest widely accepted BGM accuracy standard, which was the ISO 15197:2013 guideline.

To further reduce the risk of bias, the study was triple-blinded such that in-clinic BGM operators did not know plasma reference measurements, the reference lab did not know BGM readings, and the data analysis team worked only with BGMs designated by a code number that blinded the identity of the BGMs whose accuracy was being assessed.

Every subject had a capillary BG level reading measured by six different BGM devices (in random order), three before and three after a simultaneous reference capillary sampling was performed. Each of the three sites tested all 18 FDA-approved BGM systems by testing a new set of six BGMs across three separate studies in a round-robin fashion. 1032 of 1035 subjects completed the study, and each BGM was tested on 115 subjects, on average. The average number of strip lots per study was 2.1.

Of the 18 meters that were evaluated, only six (33%) met the predetermined accuracy standard in all three studies; five met it in only two studies; two met it in only one study. This left four FDA-cleared BGMs that did not meet the ISO 15197:2013 standard in any of the three studies.

In November 2018, the FDA released a statement promising to establish an alternative 510(k) approval "Safety and Performance Based Pathway" evaluated on objective safety and performance criteria rather than on a predicate device [15]. This new approval process could potentially give the FDA more flexibility to respond swiftly to safety concerns that arise and perform postmarket surveillance through programs such as the recently announced National Evaluation System for Health Technology [16].

Advances in analytical performance of SMBG devices

The previously mentioned 2018 review article [14] determined that newer BGMs were statistically significantly more likely to meet both the ISO 15197 2003 (OR 1.06, CI 1.06−1.07, $P < .001$, odds ratio reported for a 1-year period) and ISO 15197 2013 guidelines (OR = 1.06, CI 1.03−1.06, $P < .01$, odds ratio reported for a 1-year period). Which advances in technology best correlate temporally with this improved analytical performance in more recently developed BGMs?

Pfutzner proposed that several technological advances in BGMs contribute to the improved analytical performance in more recently developed BGMs [17]. First, device manufacturers have been able to eliminate hematocrit interference by using mathematical correction algorithms that have been refined over the course of time. In addition, changes in test strip chemistry, changes in electrode design, and other algorithmic improvements have contributed to improved accuracy in newer meters.

When analyzing the test strip enzyme in meters over time and their analytical performance, King and colleagues demonstrated that glucose dehydrogenase

(GDH)-based test strips outperformed those using glucose oxidase (GOX) for ISO 15197 2003 ($OR = 0.23$, CI 0.13–0.38, $P < .001$) and ISO 15197 2013 ($OR = 0.24$, CI 0.15–0.35, $P < .001$). The GDH enzyme, different from GOX, allows BGMs to not depend on the partial pressure of oxygen in the blood [18]. When the investigators looked deeper and compared different enzymatic cofactors for GDH, the review did not find a significant difference in accuracy.

Conclusion

The recently published findings from the DTS-BGM Surveillance Program revealed that only 6 of 18 (33%) FDA-cleared meters met the protocol-specified accuracy standard similar to current ISO and FDA guidelines for analytical performance in all three separate tests. As these 18 meters represented 90% of all commercially used meters from 2013 to 2015, many patients are depending on BGM systems that have been shown to fall short of current standards in what was the largest study ever conducted of exclusively FDA-cleared BG monitors using a protocol that was developed with input from the FDA. Contributing to the problem, the FDA lacks resources to continuously monitor manufacturing processes, many of which take place internationally. Furthermore, Medicare mail-order suppliers offer only a limited selection of brands of BGM systems and supplies. The latest Office of the Inspector General report [19] revealed that 78.8% of the SMBG systems sold to Medicare beneficiaries via mail order in Q4 2016 were made by manufacturers of systems that did not meet current standards.

Given the significant clinical role that BGMs play in the day-to-day management of diabetes, inadequate performance and accuracy may increase the risk of both hypoglycemia and hyperglycemia in patients, either of which can be costly on both a financial and clinical scale. Therefore, diabetes stakeholders such as regulators, patients, providers, and payers must seriously consider the implications of the brands of BGMs that they authorize and recommend, in light of these devices' analytical performance.

List of authors

David Ahn, MD is an endocrinologist and serves as Program Director of the Mary and Dick Allen Diabetes Center for Hoag Memorial Hospital Presbyterian in Newport Beach, California. At this institution, he is the recipient of the Dr. Kris V Iyer Endowed Chair in Diabetes Care.

David Klonoff, MD is an endocrinologist specializing in the development and use of diabetes technology. He is Medical Director of the Dorothy L. and James E. Frank Diabetes Research Institute of Mills-Peninsula Medical Center in San Mateo, California and a Clinical Professor of Medicine at UCSF.

References

[1] Choudhary P, Genovese S, Reach G. Blood glucose pattern management in diabetes: creating order from disorder. Journal of Diabetes Science and Technology 2013;7: 1575–84.

[2] Breton MD, Kovatchev BP. Impact of blood glucose self-monitoring errors on glucose variability, risk for hypoglycemia, and average glucose control in type 1 diabetes: an in silico study. Journal of Diabetes Science and Technology 2010;4:562–70.

[3] U.S. Department of Health and Human Services, U.S. Food and Drug Administration. Self-monitoring blood glucose test systems for over-the-counter use: guidance for industry and food and drug administration staff [Internet]. 2016. Available from: https://www.fda.gov/downloads/ucm380327. pdf. [Accessed 23 November 2018].

[4] International Organization for Standardization. In vitro diagnostic test systems—requirements for blood-glucose monitoring systems for self-testing in managing diabetes mellitus. ISO 15197. 2013 (E).

[5] Klonoff DC, Prahalad P. Performance of cleared blood glucose monitors. Journal of Diabetes Science and Technology 2015;9:895–910.

[6] Freckmann G, Baumstark A, Pleus S. Do the new FDA guidance documents help improving performance of blood glucose monitoring systems compared with ISO 15197? Journal of Diabetes Science and Technology 2017;11:1240–6.

[7] Ekhlaspour L, Mondesir D, Lautsch N, et al. Comparative accuracy of 17 point-of-care glucose meters. Journal of Diabetes Science and Technology 2017;11:558–66.

[8] Klonoff DC, Lias C, Beck S, et al. Development of the diabetes technology society blood glucose monitor system surveillance protocol. Journal of Diabetes Science and Technology 2016;10:697–707.

[9] Klonoff DC, Parkes JL, et al. Investigation of the accuracy of 18 marketed blood glucose monitors. Diabetes Care August 2018;41(8):1681–8.

[10] International Organization for Standardization. In vitro diagnostic test systems—requirements for blood-glucose monitoring systems for self-testing in managing diabetes mellitus. EN ISO 15197. 2003. E.

[11] U.S. Department of Health and Human Services, U.S. Food and Drug Administration. Self-monitoring blood glucose test systems for over-the-counter use: draft guidance for industry and food and drug administration staff [Internet]. 2018. Available from: https://www.fda.gov/downloads/UCM626742.pdf. [Accessed 30 December 2018].

[12] Parkes JL, Slatin SL, Pardo S, Ginsberg BH. A new consensus error grid to evaluate the clinical significance of inaccuracies in the measurement of blood glucose. Diabetes Care 2000;23:1143–8.

[13] Baumstark A, Pleus S, Schmid C, Link M, Haug C, Freckmann G. Lot-to-lot variability of test strips and accuracy assessment of systems for self-monitoring of bloo glucose according to ISO 15197. Journal of Diabetes Science and Technology 2012;6(5): 1076–86.

[14] King F, Ahn D, Hsiao V, Porco T, Klonoff DC. A review of blood glucose monitor accuracy. Diabetes Technology and Therapeutics December 2018;20(12):843–56.

[15] U.S. Department of Health and Human Services, U.S. Food and Drug Administration, Statement from FDA Commissioner, Gottlieb S, Shuren J. Director of the Center for Devices and Radiological Health, on transformative new steps to modernize FDA's 510(k) program to advance the review of the safety and effectiveness of medical devices. 2018. Available from: https://www.fda.gov/NewsEvents/Newsroom/PressAnnouncements/ucm626572.htm. [Accessed 30 December 2018].

[16] U.S. Department of Health and Human Services, U.S. Food and Drug Administration, Statement from FDA Commissioner, Gottlieb S, Shuren J. Director of the center for devices and radiological health, on FDA's updates to medical device safety action plan to enhance post-market safety. 2018. Available from: https://www.fda.gov/NewsEvents/Newsroom/PressAnnouncements/ucm626286.htm. [Accessed 30 December 2018].

[17] Pfutzner A. Advances in patient self-monitoring of blood glucose. Journal of Diabetes Science and Technology 2016;10:101–3.

[18] Baumstark A, Schmid C, Pleus S, et al. Influence of partial pressure of oxygen in blood samples on measurement performance in glucose-oxidase-based systems for self- monitoring of blood glucose. Journal of Diabetes Science and Technology 2013;7:1513–21.

[19] U.S. Department of Health and Human Services, Office of Inspector General. Medicare market shares of mail-order diabetes test strips from October through December 2016. 2017. Available from: https://oig.hhs.gov/oei/reports/oei-04-16-00473.pdf. [Accessed 23 November 2018].

CHAPTER

Clinical evaluation of SMBG systems

3

William L. Clarke, MD
Profesor, Emeritus of Pediatric Endocrinology, Department of Pediatrics, University of Virginia, Charlottesville, VA, United States

Systems for self-monitoring of blood glucose (SMBG) can be assessed in terms of their analytical or clinical accuracy. Analytical accuracy is a method to quantitatively assess how close the SMBG measurement is from a reference value; clinical accuracy qualitatively describes the clinical outcome of making treatment decisions using the measurements provided by the SMBG device. Error grid analysis (EGA) is a tool largely used to evaluate SMBG systems in terms of their clinical accuracy. EGA was originally designed as a metric to describe the clinical accuracy of blood glucose (BG) estimates by subjects with type 1 diabetes (T1D) participating in a research project where they estimated their BG level and then immediately performed SMBG [1]. Conventional statistical analyses of individual subjects' data revealed significant correlations between estimates and measurements but did not address the type of error made nor the clinical significance of estimation errors. Similarly, calculating percent deviation between estimates and actual BG levels (i.e., examining their analytical accuracy) failed to address these issues. In fact, a 30% deviation between an estimated BG of 120 mg/dL and the corresponding actual BG of 156 mg/dL would not be as clinically significant as a 30% deviation of an estimate of 80 mg/dL when the actual BG is 56 mg/dL. Thus a different type of analysis of the clinical implications of differences between estimated and actual BG levels was needed. EGA takes into account (1) the actual BG value, (2) the absolute difference between the estimate and the actual BG, and (3) the clinical implications of self-treatment decisions based on the estimated BG. EGA is always based on the individual's therapeutic or target BG range and assumes that treatment of BG values outside the treatment range will always occur.

The error grid (Fig. 3.1) is a two-dimensional graph divided into five zones of clinical accuracy. Zones A and B are considered clinically accurate and acceptable, respectively. Zones C, D, and E are clinically inaccurate or unacceptable. In zone A, estimated BG values fall within 20% of the reference BG (SMBG or laboratory determined BG) or are 70 mg/dL or less when the reference BG is also 70 mg/dL or less. The estimated BG values ("benign errors") in zone B are greater than 20% different from the reference, but would most likely not result in dangerous clinical decisions, that is, hypoglycemia or severe hyperglycemia. Zone C estimates are designated "overcorrection errors" because treatment decisions based on these

Glucose Monitoring Devices. https://doi.org/10.1016/B978-0-12-816714-4.00003-X

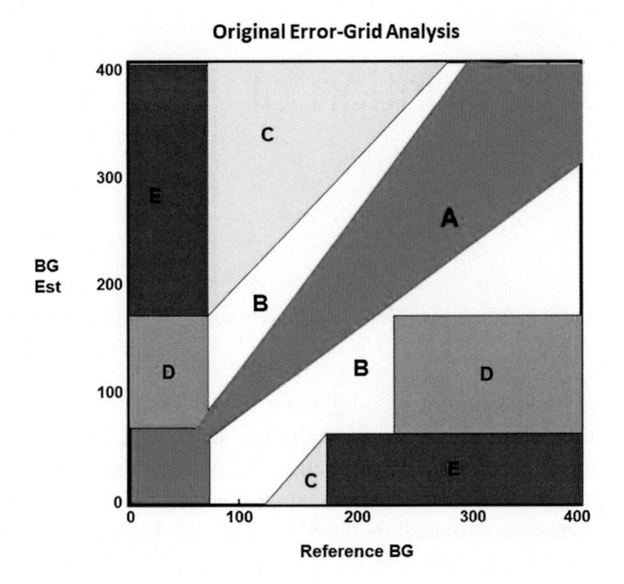

FIGURE 3.1

Error grid—zone A—clinically accurate, zone B—benign errors, zones C—overcorrection errors, zone D—failure to detect and treat errors, zones E—erroneous errors.

values might result in BG values outside of the intended therapeutic range. Zone D estimates are considered "dangerous failure to detect and treat" errors: the estimated BG is within the therapeutic range while the actual BG is either above or below that range; treatment is needed but would be omitted. Finally, zones E values are considered "erroneous treatment" errors. Estimates in these zones would suggest the need for a treatment exactly opposite to that needed, that is, giving insulin for a BG less than 70 mg/dL or giving glucose for a BG greater than 180 mg/dL.

Even before the results of the Diabetes Complications and Control Trial were known, patient-determined SMBG was becoming a routine management tool among those persons injecting insulin to manage their disease. The commercial development of self-blood glucose monitors began to grow significantly and numerous descriptions of their user-friendly attributes and statistical accuracy began to appear in the scientific literature. Early reports of accuracy (SMBG values vs. Reference values) included correlation coefficients, linear regression equations, percent deviation, precision indexes, and standard error of measurements, but no analyses of clinical accuracy [2,3]. The application of EGA to previously published statistical analyses demonstrated that clinical accuracy did not always correlate with the statistical data. In fact, no SMBG system whose data were examined using EGA produced a high percentage of clinically inaccurate or unacceptable BG readings [3]. The most common errors in these early studies (less than 5%) were zone D failures to detect glucose values outside the treatment range. Reporting clinical accuracy in performance trials of SMBG systems using EGA became more widely used and was considered a "gold standard" in industry supported publications. A recent review of analytical and clinical performance of SMBG systems revealed that accuracy was assessed most commonly using statistical bias, Bland-Altman plots, and EGA [4]. Healthcare professionals who had become familiar with EGA began using the term "clinical implications" to discuss the differences between patient SMBG and reference BG levels with their patients and colleagues.

As EGA is based on a specific treatment range, it is useful to consider the effects of different treatment goals or target ranges on clinical accuracy. For instance, some clinicians may prescribe a lower and narrower target range, 60–120 mg/dL, for obstetrical patients with T1D. This change would result in an increase in the zone B clinically acceptable range, but because the target range has been decreased, the zone D "failure to detect and treat" ranges would also decrease at the expense of an increase in the zone C "overcorrection" ranges (Fig. 3.2A). Using a target range of 100–200 mg/dL, as might be prescribed for an elderly patient, would increase the zone A clinically accurate range while decreasing the zone B clinically acceptable range. The zone D, "failure to detect and treat" range would increase as would the zone E "erroneous treatment" range (Fig. 3.2B). Current FDA guidelines for approval of new SMBG devices require that 95% and 99% of SMBG measurements be within 15% and 20% of reference values [5]. EGA based on these criteria would result in a narrower clinically accurate zone A and larger clinically acceptable zone B. Zones C, D, and E would remain unchanged unless the treatment range was changed.

Parkes et al. developed a different error grid whose zones of clinical accuracy were created by asking a random group of 100 physicians at a medical conference to characterize SMBG versus reference values based on their assumptions of risk [6]. The risk categories were similar to, but not exactly the same as, those of the EGA.

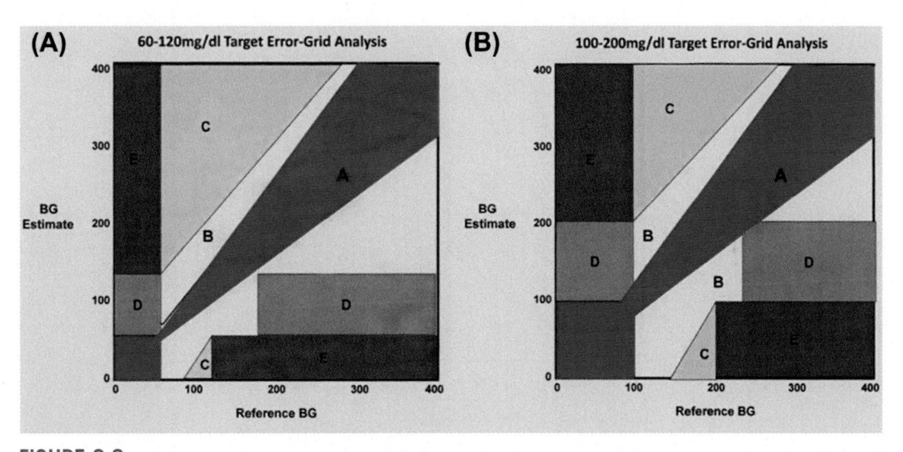

FIGURE 3.2

(A) Error grid constructed for target BG range of 60–120 mg/dL. (B) Error grid constructed for target BG range of 100–200 mg/dL.

They were as follows: zone A—no effect on clinical action; zone B—altered clinical action with little to no effect on clinical outcome; zone C—altered clinical action likely to affect clinical outcome; zone D—altered clinical action with significant clinical risk; zone E—altered clinical action with potentially dangerous consequences. Of interest, no lower zone E, that is, SMBG measurement in the low BG range when the reference was above 250 mg/dL, was created: treatment of an elevated BG with rapid-acting carbohydrate was not considered to be a dangerous clinical decision. The zones fan out from hypoglycemia, designated as BG less than 50 mg/dL, without regard to a clinical target or treatment range (Fig. 3.3). Thus while eliminating some of the discontinuous transitions from one zone to another in EGA (e.g., zone B to zone E without passing through zones C and D in some places on the grid) this consensus error grid (CEG) omits the relationship between accuracy and a specific treatment or target BG range. For example, EGA zone D "failures to detect and treat" reference BG values outside the target range are designated "altered clinical action—likely to affect clinical outcome" on the CEG.

Both EGA and the CEG are widely used to express the clinical accuracy of SMBG systems, but they have not been part of the performance data required by regulatory agencies to support applications for approval of new devices, nor have they been used by regulatory agencies and SMBG manufacturers to assess the clinical

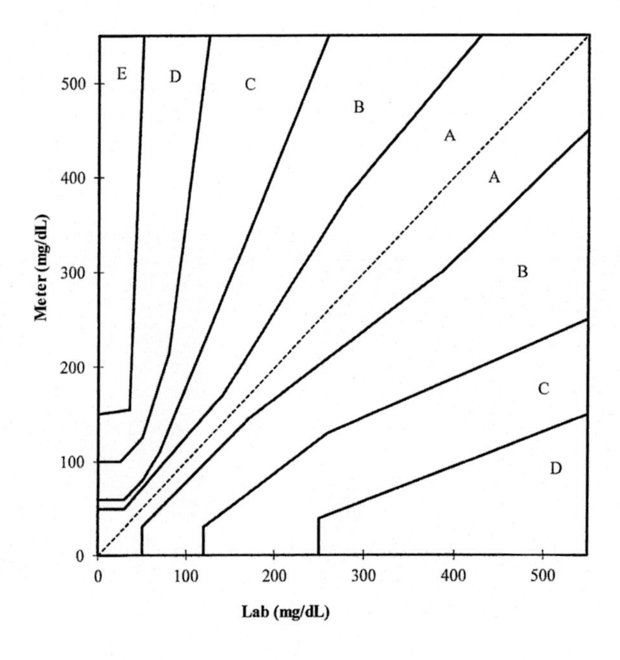

FIGURE 3.3

Consensus error grid.

significance of postmarketing SMBG system-generated clinical errors. To address this need, a group of representatives from FDA, NIH, academia, US Army, and industry convened to develop a new error grid, the surveillance error grid (SEG), which would assess the degree of clinical risk associated with clinically inaccurate SMBG systems in postmarket decision-making [7]. Diabetes clinicians were recruited to provide their assessments of the clinical risk of errors associated with four different patient scenarios [1]: T1D patient using an insulin pump [2]; type 2 diabetes (T2D) patient using insulin [3]; T2D patient not using insulin [4]; T1D patient using multiple daily injections and continuous glucose monitoring system; and supply a minimum and maximum BG range for each of the following types of clinical action: (A) emergency treatment for low BG; (B) take oral glucose; (C) no action needed; (D) take insulin; (E) emergency treatment of high BG. Scenario [3], that considered a noninsulin-treated patient, had a different description for clinical action (D), which was exercise and eat less. Clinicians were then asked to compare SMBG measurements from each of the five possible BG ranges (A—E) to actual BG concentrations from the same five ranges and select on a scale of 1—9 the magnitude of clinical risk for each possible combination. If the measured BG and the actual BG were in the same range, that square was prefilled as a no-risk outcome. Of note, each respondent used their own glucose ranges for A—E, which they had previously defined. A grid was then created for each respondent, such that each point on the grid represented a data pair consisting of reference glucose on the *x*-axis and measured glucose on the *y*-axis. Each data point was then integrated and averaged for the entire set of respondents such that each data point could be assigned a unique mean score according to the mean perception of clinical risk for that data pair. This calculation generated a clinical risk for each combination of measured and reference BG and a gradual spectrum of risk within each risk zone that was now defined by a range of risk scores. The color-coded final surveillance error grid displays a risk estimate for each SMBG—reference pair (Fig. 3.4). This analysis differs from both the EGA and the CEG where data points within the same risk zone are all assigned the same risk and data pairs that are categorized similarly may differ significantly. Comparisons of the EGA/CEG and the SEG are shown in Fig. 3.5. Using a set of 10,000 simulated reference versus SMBG data pairs, the SEG was shown to correlate significantly with either the EGA or CEG, while the correlation between the EGA and the CEG was stronger. It is unclear whether or not or how the continuous risk information provided by the SEG will be used by clinicians and diabetes educators in patient management. However, the SEG has been used to help identify clinical risk in postmarketing studies of SMBG accuracy [8].

In summary, the clinical accuracy of SMBG systems is critical to the utility of the information generated in the management of patient with diabetes. Understanding the potential risk associated with differences between reference and SMBG results can be of immediate importance to clinical decision-making. Error grids have become a useful way to both visually quantitate and present clinical accuracy.

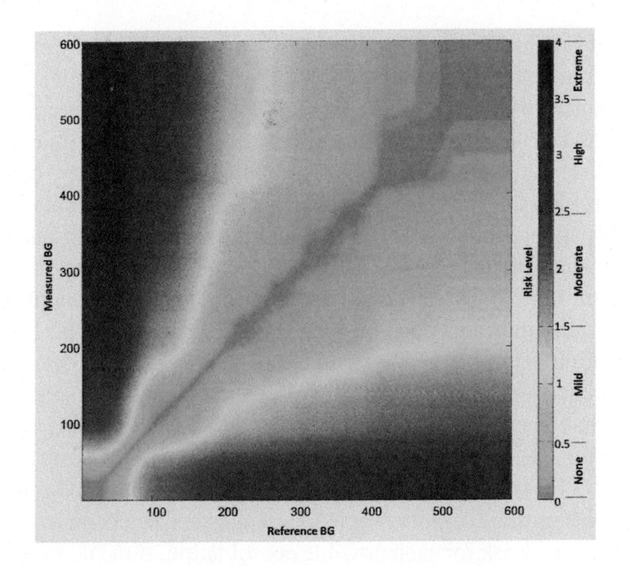

FIGURE 3.4

Surveillance error grid.

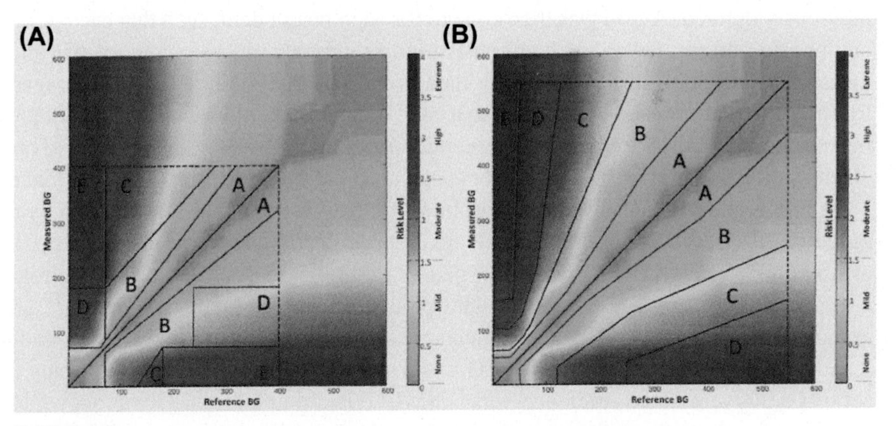

FIGURE 3.5

Surveillance error grids with EGA or CEG superimposed.

References

[1] Cox D, Clarke W, Gonder-Frederick L, Pohl S, Hoover C, Snyder A, Zimbelman L, Carter W. Accuracy of perceiving blood glucose in IDDM patients. Diabetes Care 1985;8:529–36.

[2] Clarke W. The original Clarke error grid analysis (EGA). Diabetes Technology and Therapeutics 2005;7:776–9.

[3] Clarke W, Cox D, Gonder-Frederick L, Carter W, Pohl S. Evaluating the clinical accuracy of self-blood glucose monitoring systems. Diabetes Care 1987;10:622–8.

[4] Boren S, Clarke W. Analytical and clinical performance of blood glucose monitors. Journal of Diabetes Science and Technology 2010;4:1–14.

[5] Self-monitoring blood glucose test systems for over-the-counter use: guidance for Industry and Food and Drug Administration Staff. Document issued on October 11, 2016.

[6] Parkes J, Slatin S, Pardo S, Ginsberg B. A new consensus error grid to evaluate the clinical significance of inaccuracies in the measurement of blood glucose. Diabetes Care 2000;23(8):1143–5.

[7] Klonoff D, Lias C, Vigersky R, Clarke W, Parkes J, Sacks D, Kirkman S, Kovatchev B, the Errror Grid Panel. The surveillance error grid. Journal of Diabetes Science and Technology 2014;8:658–72.

[8] Klonoff D, Parkes J, Kovatchev B, Kerr D, Bevier W, Brazg R, Christiansen M, Bailey T, Nichols J, Kahn M. Investigation of the accuracy of 18 marketed blood glucose monitors. Diabetes Care 2018;41:1681–8.

CHAPTER

Consequences of SMBG systems inaccuracy

4

Enrique Campos-Náñez, PhD [1], Kurt Fortwaengler, PMP [2]

[1] *Principal Algorithm Engineer, Research & Development, Dexcom Inc, Charlottesville, VA, United States;* [2] *Disease Modeling, Global Market Access, Roche Diabetes Care, Mannheim, Germany*

Introduction

Glucose monitoring is critical to the proper management of type 1 and type 2 diabetes. Type 1 patients must monitor their glucose multiple times a day, every day, to avoid serious complications such as severe hypoglycemia and diabetic ketoacidosis [1,2]. Although early type 2 patients may only require the sporadic use of glucose monitoring to achieve good control, as the disease progresses glucose monitoring brings incremental benefits and becomes eventually critical for insulin-dependent type 2 patients [3]. In fact, the proper and frequent use of glucose monitoring will result in enhanced glycemic control [2–5] and will lead to improved quality of life of diabetic patients [3,6].

Among the competing glucose monitoring solutions at the patient's disposal, self-monitoring of blood glucose (SMBG) by means of blood glucose monitoring (BGM) systems is a frequent choice (and sometimes only choice [7]) among type 2 patients and type 1 patients, and the benefits of its frequent use have been documented [1,2]. As patients make multiple decisions throughout the day, every day, and must rely on glucose estimates provided by these devices, many natural questions arise: to what extent do self-treatment decisions based on erroneous measurements result in under or over bolusing, or failure to correct hypoglycemia? Can sustained erroneous decisions compound over time and lead to poor glycemic outcomes and long-term complications? Do these long-term effects also translate into the loss of quality of life and increasing healthcare system costs? Commercially available glucose monitoring devices vary greatly in their technology and performance [8–14]. How do these differences between devices affect the patient's ability to manage their disease? Moreover, regulation and standards have established performance requirements for these systems. To what extent do they succeed? In other words, does compliance with a standard translate in better outcomes for a patient? Finally, when is accuracy enough? Is it clear that in the presence of metabolic, behavioral, and environmental noise a more accurate BGM system will always improve outcomes? Glucose monitoring, being the only source of ground truth for patients, takes a central piece in the management of these chronic diseases.

Glucose Monitoring Devices. https://doi.org/10.1016/B978-0-12-816714-4.00004-1

This work focuses on understanding how the accuracy characteristics of these technologies contribute to the quality of glycemic control.

This chapter surveys current efforts to characterize hypothesized relationships (illustrated in Fig. 4.1) between accuracy and outcomes. As shown in the figure, these range from the clinical to the financial, from the immediate to the long term. The approaches discussed here cover vastly different methodologies from direct clinical observations and surveys, to modeling and simulation-based studies, to meta-analysis. In the end, and despite their methodological differences, these studies deliver a clear and consistent message. Inaccurate BGM systems can negatively affect the patient's glycemic control, quality of life, and finances. Continued use of inaccurate BGM systems may result in an increased prevalence of long-term health complications affecting both the patient's quality of life and the financial health of families. When we switch our focus from an individual to the entire population, these effects aggregate over time to affect worker productivity and healthcare costs.

The chapter is organized as follows. In the first section, we set up the problem of characterizing the effect of accuracy. This requires defining how we measure the inputs (accuracy) and the outputs (clinical, quality of life, financial cost). We discuss the nuances and complexities involved that make this an interesting problem. In Section Quantifying the effect of inaccurate BGM systems, we introduce an approach to the problem based on modeling and simulation, in which we decompose the overall problem into more manageable steps. Finally, in Section Accuracy and its consequences, we present an extended illustration of our approach. The example describes how modeling and simulation can be used to estimate the effects of SMBG accuracy on patients using multiple daily injections (MDI) both in terms of short-term clinical and cost metrics and long-term clinical complications and their cost to the healthcare system using the case of the United Kingdom.

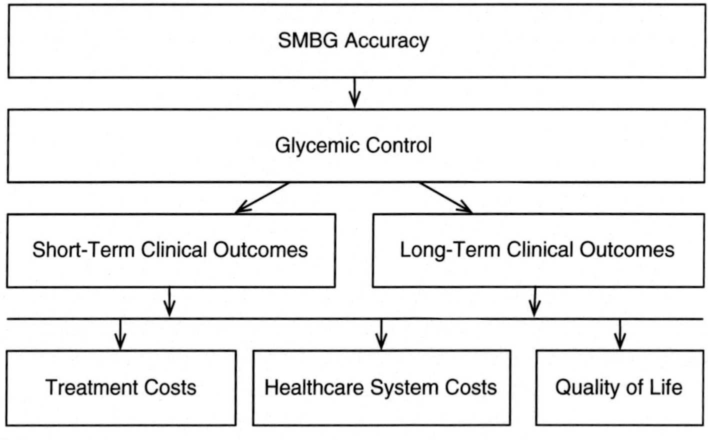

FIGURE 4.1

Short-term and long-term consequences of BGM system accuracy.

Quantifying the effect of inaccurate BGM systems

In this discussion, we focus on SMBG using BGM systems that provide a single estimate of plasma glucose using a direct capillary blood measurement. Although a patient can opt instead for a flash glucose monitor (FGM) device or a continuous glucose monitor (CGM) [15], BGM systems are still prevalent in the market and will be for years to come, particularly if we focus on type 2 diabetes. In many underdeveloped countries, the BGM system technology is the only option for patients [7,16,17]. Furthermore, understanding the effect of BGM system accuracy can be directly applicable to FGM and to CGM devices requiring frequent calibration. In the latter case, the accuracy characteristics of the BGM system used to calibrate will spill over to the CGM device affecting its performance [18,19].

The analytical and clinical accuracy of glucose monitoring devices has been a long-standing preoccupation of the community [1]. Separate efforts have developed analytical [20–22] and clinical [23–25] methods, metrics, and visualization tools. A plethora of BGM system accuracy assessments have become recently available [8,10–14,26,27], some of these conducted after device approval [9]. As a result of these efforts, many recommendations [28–33], and international standards such as ISO 15197 in its 2003 and 2013 versions [34,34] have been produced.

Despite the abundance of information, the study of the relationship between analytical and clinical accuracy and short- and long-term consequences for the patient and the healthcare system is still in its infancy and represents a challenging problem. The reasons for this are multiple and are worth discussing here.

A complex system

Fig. 4.2 represents a simplified system view of self-treatment. To illustrate this, consider the case of severe hypoglycemia. A poor glucose BGM system reading (technology) can lead to over bolus producing a severe hypoglycemia event [35]

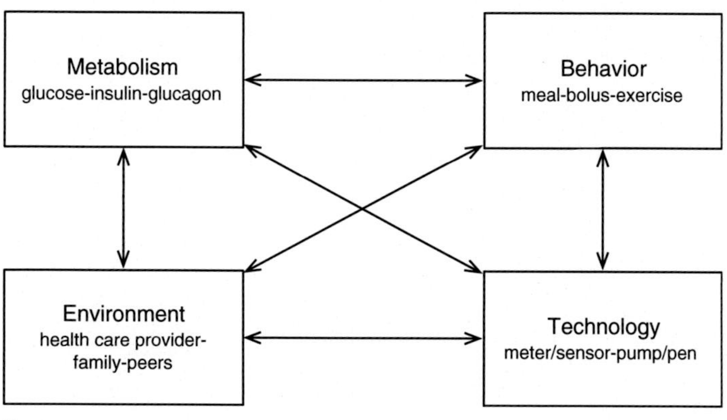

FIGURE 4.2

The diabetes management system.

(metabolism). The patient may develop a fear of hypoglycemia [36–38] and override its therapy parameters (behavior) [39]. Frequent hypoglycemia may also lead to decreased hypoglycemia awareness [40], increased glucose variability, and reduced autonomic response during counterregulation [41]. This is in addition to intraday and interday variability of the patient's metabolism, exercise, stress, and other hormonal changes. A different metabolic state will make the same BGM system error have different effects.

Patient behavior is the main driver

Behavior is a fundamental part of this control system and, in the absence of automated insulin delivery systems, it is the main driver of glycemic control [42,43]. The stated (nominal) goal of this system is to properly control glucose; nevertheless, patients have subtle, adaptive, and context-varying goals, such as reducing stress [44], social pressure [45,46], and time required to manage their disease. Variability in therapy modes, devices, drugs, and, most importantly, the natural physiological variability often require patients to continuously adjust their behavior. Despite efforts to standardize the management of diabetes, from the patient's vintage point diabetes management is still a heavy cognitive challenge requiring frequent adaptation, which involves switching therapy modes, adjusting exercise, food intake, hypoglycemia self-treatment, and insulin dosing strategies. This constantly changing environment makes it hard to isolate and adequately quantify what, if any, is the contribution of BGM system accuracy to outcomes. In Fig. 4.3 we illustrate some of these behaviors, following an approach similar to Ref. [47]. Upon entering

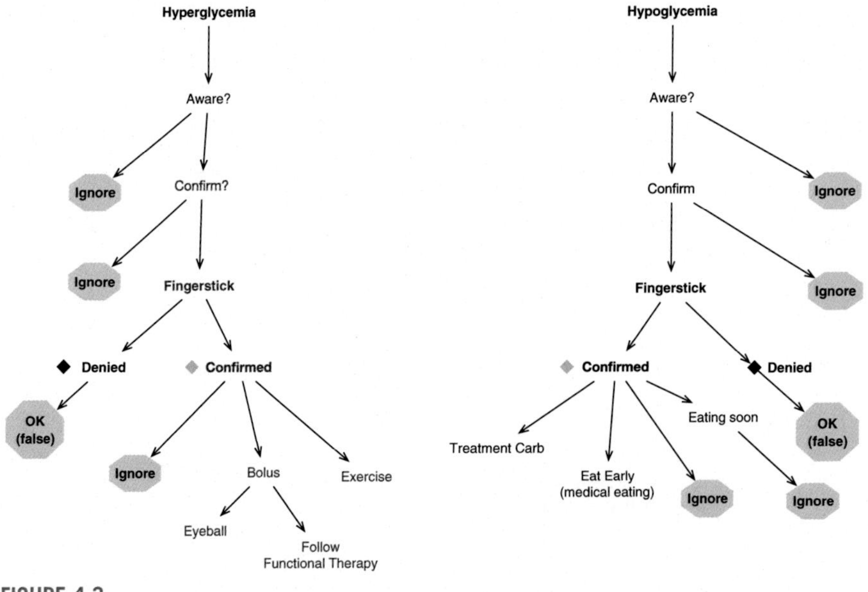

FIGURE 4.3

An example of choices around hypo/hyperglycemia.

into hyperglycemia, a patient may be aware of it (physiology), make a decision about confirming or not hyperglycemia (psychology), and upon confirmation, the patient may opt to ignore or treat that condition (behavior). The patient's treatment may include bolusing or exercising. The impact of these behaviors is very significant and, in some cases, drowns the effect of BGM system noise. More interestingly, in some forms of therapy like continuous subcutaneous insulin infusion (CSII) or sensor-augmented pump, behavioral "noise" may amplify the effect of BGM system noise. For example [42], shows that missing boluses may not only be more impactful than BGM system error but also make accuracy more critical as patients that skip boluses have less opportunities to correct.

Effects spread over time and space

Self-treatment decisions based on inaccurate estimates of a patient's plasma glucose level may result in poor glycemic control. A BGM system that incorrectly reports a high blood glucose level may drive a patient to hypoglycemia by administering an unnecessary or unnecessarily large bolus (over bolus) [18,48,49]. In the most extreme cases, measurement errors may result in severe hypoglycemia events (SHE) requiring hospitalization [50]. Similarly, underestimated glucose levels may drive the patient to hyperglycemia (under bolus). Chronic hyperglycemia in the long term will manifest itself in higher hemoglobin A1c (HbA1c), increased risk of diabetic ketoacidosis (DKA), and long-term clinical complications, such as retinopathy, cardiac disease (congestive heart failure), and renal disease [51]. Ultimately, chronic clinical conditions will result in reduced quality of life, increased cost of living, and increased cost of healthcare at a population level [52–54]. As illustrated in Fig. 4.1, the consequences of an inaccurate meter span the patient's life and may eventually involve the entire organism. A limited, by no means complete, classification of the effects comprises:

Clinical outcomes

The quality of glycemic control. We separate this into short term and long term.

Short term

Time in range, time in hypoglycemia, time in hyperglycemia; glucose variability; risk of severe SHE and severe hyperglycemia (DKA).

Long term

HbA1c, diabetes-related complications: for example, retinopathy (up to blindness), nephropathy (up to kidney failure, requiring transplantation or dialysis), diabetic foot ulcer, amputation and vascular complications (including stroke, angina, heart failure, myocardial infarction).

Quality of life

The effect of the disease in perceived quality of life. Examples are stress, anxiety, and depression related to the disease. Quality of life lost due to the disease.

Financial outcomes

Short term

Cost of therapy (e.g., insulin, meter, and strips) and acute complications (e.g., SHE and DKA); the opportunity cost of managing the disease.

Long term

Quality-adjusted life years (QALYs), cost of managing long-term clinical complications, overall healthcare system costs.

Requirements

Given the complexity of the task, it is useful to reflect on what the requirements are for a method for assessing the effects of accuracy. It is our position that assessment of the effects of accuracy need to take into account canonical elements of variability in the metabolic–behavior–device–environment system. Some of these elements include the following:

Physiological variability

A study should trace the effects of measurement errors across a representative sample of metabolic variability across the population, but also within the patient (e.g., intraday variability of insulin sensitivity). The same meter error may have different effects for two different persons or two different effects for the same person at two different times.

Behavioral variability

As argued earlier and shown in Ref. [42], accuracy effects can be attenuated or magnified to a large degree by human behavior. Accuracy studies should incorporate sufficient variability for each therapy mode and use case of a meter, for example, MDI versus CSII patients, the meter used for calibration only or also for decision-making.

Device and lot variability

As shown in Refs. [11,13,14], each device has characteristic meter errors. Errors vary from lot to lot [10,27] and even from device to device. When attempting to analyze a single device, sufficient intradevice variability should be incorporated. Most studies available pool all meter data together and present an aggregate (population) analysis. Performance variability across meters of the same brand is critical for accuracy risk characterization. When possible, accuracy data from postmarket evaluations should be included, in case manufacturing process variations have effects not present during development [55].

Therapy modes

Meters are used in multiple therapy modes and use cases. In some cases, as the only source of information, or to confirm measurements obtained by another device (when a CGM is used in adjunctive mode), or only for calibration (CGM in nonadjunctive mode). The effects of accuracy will be different in each therapy mode and use case, and this information must be clearly recorded and reported as part of the study.

Time span

A system study must somehow assess the effects of accuracy as it pertains accrued costs and more importantly the development of diabetes-related complications.

The rest of the chapter focuses on describing a systems approach that has been applied to this problem and an example of its application to pump users. First, we motivate our approach by discussing the result of direct observational studies and their limitations. We then discuss modeling and simulation as an alternative approach and how it addresses the stated requirements.

Accuracy and its consequences

The first studies relating accuracy and its clinical effects coined the term *clinical* accuracy [23,25]. The rationale here is to directly transform measurement errors into quantities that assign these errors a clinical significance. This is the approach followed in Refs. [24,56] where (measurement, reference) pairs are classified into zones in the plane. Each zone is associated with certain clinical significance. For example, meter measurements that are extremely large with respect to the reference value are labeled erroneous (D or E) as opposed to accurate or benign. The approach lends itself well to a graphical display and is an excellent communication tool. There are two potential shortcomings of the approach. On one hand, the boundaries of these regions are discrete and the same for all the population. Considering the metabolic variability of patients, it is unlikely that such boundaries are equally accurate for all. On the other hand, the clinical consequences of inaccuracy represent a static view of a dynamic system. It is conceivable that repeated errors, even mild ones, can have long-term and cascading effects.

An alternate approach is to directly observe the clinical outcomes in a population. For example, in Ref. [49], meter accuracy was assessed by comparing SMBG meter measurements to a reference such as Yellow Springs Instrument (YSI) results taken during patient visits. Measurement errors were classified into accurate, benign, or erroneous according to the error grid analysis (EGA) [23]. The study reports significant correlations between erroneous reports and HbA1c levels and the incidence of severe hypoglycemia. Other studies [57] show that the improper use of meters requiring calibration to match a specific strip code may also lead to insulin dosing errors that can have life-threatening consequences, for example, can lead to an increased number of hypoglycemia levels under 50 mg/dL. Among the limitations of the approach are the reduced number of meter brands that can be observed, and accuracy variability across lots [27]. In addition, results are aggregated making it difficult to understand the performance of individual meters.

Finally, EGA meter reading classifications have limitations that can lead to inaccurate estimations [58]. In other words, while it is possible to observe patient's use of a specific meter and even observe short-term clinical benefits [3,49,57], doing so for a long enough time to observe long-term complications would be costly. In fact, it would be near impossible to *simultaneously assess all available meters*. Such an experiment would simply be too costly and impractical. Furthermore, to understand *corner* cases in accuracy, a study may need to put a patient at risk.

An alternative to clinical studies is to resort to computer simulations or in silico studies to assess meter performance and its consequences. Computer simulation is first used to understand the relationship between meter errors and glycemic control, which can be translated into short-term clinical outcomes. Additional models are then applied to translate these clinical outcomes, such as changes in HbA1c, into long-term complications of diabetes. Finally, fingerstick and insulin costs, together with treatment costs associated with severe hypoglycemia treatment in hospitals, along with the costs of treating other diabetes-related complications is estimated using country and healthcare system-specific figures.

Modeling and simulation

In this section, we present a model and simulation-based framework to characterize the consequences of BGM systems inaccuracy. In the discussion, we will detail the steps that were required to develop the simulation framework and how the available tools allow to satisfy all the requirements discussed in Requirements section.

Metabolic models and simulators (metabolic variability)

A first step requires to connect an erroneous meter reading to the treatment decision and immediate effect on glycemic control. For this purpose, models of the glucose-insulin regulation system are needed, such as the so-called minimal model that greatly improved our understanding and measurement of glucose-insulin metabolism [59,60]. This model and variants were then extended with the availability of models of meal absorption [61], glucagon secretion [62], and insulin/glucagon transport [63−66].

Metabolic modeling efforts laid the groundwork for metabolic simulators such as the University of Virginia (UVA)/Padova [67−69] and other compartmental simulators [70]. Although originally designed with the specific goal to support the development of artificial pancreas systems [71−73], metabolic simulators have been extensively applied to other forms of insulin dosing therapy modalities, and accuracy studies in particular.

Behavioral modeling and simulation (behavioral variability and therapy modes)

Parallel efforts were made to develop behavioral models of self-treatment of hypoglycemia [39,74,75], eating, and bolusing [18,76]. These models were applied to the UVA/Padova simulator in several accuracy studies [18,76], but are still of limited variability and are applicable only to CSII-based therapy. For example, in Ref. [18], all patients are assumed to behave randomly according to a pattern consistent with observational study data. Behavior is limited to the likelihood of fingersticking under circumstances that arise in CSII treatment such as hypo and hyperglycemia awareness, fingersticking during meals, and fingersticking 2 h after a meal. Other behaviors, such as meal-related ones are assumed to be independent [39,75].

Integrated metabolic/behavioral simulation

An alternative simulation approach [77] uses a hybrid approach where compartmental estimates are complemented with data-driven (patient-specific) quantifications of metabolic and behavioral variations obtained by model inversion. These models have been extensively used in the development, evaluation, and risk assessment of insulin dosing technologies [18,48,78].

Modeling glucose monitoring devices (device and lot variability)

The measurement error of several glucose monitoring systems has been modeled extensively [35,79,80]. In particular, models for SMBG meters became available recently [18,48,81] in great measure due to the availability of and extensive array of accuracy assessments [8,10,11,13,14]. Meter and CGM models make it possible to realistically simulate not only many commercially available monitoring devices [69] but also meters/sensors with hypothetical characteristics as in Refs. [48,78] where meters/sensors with increasing noise are systematically simulated. When assessing a particular meter brand, it is important to understand variability across meters of the same brand and model, or even across test-strip lots within the same brand [27]. This can be achieved by explicitly characterizing the distribution of model parameters [79,81]. Alternatively, a separate model can be identified for each meter dataset. During a simulation, the collection of models obtained this way can be sampled randomly [18,76,78,80].

In silico accuracy studies

An in silico accuracy study puts together all the elements described earlier. As shown in Fig. 4.4, a highly accurate glucose–insulin–glucagon metabolism model can be integrated with a meter model and a behavioral model of the patient, to understand the effect of dosing decisions. This approach was originally developed to test the performance of closed-loop algorithms [60,71,72]. Building on these studies [42,48,76], simulation has been used to quantify SMBG accuracy error effects on glucose variability, risk of hypoglycemia, and average glucose. As described in Fig. 4.4, these studies simulate virtual patients making decisions according to a set of prespecified decision-making rules informed by their behavioral model as well as simulated SMBG meter measurements. The decision is fed back to the simulator to estimate metabolic dynamics until the next decision point in time. Information about glucose concentration, meter readings, meals, insulin bolus, and rescue carbohydrates are recorded throughout the simulated scenario.

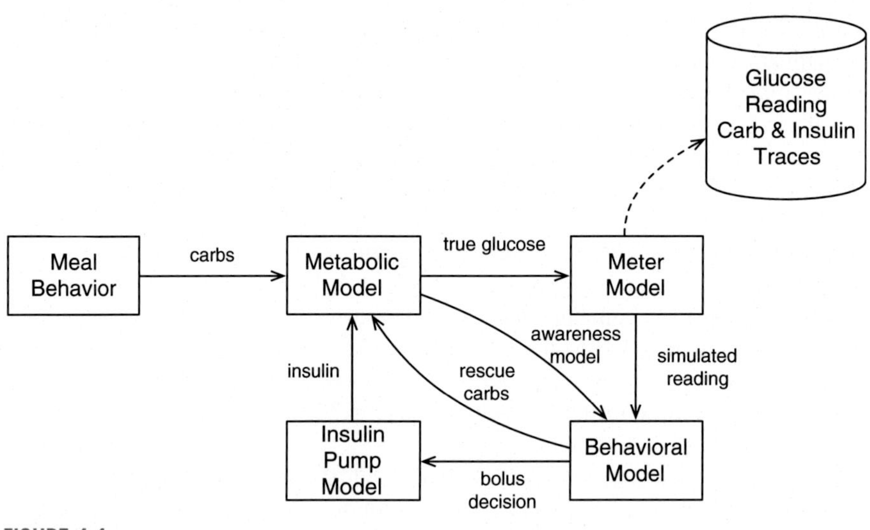

FIGURE 4.4

Structure of an in silico accuracy simulation.

Once enough data have been simulated, glycemic control outcomes can be easily computed from the observed glucose traces, such as time in normoglycemic range, time in hypo and hyperglycemia, number and duration of hypo and hyperglycemic events, and average plasma glucose concentration. In addition to fingerstick-based control, in silico accuracy studies have also been applied to assess the effect of CGM accuracy on clinical outcomes and safety of CGM for nonadjunct use [78,80], as well as to understand the interplay between SMBG and CGM accuracies [18]. In fact, several glucose manufacturers submit in silico studies as evidence to regulatory bodies.

From in silico results to short-term clinical outcomes

Computing certain clinical outcomes directly from simulated glucose traces requires additional results. For example, severe hypoglycemia cannot generally be simulated using a glucose-insulin-glucagon metabolic simulator. Instead, analysis of glucose traces can be used to estimate low blood glucose risk through the low blood glucose index (LBGI) [82]. In turn, LBGI can be used to estimate severe hypoglycemia incidence [50]. Similarly, estimating HbA1c from the simulation is not possible with the available simulators. In this case, in silico studies frequently rely on published models relating average glucose concentration and HbA1c [83].

A key advantage of an in silico approach is that it enables the exhaustive exploration of meters/sensors across their error characteristics, such as bias and noise. Only through such an exhaustive exploration, it is possible to quantitatively characterize the relationships between accuracy characteristics and clinical outcomes. For example, in Ref. [78], simulation is used to determine mean absolute relative difference (MARD) levels that are acceptable if a CGM meter is used non-adjunctively (e.g., MARD $\leq$ 10%). Similarly, simulation results have been used to understand accuracy requirements to achieve clinical performance requirements [30]. A recent study more relevant to glucose meters [76] shows that a regression model can be used to characterize relationships between SMBG bias and noise and clinical outcomes such as severe hypoglycemia events and HbA1c, total daily insulin, and fingerstick use.

Long-term health and complications

From in silico results to long-term complications

Relating quality of glycemic control and long-term clinical complications is the last step of the process, and perhaps the most difficult to perform [84].

One possible approach is to use population-level published literature relating changes in a clinical outcome metric to the incidence of long-term consequences. For example, in Ref. [54] authors relate changes in HbA1c baseline values to the incidence of complications such as retinopathy, renal failure, heart disease, and their associated costs. Fig. 4.5 shows an example of such an approach. Changes in HbA1c (x-axis) have a corresponding increase in the incidence of diabetes-related complications (y-axis). Knowing the number of events before and after the change allows us to calculate the effect on the total complication-related cost. By filtering the underlying literature for, for example, geographical region, patient age, diabetes type, or duration of the disease, the approach can be adapted to nonaverage populations.

An alternative approach [85] used the results of a health survey (EQ-5D) to develop a linear regression model associating changes in HbA1c to a measure of health (or health utility).

A third approach, used by Refs. [85–87], is to relate changes to HbA1c with changes to the progression of diabetes complications. Here, a Markov cohort modeling approach is used [88], as illustrated in Fig. 4.6. In this approach, each patient is considered to be in a state of progression, expressed by the state of the Markov chain. The transition from one state to another marks the progression of the disease and is expressed in conditional probabilities or transition rates. By analyzing the dynamics of the Markov chain, it is possible to assess the population distribution across disease progression states. Examples of this modeling approach are the Michigan Model for Diabetes [88], ECHO-T2DM [89], and the CORE model [90,91]. In Refs. [86,87], the transition probabilities in the chain are also conditional on HbA1c (illustrated as dotted lines in Fig. 4.6). For example, it is assumed that the probability of progressing to full blindness in the next period (see the bold line in Fig. 4.6) is conditional on the current state of the patient (Macular Degeneration) and its HbA1c level.

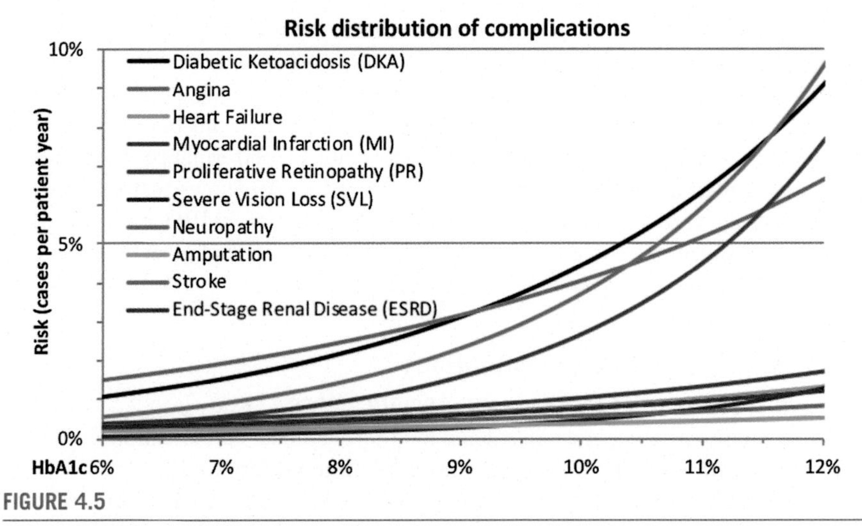

FIGURE 4.5

Effect of HbA1c on the increased risk of complications.

Based on Fortwaengler K, Parkin CG, Neeser K, Neumann M, Mast O. Description of a new predictive modeling approach that correlates the risk and associated cost of well-defined diabetes-related complications with changes in glycated hemoglobin (HbA1c). Journal of Diabetes Science and Technology 2017;11(2):315–323.

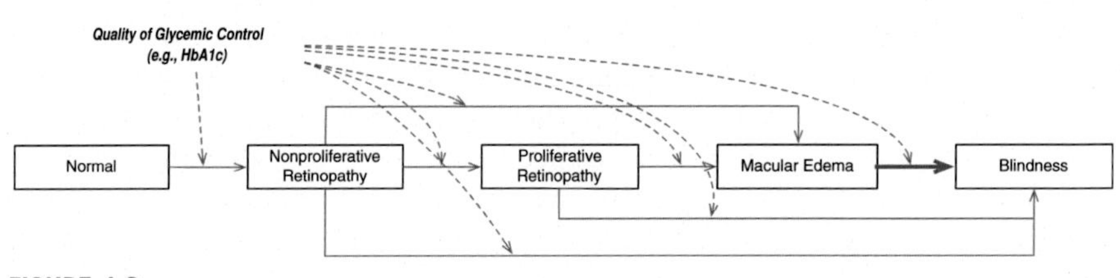

FIGURE 4.6

Markov chain model of progression of retinopathy. *Dotted lines* represent how HbA1c affects the likelihood of a patient transitioning to a more advanced stage of the disease.

Adapted from Zhou H, et al., A computer simulation model of diabetes progression, quality of life, and cost. Diabetes Care 2005;28(12):2856–63.

From clinical performance to financial outcomes

Once the incidence of short- and long-term complications has been established, attaching a financial cost is straightforward. In Refs. [86,87,92], the overall economic value of a certain level of clinical performance (measured as changes to HbA1c) is reported for Canada [86] and England [87,92]. In addition to direct financial figures, the impact of accuracy can also be expressed in QALYs [93]. In addition to relating meter accuracy, short-term clinical outcomes, such as HbA1c, and long-term outcomes and costs, in silico studies have also looked at the relationship between compliance with ISO accuracy standards (ISO 15197:2003 and 15197:2013) and costs [92].

An extended illustration

In this section, we discuss the design and results of two previously published studies [18,54,76,92] that illustrate the approach introduced in Accuracy and its consequences section. The goals of the project were (1) to assess how different aspects of a BGM system accuracy affect HbA1c, severe hypoglycemia, insulin, and fingerstick use (and costs); and (2) understand how the BGM system accuracy affects long-term financial costs to the patient and the health system. A BGM system includes the meter as well as the strips (both can cause inaccuracy). The project was constrained to type 1 diabetes patients using CSII. The general approach followed is described in Fig. 4.7.

The motivation of the study (first level of Fig. 4.7) is to understand how BGM system accuracy contributes to clinical outcomes and ultimately to healthcare costs. In our approach (second level of Fig. 4.7), we needed to simulate patients with different brands of BGM systems, observe how their glycemic control and clinical outcomes were affected, and ultimately assess whether the systems were resulting in increased costs for the health system. To make this possible a set of tools and methods were applied (third level in the figure). First, we created an in silico evaluation platform based on the UVA/Padova simulator [69] where models of multiple BGM systems were coded along with a behavioral model. Second, using the glucose traces resulting from our in silico studies as basis, severe hypoglycemia events per person per year and HbA1c levels were estimated. Simulation outputs were also used to monitor insulin and fingerstick use. Next, the HbA1c Translator was used to produce an estimate of diabetes-related complications due to changes in HbA1c. Finally, using treatment costs for each complication from the United Kingdom, the total costs to the healthcare system were computed. In the following sections, we describe each of the stages.

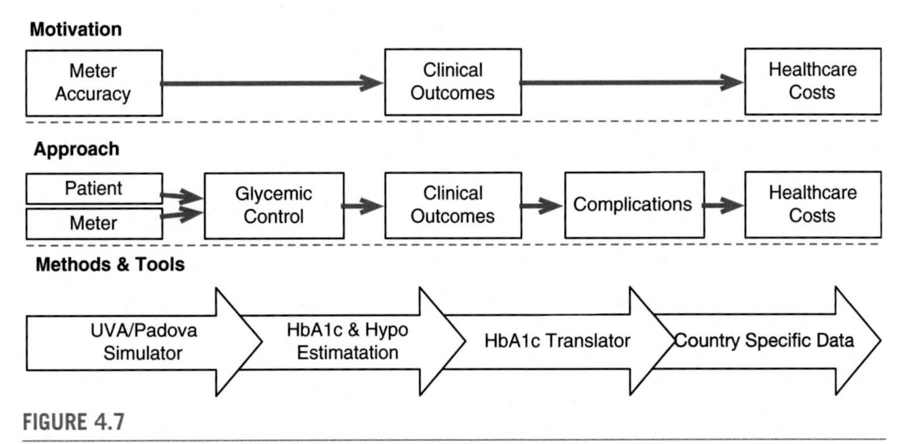

FIGURE 4.7

Approach followed in our illustration.

The in silico study

As described in Fig. 4.8, we built a simulation model for each of the 43 commercial BGM systems reported in Ref. [13] (a later study included an additional nine BGM systems reported in Ref. [10]). In parallel, we used archived deidentified data collected during the project funded by the National Institutes of Health/National Institute of Diabetes and Digestive and Kidney Diseases grant RO1 DK 085623 (see clinicaltrial.gov for clinical trial registration number NCT01434030) to characterize patients' behaviors. Device and behavioral models were integrated into a 30-day simulation based on the UVA/Padova Simulator [69]. The following sections describe with a bit more detail these calculations.

Meter models

As mentioned earlier, this study included separate models for 43 BGM systems commercially available in 2012 [13] and 9 that were available later in 2015 [10,13]. The BGM systems models used relied on separate statistical descriptions of the noise on upper (>100 mg/dL) and lower (<100 mg/dL) glucose ranges, an approach also used in Ref. [81].

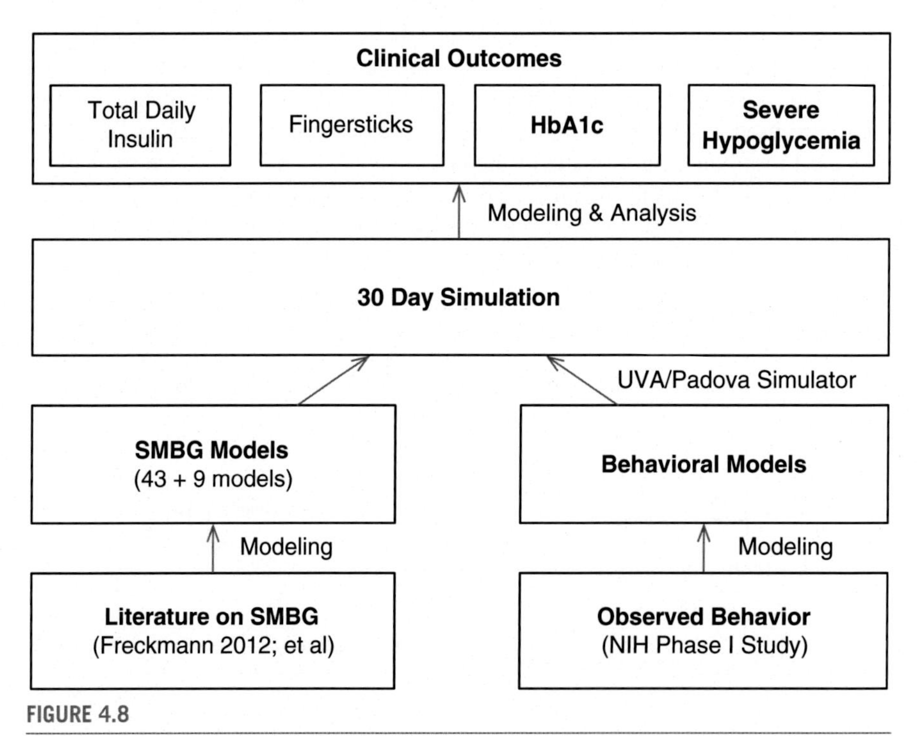

FIGURE 4.8

Design of the in silico study for this illustration.

Behavioral models

For the purpose of the study, behaviors such as self-treatment of hypoglycemia, meal, and bolusing were modeled as independent behaviors. The prospective models were parameterized using clinical data. The models were statistically validated against observed behavior. The main behavioral modules are depicted in Fig. 4.9. A more detailed description of the models is contained in Ref. [76].

In our approach, meals are generated using a Markov model that responds only to time of day, and the time and size of the previous meal. The implicit assumption is that there is no medical eating (e.g., a meal is taken early to avoid hypoglycemia). Meal boluses are taken around the time of the meal (a small random shift of bolus time is applied). Correction bolus and hypoglycemia treatments occur in response to fingersticks indicating hyper or hypoglycemia, respectively. In turn, fingersticks are initiated in response to hypo or hyperglycemia awareness models (the patient perceives symptoms). Each step in these perception–confirmation–action chains is described by a simple Bernoulli probabilistic model with parameters estimated from observational studies mentioned earlier.

Clinical and financial outcomes

For the purpose of this analysis, we focused on two meters' characteristics: error and bias [76]. We define an error as the fraction of meter measurements whose absolute relative difference (ARD) with respect to true plasma glucose exceeds 5%. On the other hand, bias is defined as the average difference between meter measurement and true plasma glucose concentration.

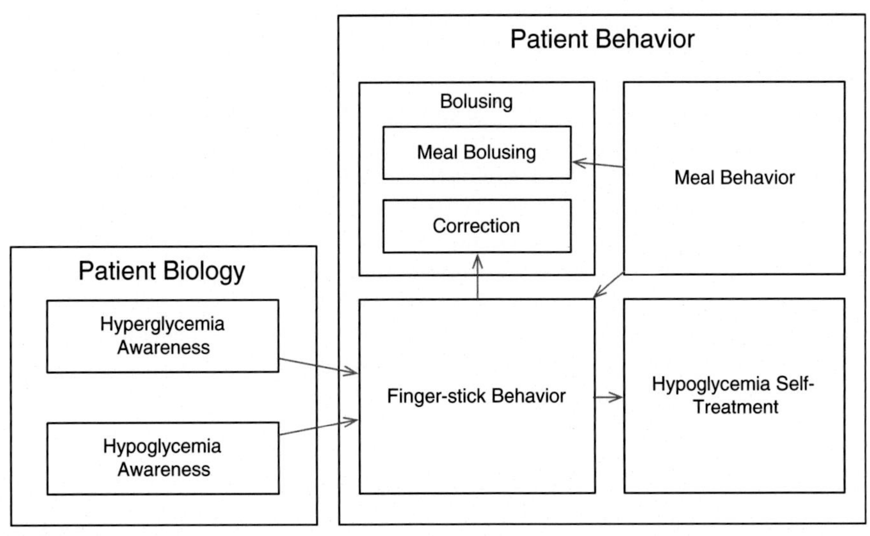

FIGURE 4.9

Behavioral components and their interaction with other behavioral and biological components.

We assessed the influence of error and bias on the clinical outcomes derived from the simulation results; these included HbA1c, severe hypoglycemia, insulin, and fingerstick use. Then, we proceeded to estimate the impact of HbA1c changes on the incidence of DKA and long-term complications. Finally, we estimated the impact of the observed inaccuracy on the overall financial costs. This included the costs related to long-term complications, DKA and severe hypoglycemic events, as well as the costs of insulin and fingerstick use.

Results: clinical outcomes

The results of the study are summarized in Fig. 4.10. Each dot in the figure corresponds to the average outcomes of the entire in silico population using one particular BGM system and represents four values: BGM system noise and bias (respectively *x*-axis and dot *color*), and associated HbA1c and incidence of severe hypoglycemia (respectively *y*-axis and dot *size*).

For example, consider the scenario where patients use an ideal sensor, that is, one reporting the exact plasma glucose value. The *x*-coordinate value is 0 (no noise), and the dot has a green color (no bias). The use of perfect information results in an estimated HbA1c of almost 8.8% with an incidence in severe hypoglycemia of around 1 event every 6 months (as per the size of the dot). Now consider the top-right small dark blue dot. This corresponds to an erratic BGM system (one with more than 80% of its measurements with errors greater than 5% of the true value). This BGM system also has a large negative bias (−20 mg/dL), hence the dark blue color. Using this BGM system will drive the in silico population to a reduction of hypoglycemia with respect to the ideal case, but at the cost of an increase of HbA1c of 0.5%, up to a 9.2% level. An alternate example is to consider the big yellow dot at the bottom right of Fig. 4.10. This BGM system is noisy with an *x*-axis of 70 meaning that 70% of its readings have an ARD of more than 5%. It is deep yellow because of its large positive bias. As a result of this bias (the BGM system overreports glucose levels), patients using the meter will experience a lower HbA1 (*y*-axis around 8.4% down from 8.8% HbA1c) at the cost of increased hypoglycemia (size of the bubble $\approx$ 2 hypoglycemic events every 6 months).

Overall, the results show a clear inverse relationship between bias (color) and HbA1c. Note that the same effect is apparent in the incidence of severe hypoglycemia, as BGM systems with large positive bias will more frequently drive patients to over bolus and severe hypoglycemia. Our error metric alone seems to only very mildly affect severe hypoglycemia. In this next section, we present regression models that formalize and quantify these relationships.

Accuracy and clinical outcomes: a regression model

In an effort to further clarify the accuracy-to-outcome relationship, a set of linear regression models was developed. The models explain HbA1c, severe hypoglycemia, total daily insulin, and fingerstick count as a function of error and bias [76].

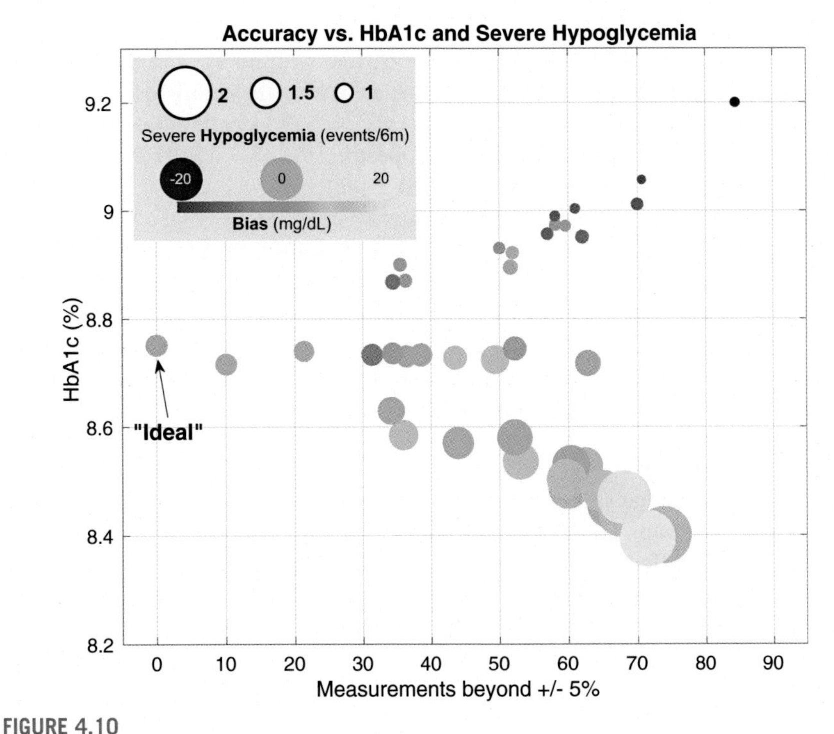

FIGURE 4.10

Clinical outcomes from the in silico study.

They showed a very good fit and quantify those relationships. This made it possible to understand how (nonexisting) BGM systems with given error and bias will perform clinically.

The in silico study shows a clear relationship between BGM system accuracy and quality control, measured in this case in terms of HbA1c and severe hypoglycemia incidence. Meters with systematic bias will affect both HbA1c (inversely) and hypoglycemia (proportionally). This agrees with the results reported in Ref. [49], which shows that consistently positive bias may increase episodes of hypoglycemic coma.

Determining financial impact

To understand the effect of accuracy in terms of available standards, meters were classified according to their compliance with ISO standards for accuracy. Their classification is represented in Fig. 4.11. Out of the 43, seven failed to comply with ISO 15197:2003. Out of the remaining meters, 14 satisfied ISO 15197:2003, but were not ISO 15197:2013 compliant. The final 22 m, compliant with ISO 15197:2013 were further split by the middle according to their level of performance in the 10/10, that is, ± 10 mg/dL when the references are >100 mg/dL, and $\pm 10\%$ when the reference is >100 mg/dL.

Cost compared to an ideal

Compared to an "ideal" BGM system, that is, one with no error and no bias, the average additional cost of inaccuracy associated with the entire group of BGM systems was £155 per patient year (PPY) (£95 to £219, depending on the population). The average additional cost of not meeting the ISO 15197:2003 standard [34] is approximately £306 PPY (£169 to £446), which is £178 PPY (£85 to £270) more than the additional cost for the average ISO 15197:2003 compliant BGM system. It is also £235 PPY (£113 to £357) more expensive than systems compliant with the ISO 15197:2013 [34]. Complying with the ISO 15197:2003 but failing to comply with the ISO 15197:2013 results in an additional cost of £216 PPY (£123 to £311). Splitting BGM systems according to 10/10 performance shows that the bottom results in a higher additional cost of £79 compared to the top-performing group (£64). The left panel of Fig. 4.12 illustrates these effects.

Worst-case costing

Alternative, and perhaps as a measure of risk, one can focus on the worst BGM system in each group. The result illustrated in the right panel of Fig. 4.12, shows the same relative effects. The overall worst system, that is, the most expensive not compliant to ISO 15197:2003, has an additional cost of £597 compared to the ideal system. Complying with ISO 15197:2003, but not with the 2013 version results in a maximum of £440 in additional costs. The worst system complying with ISO 15197: 2013 has an additional cost of £278 compared to the ideal BGM system.

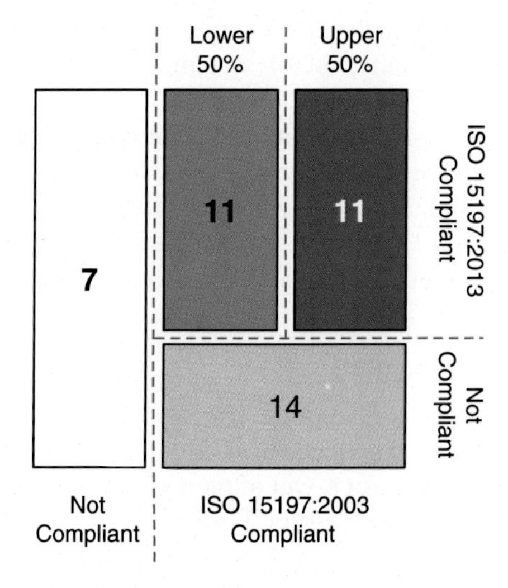

FIGURE 4.11

Classification of 43 commercial BGM systems according to their compliance with accuracy standards.

Observations from the illustrative example

Based on our approach, we showed that there is a clear relationship between BGM system accuracy and clinical outcomes, with the highest costs being associated with meter systems not meeting the ISO 15197:2003 standard. Our analysis makes it clear that BGM systems compliant with the ISO 15197:2013 standard have a limited effect on HbA1c, severe hypoglycemia incidence, total daily insulin, and fingerstick frequency. For the top 50% BGM systems complying with ISO 15197:2013, the worst observed effect on the HbA1c values was 0.12% (absolute) while the worst increase in SHE cases could reach 0.36 cases PPY. Compare this to the results for the group of BGM systems, not compliant with any ISO standard: for the HbA1c values, the increases could reach approximately 0.47% while the number of annual SHE could increase by up to 1.70 cases per year (it is important to understand that the system leading to an increase in HbA1c of 0.47% is not the same as the one increasing the SHE by 1.70 cases PPY!). In addition, insulin consumption could increase by up to 5.5 units/day and the number of fingersticks by up to 1.0 tests/day.

Lower costs are associated with systems meeting the ISO 15197:2013 standard. Using BGM systems that meet the system accuracy criteria of the ISO 15197:2013 standard can help reduce the clinical and financial consequences associated with the inaccuracy of BGM devices. The worst-case analysis shows that compliance with stricter accuracy standards reduces clinical and financial risk for patients and healthcare systems.

Bias is the accuracy characteristic that has the biggest influence on HbA1c, and both BGM system error and bias have a significant effect on the incidence of severe hypoglycemia.

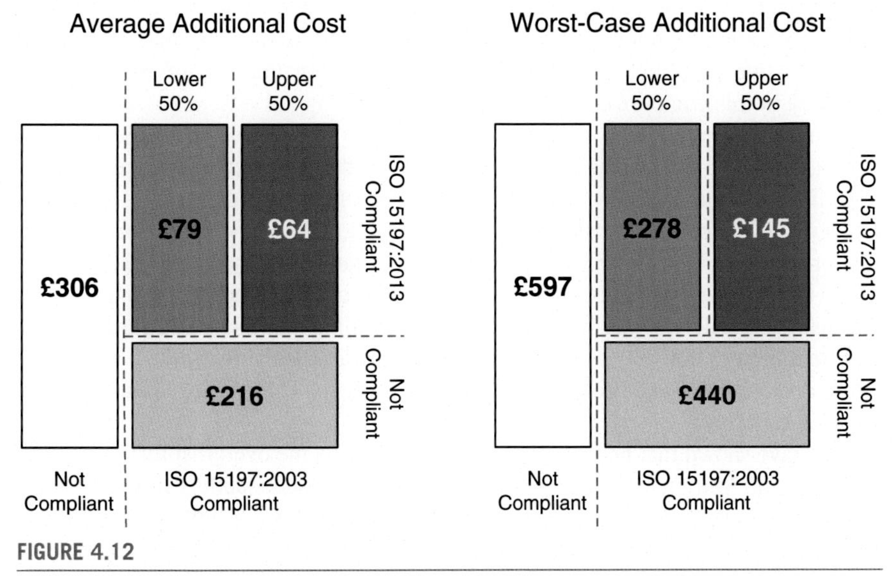

FIGURE 4.12

Costs associated with compliance with ISO standards of accuracy [34].

Patient behavior plays a critical role in these assessments. Variations of the in silico study presented here where patients are always compliant (bolus correctly and on-time) are needed. Here, we see that the effect of error disappears, likely due to a law of large numbers: in the absence of systematic bias, errors tend to cancel each other out. Alternatively, one can conclude that noncompliant behavior increases the effect of the inaccuracy: noncompliant patients have less opportunities to correct, highlighting the importance of accurate measurements during those times.

Limitations

The in silico study does not accommodate any long-term behavioral adaptations. Although short-term behavior (meals, bolus, etc.) were considered in the study, it is unlikely that a patient that experiences frequent hypoglycemic events will not adapt. These adaptations could include switching BGM systems, adjusting their insulin therapy, modify eating and exercise behaviors. In addition, our study assumes that BGM system accuracy remains constant throughout the progression of a patient's disease. This ignores developments that might happen during the lifetime of the patient. Finally, the results were limited to CSII. It is likely that the results can be extended to patients using MDI therapy, at least at a qualitative level.

Conclusions and future work

Assessing the effects of BGM system accuracy is a challenging task. On one hand, the effects of poor decisions span a long time, from the immediate to the very long term, affecting at the same time many aspects of a patient's life. To complicate things, behavior, technology, and environmental considerations impose limits on how effective glycemic control can become. Isolating the role of BGM system accuracy in this complex environment is nontrivial.

Simulation and systems modeling can shed light into this process. The ability to stimulate metabolism, behavior, and technology as they interact in practice is invaluable in understanding how this complex system interacts to produce outcomes. However, many challenges still exist. Better metabolic models, particularly models that properly account for long-term metabolic variations are needed. This is particularly true in type 2 diabetes where poor glycemic control leads to the progression of the disease. Models of the interplay between a failing glucose-insulin metabolism and treatment options for a type 2 patient are still in their infancy. The amount of treatment options and combinations make this a challenging combinatorial modeling problem.

We have shown that behavior has a strong effect on the overall ability to achieve good control. More work is necessary to advance our understanding of the patient's behavior. What are the patient's goals? What is the best use of information to help this patient achieve his/her goals? Recent reports [94] show that little progress has been made, despite clear improvements in accuracy, insulin formulations, and therapeutic choices. To better understand the effect and relevance of accuracy, it is necessary to understand how patients behave in response to the information that is available to them.

Commonly used definitions and metrics of clinical accuracy have been critical in guiding the community toward better systems. Tools like the EGA, in its various incarnations, have provided the first line of defense in understanding consequences of inaccuracy. Unfortunately, these tools have the limitation of not accounting for the environment in which a certain error is committed. The question that perhaps should be asked is: what is the best use of noisy information, given the patient's unique behavioral traits and particular goals? In other words, we need to understand how contextual and behavioral inaccuracy interacts with BGM system inaccuracy to create a true clinical or therapeutical accuracy.

One final comment was related to the almost exclusive use of HbA1c as a surrogate for glycemic control. In understanding the long-term effects of inaccuracy, other metrics are necessary. For example, there are indications of the relationship of glucose variability on retinopathy [95,96]. It is time for the industry to move to a more comprehensive assessment of glycemic control, which includes indices of glycemic variabilities, such as high and low blood glucose risk indices. Characterizing the relationship between BGM system accuracy and these additional glucose control metrics, as well as the relationship between these metrics and long-term complications is a necessary next step.

Disclosure

E. Campos-Náñez received no financial or resource support for the production of this manuscript as an employee of Dexcom, Inc. The views expressed here are exclusively his own.

References

[1] Hirsch IB, et al. Self-Monitoring of Blood Glucose (SMBG) in insulin- and non−insulin-using adults with diabetes: consensus recommendations for improving SMBG accuracy, utilization, and research. Diabetes Technology and Therapeutics 2008; 10(6):419−39.

[2] Klonoff DC, et al. Consensus report: the current role of self-monitoring of blood glucose in non-insulin-treated type 2 diabetes. Journal of Diabetes Science and Technology 2011;5(6):1529−48.

[3] Barnard KD, Young AJ, Waugh NR. Self monitoring of blood glucose - a survey of diabetes UK members with type 2 diabetes who use SMBG. BMC Research Notes Nov. 2010;3(1):318.

[4] Poolsup N, Suksomboon N, Jiamsathit W. Systematic review of the benefits of self-monitoring of blood glucose on glycemic control in type 2 diabetes patients. Diabetes Technology and Therapeutics 2008;10(s1). S-51−S-66.

[5] Poolsup N, Suksomboon N, Rattanasookchit S. Meta-analysis of the benefits of self-monitoring of blood glucose on glycemic control in type 2 diabetes patients: an update. Diabetes Technology and Therapeutics 2009;11(12):775–84.
[6] Oki JC, Flora DL, Isley WL. Frequency and impact of SMBG on glycemic control in patients with NIDDM in an urban teaching hospital clinic. The Diabetes Educator 1997;23(4):419–24.
[7] Bassili A, Omar M, Tognoni G. The adequacy of diabetic care for children in a developing country. Diabetes Research and Clinical Practice 2001;53(3):187–99.
[8] Klonoff DC, et al. Investigation of the accuracy of 18 marketed blood glucose monitors. Diabetes Care 2018;41(8):1681–8.
[9] Klonoff DC, Prahalad P. Performance of cleared blood glucose monitors. Journal of Diabetes Science and Technology 2015;9(4):895–910.
[10] Freckmann G, Link M, Schmid C, Pleus S, Baumstark A, Haug C. System accuracy evaluation of different blood glucose monitoring systems following ISO 15197:2013 by using two different comparison methods. Diabetes Technology and Therapeutics 2015;17(9):635–48.
[11] Freckmann G, Baumstark A, Schmid C, Pleus S, Link M, Haug C. Evaluation of 12 blood glucose monitoring systems for self-testing: system accuracy and measurement reproducibility. Diabetes Technology and Therapeutics 2014;16(2):113–22.
[12] Brazg RL, Klaff LJ, Parkin CG. Performance variability of seven commonly used self-monitoring of blood glucose systems: clinical considerations for patients and providers. Journal of Diabetes Science and Technology 2013;7(1):144–52.
[13] Freckmann G, Schmid C, Baumstark A, Pleus S, Link M, Haug C. System accuracy evaluation of 43 blood glucose monitoring systems for self-monitoring of blood glucose according to DIN EN ISO 15197. Journal of Diabetes Science and Technology 2012; 6(5):1060–75.
[14] Freckmann G, et al. System accuracy evaluation of 27 blood glucose monitoring systems according to DIN EN ISO 15197. Diabetes Technology and Therapeutics 2010; 12(3):221–31.
[15] Heinemann L, Freckmann G. CGM versus FGM; or, continuous glucose monitoring is not flash glucose monitoring. Journal of Diabetes Science and Technology 2015;9(5): 947–50.
[16] El-Shazly M, et al. Health care for diabetic patients in developing countries. Public Health 2000;114(4):276–81.
[17] Escalante M, Gagliardino JJ, Guzmán JR, Tschiedel B. Call-to-action: timely and appropriate treatment for people with type 2 diabetes in Latin America. Diabetes Research and Clinical Practice 2014;104(3):343–52.
[18] Campos-Náñez E, Breton MD. Effect of BGM accuracy on the clinical performance of CGM: an in-silico study. Journal of Diabetes Science and Technology 2017;11(6): 1196–206.
[19] Heinemann L. Continuous Glucose Monitoring (CGM) or Blood Glucose Monitoring (BGM): interactions and implications. Journal of Diabetes Science and Technology 2018;12(4):873–9.
[20] Chen ET, Nichols JH, Duh S-H, Hortin G. Performance evaluation of blood glucose monitoring devices. Diabetes Technology and Therapeutics 2003;5(5):749–68.
[21] Bergenstal RM. Evaluating the accuracy of modern glucose meters. Insulin 2008;3(1): 5–14.

[22] Freckmann G, Pleus S, Grady M, Setford S, Levy B. Measures of accuracy for continuous glucose monitoring and blood glucose monitoring devices. Journal of Diabetes Science and Technology 2019;13(3). https://doi.org/10.1177/1932296818812062.
[23] Clarke WL, Cox D, Gonder-Frederick LA, Carter W, Pohl SL. Evaluating clinical accuracy of systems for self-monitoring of blood glucose. Diabetes Care 1987;10(5): 622–8.
[24] Clarke WL. The original Clarke Error Grid Analysis (EGA). Diabetes Technology and Therapeutics 2005;7(5):776–9.
[25] Kovatchev B, Anderson S, Heinemann L, Clarke W. Comparison of the numerical and clinical accuracy of four continuous glucose monitors. Diabetes Care 2008;31(6): 1160–4.
[26] Kilo Charles S, et al. Evaluation of a new blood glucose monitoring system with autocalibration for home and hospital bedside use. Diabetes Research and Clinical Practice 2006;74(1):66–74.
[27] Baumstark A, Pleus S, Schmid C, Link M, Haug C, Freckmann G. Lot-to-lot variability of test strips and accuracy assessment of systems for self-monitoring of blood glucose according to ISO 15197. Journal of Diabetes Science and Technology 2012;6(5): 1076–86.
[28] Boyd JC, Bruns DE. Quality specifications for glucose meters: assessment by simulation modeling of errors in insulin dose. Clinical Chemistry 2001;47(2):209–14.
[29] Boyd JC, Bruns DE. Chapter 16 Monte Carlo simulation in establishing analytical quality requirements for clinical laboratory tests: meeting clinical needs. Methods in Enzymology 2009;467:411–33. Academic Press.
[30] Karon BS, Boyd JC, Klee GG. Glucose meter performance criteria for tight glycemic control estimated by simulation modeling. Clinical Chemistry 2010;56(7):1091–7.
[31] U. Food and D. Administration. Self-monitoring blood glucose test systems for over-the counter use—draft guidance for industry and food and drug administration staff. Food; Drug Administration; 2015.
[32] Krouwer JS. Why the details of glucose meter evaluations matters. Journal of Diabetes Science and Technology 2019;13(3):559–60. https://doi.org/10.1177/1932296818803113.
[33] Grino M, Alitta Q, Oliver C. Response to 'blood glucose monitoring data should be reported in detail when studies about efficacy of continuous glucose monitoring systems are published'. Journal of Diabetes Science and Technology 2019;13(1):154–5. https://doi.org/10.1177/1932296818805749.
[34] for Standardization IO. In vitro diagnostic test systems—requirements for blood-glucose monitoring systems for self-testing in managing diabetes mellitus. International Organization for Standardization; 2003 and 2013.
[35] Breton M, Kovatchev B. Analysis, modeling, and simulation of the accuracy of continuous glucose sensors. Journal of Diabetes Science and Technology 2008;2(5):853–62.
[36] Cox DJ, Irvine A, Gonder-Frederick L, Nowacek G, Butterfield J. Fear of hypoglycemia: quantification, validation, and utilization. Diabetes Care 1987;10(5):617–21.
[37] Wild D, von Maltzahn R, Brohan E, Christensen T, Clauson P, Gonder-Frederick L. A critical review of the literature on fear of hypoglycemia in diabetes: implications for diabetes management and patient education. Patient Education and Counseling 2007;68(1):10–5.
[38] Anderbro T, et al. Fear of hypoglycaemia in adults with type 1 diabetes. Diabetic Medicine 2010;27(10):1151–8.

[39] Breton M, Clarke WL, Farhy L, Kovatchev B. A model of self-treatment behavior, glucose variability, and hypoglycemia-associated autonomic failure in type 1 diabetes. Journal of Diabetes Science and Technology 2007;1(3):331–7.
[40] Bolli GB. Hypoglycaemia unawareness. Diabetes and Metabolism 1997;23(Suppl. 3): 29–35.
[41] Fabris C, Farhy LS, Anderson SM, Nass RM, Kovatchev BP, Breton MD. Recent exposure to hypoglycemia increases glucose variability following a hyper/hypoglycemic metabolic challenge in T1D. Journal of Diabetes Science and Technology 2018; 12(2):311–7.
[42] Campos-Náñez E, Kovatchev BP. Impact of meal constituents on artificial pancreas algorithms. Diabetes Technology and Therapeutics 2016;18(10):607–9.
[43] Kovatchev B. Automated closed-loop control of diabetes: the artificial pancreas. Bioelectronic Medicine Nov. 2018;4(1):14.
[44] Herzer M, Hood KK. Anxiety symptoms in adolescents with type 1 diabetes: association with blood glucose monitoring and glycemic control. Journal of Pediatric Psychology 2010;35(4):415–25.
[45] Thomas AM, Peterson L, Goldstein D. Problem solving and diabetes regimen adherence by children and adolescents with IDDM in social pressure situations: a reflection of normal development. Journal of Pediatric Psychology 1997;22(4):541–61.
[46] Pendley JS, Kasmen LJ, Miller DL, Donze J, Swenson C, Reeves G. Peer and family support in children and adolescents with type 1 diabetes. Journal of Pediatric Psychology 2002;27(5):429–38.
[47] Kovatchev B, Cox D, Gonder-Frederick L, Schlundt D, Clarke W. Stochastic model of self-regulation decision making exemplified by decisions concerning hypoglycemia. Health Psychology 1998;17(3):277–84.
[48] Breton MD, Kovatchev BP. Impact of blood glucose self-monitoring errors on glucose variability, risk for hypoglycemia, and average glucose control in type 1 diabetes: an in silico study. Journal of Diabetes Science and Technology 2010;4(3):562–70.
[49] Boettcher C, et al. Accuracy of blood glucose meters for self-monitoring affects glucose control and hypoglycemia rate in children and adolescents with type 1 diabetes. Diabetes Technology and Therapeutics 2015;17(4):275–82.
[50] Kovatchev BP, Cox DJ, Gonder-Frederick LA, Young-Hyman D, Schlundt D, Clarke W. Assessment of risk for severe hypoglycemia among adults with IDDM: validation of the low blood glucose index. Diabetes Care 1998;21(11):1870–5.
[51] Huang ES, Laiteerapong N, Liu JY, John PM, Moffet HH, Karter AJ. Rates of complications and mortality in older patients with diabetes mellitus: the diabetes and aging study. Journal of the American Medical Association Internal Medicine 2014;174(2): 251–8.
[52] Hammer M, Lammert M, Mejias SM, Kern W, Frier BM. Costs of managing severe hypoglycaemia in three european countries. Journal of Medical Economics 2009; 12(4):281–90.
[53] Foos V, et al. Economic impact of severe and non-severe hypoglycemia in patients with type 1 and type 2 diabetes in the United States. Journal of Medical Economics 2015; 18(6):420–32.
[54] Fortwaengler K, Parkin CG, Neeser K, Neumann M, Mast O. Description of a new predictive modeling approach that correlates the risk and associated cost of well-defined diabetes-related complications with changes in glycated hemoglobin (HbA1c). Journal of Diabetes Science and Technology 2017;11(2):315–23.

[55] Freckmann G, Schmid C, Baumstark A, Rutschmann M, Haug C, Heinemann L. Analytical performance requirements for systems for self-monitoring of blood glucose with focus on system accuracy: relevant differences among ISO 15197:2003, ISO 15197:2013, and current FDA recommendations. Journal of Diabetes Science and Technology 2015;9(4):885–94.

[56] Parkes JL, Slatin SL, Pardo S, Ginsberg BH. A new consensus error grid to evaluate the clinical significance of inaccuracies in the measurement of blood glucose. Diabetes Care 2000;23(8):1143–8.

[57] Raine ICH, Pardo S, Parkes JL. Predicted blood glucose from insulin administration based on values from miscoded glucose meters. Journal of Diabetes Science and Technology 2008;2(4):557–62.

[58] Kollman C, Wilson DM, Wysocki T, Tamborlane WV, Beck RW. Limitations of statistical measures of error in assessing the accuracy of continuous glucose sensors. Diabetes Technology and Therapeutics 2005;7(5):665–72.

[59] Bergman RN. Toward physiological understanding of glucose tolerance: minimal-model approach. Diabetes 1989;38(12):1512–27.

[60] Cobelli C, Man CD, Pedersen MG, Bertoldo A, Toffolo G. Advancing our understanding of the glucose system via modeling: a perspective. IEEE Transactions on Biomedical Engineering May 2014;61(5):1577–92.

[61] Man CD, Camilleri M, Cobelli C. A system model of oral glucose absorption: validation on gold standard data. IEEE Transactions on Biomedical Engineering December, 2006; 53(12):2472–8.

[62] Dobbins RL, Davis SN, Neal DW, Cobelli C, Jaspan J, Cherrington AD. Compartmental modeling of glucagon kinetics in the conscious dog. Metabolism 1995;44(4):452–9.

[63] Nucci G, Cobelli C. Models of subcutaneous insulin kinetics. A critical review. Computer Methods and Programs in Biomedicine 2000;62(3):249–57.

[64] Lv D, Breton MD, Farhy LS. Pharmacokinetics modeling of exogenous glucagon in type 1 diabetes mellitus patients. Diabetes Technology and Therapeutics 2013; 15(11):935–41.

[65] Lv D, et al. Pharmacokinetic model of the transport of fast-acting insulin from the subcutaneous and intradermal spaces to blood. Journal of Diabetes Science and Technology 2015;9(4):831–40.

[66] Visentin R, et al. Improving efficacy of inhaled technosphere insulin (Afrezza) by postmeal dosing: in-silico clinical trial with the University of Virginia/Padova type 1 diabetes simulator. Diabetes Technology and Therapeutics 2016;18(9):574–85.

[67] Dalla Man C, Rizza R, Cobelli C. Meal simulation model of the glucose-insulin system. IEEE Transactions on Bio-Medical Engineering Nov. 2007;54:1740–9.

[68] Man CD, Micheletto F, Lv D, Breton M, Kovatchev B, Cobelli C. The UVA/PADOVA type 1 diabetes simulator: new features. Journal of Diabetes Science and Technology 2014;8(1):26–34.

[69] Visentin R, et al. The UVA/PADOVA type 1 diabetes simulator goes from single meal to single day. Journal of Diabetes Science and Technology 2018;12(2):273–81.

[70] Wilinska ME, Chassin LJ, Acerini CL, Allen JM, Dunger DB, Hovorka R. Simulation environment to evaluate closed-loop insulin delivery systems in type 1 diabetes. Journal of Diabetes Science and Technology 2010;4(1):132–44.

[71] Magni L, et al. Model predictive control of type 1 diabetes: an in silico trial. Journal of Diabetes Science and Technology 2007;1(6):804–12.

[72] Kovatchev BP, Breton M, Man CD, Cobelli C. In silico preclinical trials: a proof of concept in closed-loop control of type 1 diabetes. Journal of Diabetes Science and Technology 2009;3(1):44–55.
[73] Wilinska ME, et al. Overnight closed-loop insulin delivery with model predictive control: assessment of hypoglycemia and hyperglycemia risk using simulation studies. Journal of Diabetes Science and Technology 2009;3(5):1109–20.
[74] Gonder-Frederick L, Cox D, Kovatchev B, Schlundt D, Clarke W. A biopsychobehavioral model of risk of severe hypoglycemia. Diabetes Care 1997; 20(4):661–9.
[75] Shepard JA, Gonder-Frederick L, Vajda K, Kovatchev B. Patient perspectives on personalized glucose advisory systems for type 1 diabetes management. Diabetes Technology and Therapeutics 2012;14(10):858–61.
[76] Campos-Náñez E, Fortwaengler K, Breton MD. Clinical impact of blood glucose monitoring accuracy: an in-silico study. Journal of Diabetes Science and Technology 2017; 11(6):1187–95.
[77] Patek SD, et al. Empirical representation of blood glucose variability in a compartmental model. In: Kirchsteiger H, Jørgensen JB, Renard E, del Re L, editors. Prediction methods for blood glucose concentration: design, use and evaluation. Cham: Springer International Publishing; 2016. p. 133–57.
[78] Kovatchev BP, Patek SD, Ortiz EA, Breton MD. Assessing sensor accuracy for non-adjunct use of continuous glucose monitoring. Diabetes Technology and Therapeutics 2015;17(3):177–86.
[79] Facchinetti A, Favero SD, Sparacino G, Castle JR, Ward WK, Cobelli C. Modeling the glucose sensor error. IEEE Transactions on Biomedical Engineering 2014;61(3): 620–9.
[80] Breton MD, Hinzmann R, Campos-Nañez E, Riddle S, Schoemaker M, Schmelzeisen-Redeker G. Analysis of the accuracy and performance of a continuous glucose monitoring sensor prototype: an in-silico study using the UVA/PADOVA type 1 diabetes simulator. Journal of Diabetes Science and Technology 2017;11(3):545–52.
[81] Vettoretti M, Facchinetti A, Sparacino G, Cobelli C. A model of self-monitoring blood glucose measurement error. Journal of Diabetes Science and Technology 2017;11(4): 724–35.
[82] Fabris C, Patek SD, Breton MD. Are risk indices derived from CGM interchangeable with SMBG-based indices? Journal of Diabetes Science and Technology 2016;10(1): 50–9.
[83] Nathan DM, Kuenen J, Borg R, Zheng H, Schoenfeld D, Heine RJ. Translating the A1c assay into estimated average glucose values. Diabetes Care 2008;31(8):1473–8.
[84] Fritzen K, Heinemann L, Schnell O. Modeling of diabetes and its clinical impact. Journal of Diabetes Science and Technology 2018;12(5):976–84.
[85] McQueen RB, et al. Association between glycated hemoglobin and health utility for type 1 diabetes. The Patient: Patient-Centered Outcomes Research Jun. 2014;7(2): 197–205.
[86] McQueen RB, Breton MD, Ott M, Koa H, Beamer B, Campbell JD. Economic value of improved accuracy for self-monitoring of blood glucose devices for type 1 diabetes in Canada. Journal of Diabetes Science and Technology 2016;10(2):366–77.
[87] McQueen RB, et al. Economic value of improved accuracy for self-monitoring of blood glucose devices for type 1 and type 2 diabetes in England. Journal of Diabetes Science and Technology 2018;12(5):992–1001.

[88] Zhou H, et al. A computer simulation model of diabetes progression, quality of life, and cost. Diabetes Care 2005;28(12):2856–63.
[89] Willis M, Johansen P, Nilsson A, Asseburg C. Validation of the economic and health outcomes model of type 2 diabetes mellitus (ECHO-T2DM). PharmacoEconomics 2017;35(3):375–96.
[90] Palmer AJ, et al. The core diabetes model: projecting long-term clinical outcomes, costs and cost effectiveness of interventions in diabetes mellitus (types 1 and 2) to support clinical and reimbursement decision-making. Current Medical Research and Opinion 2004;20(Suppl. 1):S5–26.
[91] McEwan P, Foos V, Palmer JL, Lamotte M, Lloyd A, Grant D. Validation of the IMS CORE diabetes model. Value in Health 2014;17(6):714–24.
[92] Fortwaengler K, Campos-Náñez E, Parkin CG, Breton MD. The financial impact of inaccurate blood glucose monitoring systems. Journal of Diabetes Science and Technology 2018;12(2):318–24.
[93] Fanshel S, Bush JW. A health-status index and its applications to health-services outcomes. Operations Research 1970;18(6):967–1235.
[94] Rodbard D. State of type 1 diabetes care in the United States in 2016–2018 from T1d exchange registry data. Diabetes Technology and Therapeutics 2019;21(2):62–5.
[95] Kilpatrick ES, Rigby AS, Atkin SL. The effect of glucose variability on the risk of microvascular complications in type 1 diabetes. Diabetes Care 2006;29(7):1486–90.
[96] Sartore G, Chilelli NC, Burlina S, Lapolla A. Association between glucose variability as assessed by Continuous Glucose Monitoring (CGM) and diabetic retinopathy in type 1 and type 2 diabetes. Acta Diabetologica 2013;50(3):437–42.

CHAPTER

Modeling the SMBG measurement error

5

Martina Vettoretti, PhD
Department of Information Engineering, University of Padova, Padova, Italy

SMBG measurement error

Measurements collected by portable self-monitoring of blood glucose (SMBG) devices approximate blood glucose (BG) concentrations. The difference between SMBG measurement and true BG concentration represents the SMBG measurement error (or observational error). SMBG measurement error is typically assessed by comparing SMBG measurements to temporally matched reference measurements collected by a laboratory measurement system known to have small measurement errors. As all measurement instruments, SMBG devices are affected by both systematic and random errors. The systematic error is the predictable error component, which influences observations consistently in one direction. For example, a glucose meter that consistently under/overestimate glucose concentration is affected by a systematic error. Commonly, systematic errors are quantitatively assessed by calculating the average absolute or relative difference between SMBG measurements and reference values, also called bias. The magnitude of the systematic error defines the *accuracy* of the measurement instrument. The random error is the unpredictable error component, which leads to inconsistent values when repeated measurements of the same quantity are performed. Random errors are usually quantified by assessing the standard deviation (SD) or the coefficient of variation (CV) of the differences between SMBG measurements and reference values. The magnitude of the random error defines the *precision* of the measurement instrument.

Several studies have assessed the accuracy and precision of SMBG devices on the market, by comparing SMBG measurements with reference samples collected by high-accuracy and precision laboratory instruments [1–8]. These studies show that performance in terms of accuracy and precision can be very different for different devices. For example, Freckmann et al. [2] compared the accuracy and precision of 43 SMBG devices and demonstrated how the error's bias and CV varied significantly with the considered device (see, e.g., Fig. 2 in Ref. [2]). A similar result was obtained in a recent work by Klonoff et al. [8] (see, e.g., Table 4 in Ref. [8]). Besides the type of device used, the characteristics of the SMBG measurement error may be influenced by many other factors, such as the hematocrit level [7,9] and ambient conditions [10,11], and can vary when different lots of test strips are used [12]. Literature studies also suggest that, even when these factors are fixed

Glucose Monitoring Devices. https://doi.org/10.1016/B978-0-12-816714-4.00005-3

and measurement conditions are stable, the characteristics of SMBG measurement error can vary for different BG concentrations [13,14]. This is visible, for example, in the scatter plots of SMBG error versus reference glucose reported by Freckmann et al. for 43 different SMBG devices [2] (see Fig. 1A–C in Ref. [2]).

To receive approval from regulatory agencies and enter the market, SMBG devices must satisfy accuracy standards. For example, the International Organization for Standardization (ISO) requires that at least 95% of differences in the low glucose range and relative differences in the high glucose range (calculated by comparing SMBG measurements with high-accuracy reference measurements) not exceed a prefixed range. In particular, standard ISO 15197:2003 [15] requires that when BG concentration is < 75 mg/dL at least 95% of readings deviate less than 15 mg/dL from the reference, whereas when BG concentration is ≥ 75 mg/dL at least 95% of readings deviate less than 20% from the reference. The 2013 revision of the standard, ISO 15197:2013 [16], imposes tighter accuracy requirements: if $BG < 100$ mg/dL at least 95% of readings must deviate less than 15 mg/dL; if $BG \geq 100$ mg/dL at least 95% of readings must deviate less than 15%. A recent FDA guidance imposes that SMBG measurements must deviate no more than 15% from the reference BG values independently on BG, that is, a stricter requirement is imposed in the low glucose range [17]. Many studies assessing the performance of SMBG devices according to the accuracy standards can be found in the literature. These studies show that not all the SMBG devices on the market meet the standard accuracy requirements. For instance, only 1 of the 7 systems tested by Brazg et al. [18] satisfies standard ISO 15197:2013.

The SMBG measurement error can negatively impact glycemic control, especially in subjects with type 1 diabetes (T1D), who rely on SMBG measurements for making treatment decisions, for example, insulin bolus calculation and hyper/hypoglycemia treatment. For example, an underestimation of BG concentration can lead to a too-small insulin dose that may result in hyperglycemia, which, when prolonged and frequently repeated in time, can lead to long-term complications like cardiovascular diseases, neuropathy, nephropathy, and retinopathy. Conversely, an overestimation of BG concentration can lead to a too large insulin dose that exposes the patient to the risk of hypoglycemia, which can lead, in the short-term, to risky conditions like seizure and coma, and even to death. Particularly important is the SMBG accuracy in the low glucose range [19], where positively biased glucose measurements would not allow the patient to detect dangerous hypoglycemic events. Recently, it has also been shown that SMBG devices currently on the market present variable performance in the range of hypoglycemia (in Ref. [20], the mean absolute relative error in hypoglycemia ranged between 4.4% and 13.5%).

Why modeling the SMBG measurement error?

A model of the SMBG measurement error is a mathematical or statistical description of the SMBG measurement error that allows on one hand to summarize the error characteristics and on the other hand to generate synthetic SMBG data that can be used for the so-called in silico clinical trials. In silico clinical trials (ISCTs) are defined as "The use of individualized computer simulations in the development or regulatory evaluation of a medicinal product, medical device, or medical intervention" [21]. The idea is to recreate the concept of in vivo clinical trials (i.e., trial performed on real patients) in a simulation environment. The strength of ISCT is that they can overcome some limitations of in vivo clinical trials, such as long duration, elevated costs and, as a consequence, low numerosity. In fact, given the low cost and time required to run computer simulations, ISCT can be performed in an incredibly large number of subjects, that would be impossible to enroll in an in vivo clinical trial, because too expensive and time consuming. Another advantage of ISCT is the possibility of testing high-risk situations related to the occurrence of rare events, not observable in in vivo clinical trials because of their limited size and duration. ISCTs are thus unique procedures to test the safety of treatments based on drugs and medical devices under extreme conditions, without exposing human patients to any risk and can be used to reduce, refine, or even replace in vivo clinical trials.

Of course, an essential requirement to perform ISCT is the availability of a model of the patient's physiological response to the drug or medical device-based treatment under test, with parameterization that accounts for the interindividual variability and is, therefore, able to describe a large number of virtual subjects. Diabetes has been an area of intense modeling development in these last 20 years [22–24]. One of the most popular models of T1D patients' physiology is the UVA/Padova T1D simulator, jointly developed by the University of Padova and the University of Virginia [25–27]. However, a physiology model alone is not sufficient to realize such ISCT, because, besides the physiological response of T1D subjects to meals and insulin doses, other fundamental components, such as glucose measurements collected by glucose monitoring devices, need to be simulated to obtain realistic in silico scenarios.

The simplest strategy that could be used to generate in silico realizations of SMBG measurement error is bootstrapping from a dataset of SMBG error realizations derived from real data. However, a limitation of this approach is that the number of different SMBG error realizations that can be simulated is finite (and, in particular, equal to the cardinality of the available dataset). This represents a problem for large-scale in silico applications, for example, an in silico clinical trial with long study duration and/or a large number of virtual subjects involved. This limitation can be overtaken by using a model of SMBG measurement error, which allows to generate an unlimited number of virtual SMBG error realizations.

Therefore models of the SMBG measurement error allowing the simulation of synthetic SMBG data are very useful tools to integrate within T1D simulation platforms, like the UVA/Padova T1D simulator, that could then be used to perform ISCT, for example, to assess the impact of SMBG error on the safety and effectiveness of SMBG-based treatments.

Literature models of SMBG measurement error

Many studies assessed the accuracy and precision of SMBG devices, but only a few studies attempted to model the SMBG measurement error [32–38,40]. All these studies focused on modeling the SMBG measurement error distribution by one or more probability density function (PDF) models and considered the SMBG measurements as independently sampled from such a PDF model, that is, assuming uncorrelation between consecutive SMBG measurements. Such uncorrelation assumption is reasonable for SMBG samples because of the sparseness of these measurements, typically collected 3–5 times per day (as opposed to continuous glucose monitoring (CGM) measurements, whose quasicontinuous nature allows descriptions of the measurement error autocorrelation by using first or second-order autoregressive models [28–31]).

A first simple model of the SMBG measurement error distribution was proposed in a work by Boyd and Bruns [32] and consisted in modeling the SMBG relative error, E^{rel}, by a Gaussian distribution whose mean and SD describe the analytical bias and imprecision of the glucose meter:

$$E^{rel} = \frac{X - R}{R} \sim N(\text{Bias, SD}) \tag{5.1}$$

where X and R represent the SMBG measurement and the reference glucose value, respectively. This model was used for several applications in the literature, because it is simple and easy to apply. Boyd and Bruns used the Gaussian model of Eq. (5.1) to simulate different levels of accuracy and precision in SMBG measurements and assessed their impact on insulin dosing errors in patients who use a sliding scale for correcting insulin boluses according to SMBG measurements [32]. The same model was applied in other studies to assess the effect of SMBG analytical errors on insulin dosing errors in intensive care unit patients on tight glycemic control [33–35] and in the presence of carb-counting errors [36].

However, none of these studies validated the simple Gaussian model against empirical data. Moreover, literature evidences suggest that this simple model may represent a suboptimal, too simplistic, description of the SMBG measurement error. In particular, a critical point of this model concerns the use of a single canonical PDF over the entire BG range. In fact, scatter plots reported in Refs. [1–3,6] show that neither absolute nor relative error of SMBG data presents constant mean and SD over the entire BG range.

To deal with this issue, in a work by Breton and Kovatchev [37] a Gaussian PDF model with zero mean and SD dependent on the BG value was adopted. In particular, the relationship between SD and BG was tuned to simulate the performance of a meter satisfying the ISO 15197:2003 requirements, although Breton and Kovatchev did not explicitly report the SD-BG relationship in their publication [37]. In a work by Pretty et al. [38], a nonparametric approach was adopted, in which a bivariate kernel density model of the joint distribution of SMBG measurements and reference values was derived for the Abbott Optium Xceed (Abbott Diabetes Care, Alameda,

CA) device. Nevertheless, none of these models was validated, for example, by comparison of error realizations generated by the model versus an independent set of observations.

Another critical point of the simple Gaussian PDF model concerns the symmetry assumption of SMBG measurement error distribution. In fact, the histograms reported in Chan et al. [39] show that some SMBG devices present an asymmetric error distribution, calling for the use of PDF models allowing for nonzero skewness. This is visible also in Fig. 5.1, where the SMBG relative error distribution is reported for a sample collected with the One Touch Ultra 2 (Lifescan Inc., Milpitas, CA) device.

To overcome the limitations of the simple Gaussian model, our research group at the University of Padova proposed a new methodology to model the SMBG measurement error's PDF, which was presented in Vettoretti et al. [40]. Specifically, this method (i) deals with the variability of SMBG error characteristics with BG by using multiple PDF models in different zones of the glucose range and (ii) takes into account the asymmetry of the SMBG error distribution by using PDF models allowing for nonzero skewness. A similar approach was used in a recent work by Campos-Nañez et al. [42], where the error of 43 commercial SMBG devices was modeled using two Johnson distribution PDF models, one to describe the absolute

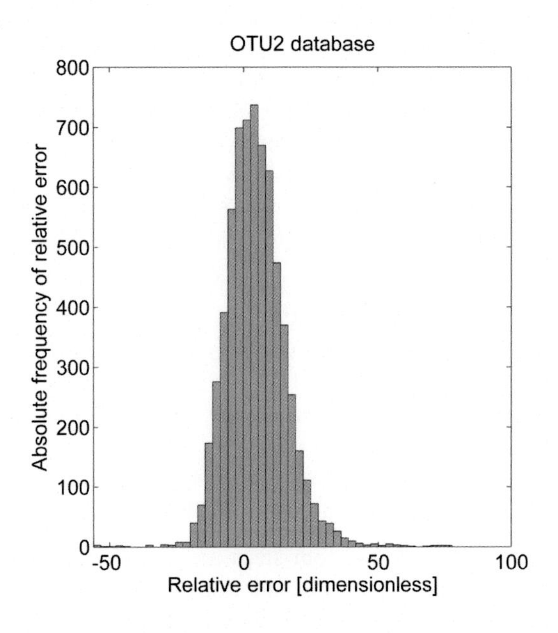

FIGURE 5.1

Histogram (absolute frequencies) of the SMBG relative error (in percentage) for a sample collected with the One Touch Ultra 2 device.

Adapted from Vettoretti M, Facchinetti A, Sparacino G, Cobelli C. A model of self-monitoring blood glucose measurement error. Journal of Diabetes Science and Technology 2017;11(4):724–735.

error in the low glucose range (BG < 100 mg/dL) and one to describe the relative error in the high glucose range (BG $\geq$ 100 mg/dL). The methodology by Vettoretti et al. was applied to two literature datasets of SMBG data, collected by One Touch Ultra 2 and Bayer Contour Next USB (Bayer HealthCare LLC, Diabetes Care, Whippany, NJ), and reference BG samples, measured by the YSI (YSI Inc., Yellow Spring, OH) laboratory equipment [40,41]. A two-zone SMBG error model employing skew-normal distributions was derived for each of the two considered datasets [40]. The models were validated by goodness-of-fit tests, which showed the superiority of the two-zone skew-normal PDF model compared to the simple Gaussian PDF model previously employed in the literature.

In the following section, the method proposed by Vettoretti et al., which currently represents the state-of-the-art approach for modeling the SMBG error distribution, is described in detail.

The state-of-the-art modeling method by Vettoretti et al.

To create a model of the SMBG measurement error PDF, two strategies are possible. The first is to resort to a parametric approach in which it is assumed that the PDF we want to describe presents a known shape characterized by a fixed, and usually small, number of parameters. The disadvantage of this approach is that it requires some assumptions. The second possibility is to use nonparametric approaches, but coping with the risk of overfitting (i.e., to include dataset-specific phenomena in the model) can be difficult. Moreover, nonparametric methods do not allow to obtain any information on the PDF's properties, for example, they do not allow to establish if the distribution is unimodal, it presents a positive skewness, etc.

Recently, our research group at the University of Padova proposed a new methodology to develop and validate parametric models of the SMBG error PDF [40], which consists of four steps:

A. definition of a training and a test set;
B. identification of BG zones of the training set in which the errors present a constant-SD distribution;
C. maximum-likelihood (ML) fitting of a PDF model to errors in each identified zone of the training set;
D. model validation by comparing the distribution of random samples simulated by the model with the distribution of data in the test set.

In the following sections, these four steps are described in detail. Of note, the method by Vettoretti et al. is general and can be easily applied to any dataset containing SMBG datapoints and matched BG references. Later in this chapter, the method will be applied to two case studies to develop a model of SMBG measurement error for two commercially available glucose meters (One Touch Ultra 2 and Bayer Contour Next USB).

Definition of training and test sets

Let us suppose we have a dataset containing n_{tot} SMBG measurements, x_i, $i = 1, \ldots, n_{tot}$, and n_{tot} reference samples, r_i, $i = 1, \ldots, n_{tot}$, collected by high-precision and accuracy laboratory instruments. For each pair (x_i; r_i), the SMBG absolute error, e_i^{abs}, and the SMBG relative error, e_i^{rel}, are calculated as follows:

$$e_i^{abs} = x_i - r_i, \quad e_i^{rel} = \frac{x_i - r_i}{r_i} \cdot 100 \tag{5.2}$$

Note that here we refer to absolute error and relative error as the signed difference and the signed percent difference between the SMBG measurement and its reference sample, respectively.

Error data are then divided into two parts. The first part, with cardinality $n_{training}$, is used as training set to derive the model of SMBG error PDF in the following steps B and C. The second part, with cardinality n_{test}, is used as test set to validate the model in step D. Absolute and relative errors of training set can be displayed in a scatter plot versus reference glucose to visually assess if the characteristics of the error distribution (e.g., mean and dispersion) significantly vary across the reference glucose range.

Constant-SD zones identification

Changes in the dispersion of absolute and relative errors with reference glucose are quantified in the training set by analyzing the sample SD. In particular, first a uniform grid g_i, $i = 1, \ldots, n_g$, where n_g is the number of points in the grid, is defined in the glucose range with step S (e.g., S = 5 mg/dL). Then, intervals centered at points g_i, $i = 1, \ldots, n_g$, with half-width L (e.g., $L = 15$ mg/dL) are defined. Finally, the sample SD of absolute and relative errors in each interval $g_i \pm L$ is calculated, which approximates the error SD (absolute or relative) at the glucose point g_i. The plot of sample SD values versus glucose points g_i, $i = 1, \ldots, n_g$, allows to visualize how the error SD (absolute or relative) varies across the glucose range and identify zones of the glucose range in which either absolute or relative error presents an approximately constant-SD distribution.

Maximum-likelihood fitting

In each constant-SD zone, the distribution of the error (absolute or relative) in the training set, here represented by the continuous random variable Y, is fitted by ML using a certain PDF model. In particular, we recommend first to test the error distribution for normality in each zone (e.g., using the Lilliefors test). Then, if the test cannot reject the hypothesis of normality, the Gaussian PDF model can be adopted, as there is no sufficient evidence to justify the use of a more complex model. Conversely, if the normality test rejects the normality hypothesis, a different PDF model should be used. A convenient choice is the skew-normal PDF model, an extension of the normal distribution allowing for nonzero skewness [43]. The skew-normal PDF is described by the following equation:

$$f_Y(y) = \frac{2}{\omega} \cdot \phi\left(\frac{y-\xi}{\omega}\right) \cdot \Phi\left(\alpha \cdot \frac{y-\xi}{\omega}\right) \tag{5.3}$$

where $\phi(\cdot)$ and $\Phi(\cdot)$ are the PDF and the cumulative distribution function of the standard Gaussian random variable, respectively, while the scalars ξ, ω, and α are the location, scale, and skewness parameters of the skew-normal PDF. The skew-normal PDF model allows to describe both positively ($\alpha > 0$) or negatively ($\alpha < 0$) skewed PDFs. Of note, when α is equal to zero, the skew-normal PDF becomes the Gaussian PDF.

Let us define $\mathbf{y} = [y_1, \ldots, y_n]^T$ as the vector storing the n samples of the error (absolute or relative), that is, n realizations of Y, falling within one of the identified constant-SD glucose zones. Then, the parameters of the skew-normal PDF that best reproduces the distribution of the data $\mathbf{y}$ are estimated by maximizing the log-likelihood function:

$$l(\xi, \omega, \alpha) = -n \cdot \log(\omega) - \frac{1}{2} z^T z - \sum_{i=1}^{n} \log(2\Phi(\alpha z_i)) \tag{5.4}$$

where z_i, $i = 1, \ldots, n$, are the n components of $\mathbf{z} = (\mathbf{y} - \xi \cdot \mathbf{1_n})/\omega$, with $\mathbf{1_n}$ representing the nx1 unitary vector.

If the error distribution does not present a significant number of left or right outliers, that is, values in the left or right tails of the error histogram with a frequency significantly greater than zero, the model of Eq. (5.3) should be sufficiently accurate to describe the error PDF. On the contrary, if the SMBG error presents nonnegligible left or right outliers, it is more convenient to use a composite model in which the distributions of nonoutliers and outliers are described by different PDF models. For example, a model obtained combining the skew-normal PDF and the exponential PDF can be used. To derive such a model, two thresholds must be identified, T_1 and T_2, to separate the nonoutlier region, $T_1 \le y \le T_2$, from the left outlier region, $y < T_1$, and the right outlier region, $y > T_2$. Let us define a discrete random variable V that can assume three values: 0, with probability $p_V(0)$, when the SMBG error y is not an outlier; 1, with probability $p_V(1)$, when y is a left outlier; 2, with probability $p_V(2)$, when y is a right outlier. Then, the model defined for the PDF of Y, $f_Y(y)$, is conditioned by the value of V. In particular, when $V = 0$ the conditional PDF of Y given V, $f_{Y|V}(y|0)$, is the skew-normal PDF model of Eq. (5.3), whose parameters are identified by ML as described earlier. When $V = 1$, $f_{Y|V}(y|1)$ is described by an exponential PDF model reversed with respect to the ordinate axis and shifted left of T_1:

$$f_{Y|V}(y|1) = \begin{cases} \lambda_1 e^{\lambda_1 (y+T_1)} & y \le T_1 \\ 0 & y > T_1 \end{cases} \tag{5.5}$$

When $V = 2, f_{Y|V}(y|2)$ is described by an exponential PDF model shifted right of T_2:

$$f_{Y|V}(y|2) = \begin{cases} \lambda_2 e^{-\lambda_2(y-T_2)} & y \geq T_2 \\ 0 & y < T_2 \end{cases} \tag{5.6}$$

Parameters λ_1 and λ_2 are estimated by ML by inverting the sample mean of left or right outliers properly shifted and reversed. Finally, the PDF of Y is obtained as follows:

$$f_Y(y) = f_{Y|V}(y|0) \cdot p_V(0) + f_{Y|V}(y|1) \cdot p_V(1) + f_{Y|V}(y|2) \cdot p_V(2) \tag{5.7}$$

where the values $p_V(1)$ and $p_V(2)$ are estimated by dividing the number of observed left or right outliers by the total number of observations in the specific constant-SD zone, while $p_V(0)$ is found as follows:

$$p_V(0) = 1 - p_V(1) - p_V(2) \tag{5.8}$$

Remark. We propose for step C the use of canonical PDF models, like the skew-normal and the exponential, as they can be easily used to generate random samples in in silico applications. Conversely, the use of a noncanonical model, for example, a spline model, would require much more complex procedures to generate random samples from the model.

Model validation

The model identified in each constant-SD zone for absolute or relative error can be validated by comparing the distribution of random samples simulated by the model with the distribution of the test set data. Model validation is an important step to perform before using the SMBG error model for simulation applications because it allows to check if the model is able to generate realistic realizations of the SMBG error. Such a validation can be performed by calculating the mean absolute difference (MAD) between the empirical distribution function of random samples simulated by the model and the empirical distribution function of the test set data, and by the application of goodness-of-fit tests, like the two-sample Kolmogorov—Smirnov (KS) and Cramér—von Mises (CvM) tests.

For such a purpose, M random samples are simulated by drawing n_{test} values from the model identified in each zone, being n_{test} the cardinality of the test set. In particular, SMBG error values are drawn from the exponential PDF of left outliers with probability $p_V(1)$, from the exponential PDF of right outliers with probability $p_V(2)$ and from the skew-normal PDF of nonoutliers with probability $p_V(0)$. If the model does not include outliers' description, then $p_V(1) = p_V(2) = 0$ and all the n_{test} SMBG error values are drawn from the skew-normal PDF. To sample a random number y from a skew-normal PDF, with generic parameters ξ, ω, and α, a three-step method is provided in Ref. [43]. First, two values u_0 and u_1 having marginal PDF N(0,1) and correlation $\delta = \alpha \big/ \sqrt{1 + \alpha^2}$ are generated by sampling u_0 and r from independent N(0,1) random variables and defining $u_1 = \delta \cdot u_0 + \sqrt{1-\delta^2} \cdot r$. Then, in the second step, a random number z sampled from a skew-normal PDF with parameters $\xi = 0$, $\omega = 1$ and $\alpha \neq 0$ is obtained as follows:

$$z = \begin{cases} u_1 & \text{if } u_0 \geq 0 \\ -u_1 & \text{otherwise} \end{cases} \tag{5.9}$$

As the third step, the realization sampled from the skew-normal PDF with parameters ξ, ω, and α is finally obtained setting $y = \xi + \omega \cdot z$.

For each of the M simulated samples, first, the empirical distribution function (EDF) is calculated, which represents an estimate of the cumulative distribution function. In particular, given a generic random sample, Y_j, $j = 1, \ldots n$, its EDF is defined as follows:

$$\widehat{F}(y) = \frac{1}{n} \sum_{j=1}^{n} I_{[-\infty\ y]}(Y_j) \tag{5.10}$$

where $I_{[-\infty\ y]}$ is 1 if $Y_j \leq y$ and 0 otherwise. Then, the MAD between the EDF of each simulated sample and the EDF of the test set is calculated, and its average value across the M simulated samples is obtained.

Finally, each of the M simulated samples is compared to the test set error data by performing, with significance level β, the two-sample KS and CvM tests, that is, nonparametric tests for the null hypothesis $H_0 =$ "the two samples are drawn from the same distribution" = based on a measure of distance between the EDFs of the two samples. The percentage of simulated samples for which KS and CvM tests reject H_0 is calculated, which should be small if the identified model of SMBG error PDF is accurate.

To avoid the results of the validation being dependent on the particular realization of random samples, we recommend repeating these validation steps N times (e.g., $N = 100$). Specifically, the average MAD and the percentage of samples for which KS and CvM tests reject H_0 can be obtained for the N groups of M random samples and finally their mean, minimum, and maximum values can be calculated.

Derivation of a model of SMBG error distribution for two commercial devices

Case study 1: modeling the One Touch Ultra 2 measurement error

Dataset

The One Touch Ultra 2 (OTU2) dataset was obtained from a larger dataset collected as part of a multicenter study conducted in 2011 (with the original specific aim being to assess the accuracy of a CGM sensor) [44]. For our purpose, in particular, it is relevant to report that 72 subjects (60 with T1D, 12 with type 2 diabetes [T2D]) participated in three clinical sessions in which SMBG measurements were collected twice per hour for a 12-h period using the OTU2 meter. A total of 6906 SMBG samples were collected (on average 95 samples per subject). In parallel to SMBG, highly accurate reference BG measurements were obtained every 15 min by YSI. During the clinical sessions, carbohydrate and insulin administrations were managed to make patients' BG concentration vary in a wide range. Fig. 5.2 shows SMBG and YSI samples collected in a representative subject.

Preprocessing of YSI and SMBG-YSI matching

To obtain realizations of the SMBG measurement error, required to develop a model of its PDF, each SMBG sample should be matched to a corresponding reference value, that is, ideally a YSI value collected at the same time. Unfortunately, the dataset considered for the analysis was originally collected to assess the accuracy of CGM sensors, thus with a scope different from ours. For this reason, as visible in Fig. 5.2, the SMBG data are in most cases not temporally aligned with the YSI samples. Consequently, the two measurements cannot be directly compared. Karon et al. [35] circumvented the problem by retaining in the analysis only the SMBG samples with a reference value collected within 5 min from the SMBG sampling time. Here, we propose a different approach that allows to reduce the number of SMBG data discarded because too far from reference samples, and to reduce the temporal matching error from 5 min to 30 s.

Specifically, YSI sequences are preprocessed according to the following two-step procedure. First, YSI measurements considered not compatible with the BG pattern are removed from the analysis (see, e.g., the full red circle in

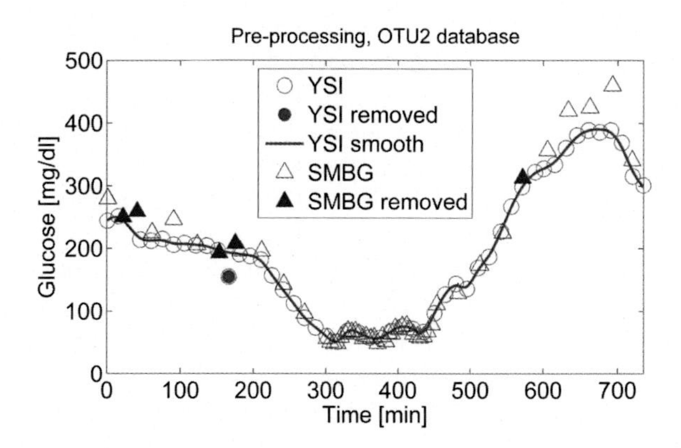

FIGURE 5.2

Preprocessing performed on a 12-h YSI sequence of the OTU2 dataset. The continuous line represents the quasicontinuous-time smoothed profile obtained from YSI measurements (*empty circles*) after removing inaccurate samples (*full circles*). Triangles represent all available SMBG measurements. SMBG samples within YSI gaps are removed from the analysis (*full triangles*).

Adapted from Vettoretti M, Facchinetti A, Sparacino G, Cobelli C. A model of self-monitoring blood glucose measurement error. Journal of Diabetes Science and Technology 2017;11(4):724–735.

Fig. 5.2). In particular, we adopt an empirical rule according to which a YSI sample is considered not compatible with the BG pattern if it deviates upward/downward more than a certain quantity θ from the previous and the following YSI samples, and the following YSI sample is in line with the previous YSI sample, that is, it does not deviate upward/downward more than ε from the previous YSI sample. More specifically, the jth YSI sample (empty red circles in Fig. 5.2), YSI_j, is removed if $YSI_j < YSI_{j-1} - \theta(YSI_{j-1})$, $YSI_j < YSI_{j+1} - \theta(YSI_{j+1})$ and $YSI_{j+1} > YSI_{j-1} - \varepsilon$, or $YSI_j > YSI_{j-1} + \theta(YSI_{j-1})$, $YSI_j > YSI_{j+1} + \theta(YSI_{j+1})$ and $YSI_{j+1} < YSI_{j-1} + \varepsilon$. In particular, we set ε to be 5 mg/dL and $\theta(x)$ to be 15 mg/dL when $x \leq 100$ mg/dL and $15\% \cdot x$ when $x > 100$ mg/dL.

In the second step, from each 12-h YSI sequence, an estimate of the BG profile is reconstructed on a quasicontinuous time grid, for example, with a 1-min step, by the nonparametric smoothing stochastic approach described in Ref. [45], which allows to compensate for the unavoidable presence of measurement error in YSI samples (zero-mean, uncorrelated and with constant CV equal to 2%, according to Ref. [46]). Notably, as shown in Ref. [45], the use of an ML smoothing criterion in this method limits the risk of introducing distortion/bias in the BG profile reconstructed from the YSI samples (as visible in the representative example of Fig. 5.2).

Samples of the BG profile reconstructed on the 1-min step virtual grid (continuous line in Fig. 5.2) can be used as references for the calculation of the SMBG measurement error. For such a scope, each SMBG sample (empty blue triangles in Fig. 5.2) was matched to the nearest (in time) sample of the reconstructed BG profile. Hence, the temporal distance between the SMBG sample and the reference sample was not greater than 30 s.

Of note, when one or more YSI samples are missing, and thus two consecutive YSI measurements are 30 min or more apart, the reconstruction of the BG profile over the gaps could be not reliable. Therefore SMBG measurements falling inside these gaps, that is, 4.37% of all the available SMBG measurements, were excluded from the analysis (full blue triangles in Fig. 5.2).

After preprocessing the data as described, the SMBG samples selected for the analysis resulted well distributed in the glycemic range, with a number of samples in hypoglycemia and hyperglycemia sufficiently high to allow for an accurate description of the SMBG measurement error also in these extreme conditions. In particular, the SMBG samples selected for the analysis were distributed in the glycemic range as follows: 356 samples were in hypoglycemia below or equal to 50 mg/dL, 930 in hypoglycemia above 50 mg/dL and below or equal to 70 mg/dL, 2958 in euglycemia between 70 mg/dL and 180 mg/dL, 1479 in hyperglycemia equal to or above 180 mg/dL and below 250 mg/dL, and 881 in hyperglycemia equal to or above 250 mg/dL.

Model development

The obtained SMBG-BG matched pairs were used to calculate SMBG absolute and relative errors as in Eq. (5.2). Then, the whole set of SMBG error data, with cardinality $n_{tot} = 6604$, was divided into a training set with cardinality $n_{training} = 2/3 \cdot n_{tot}$

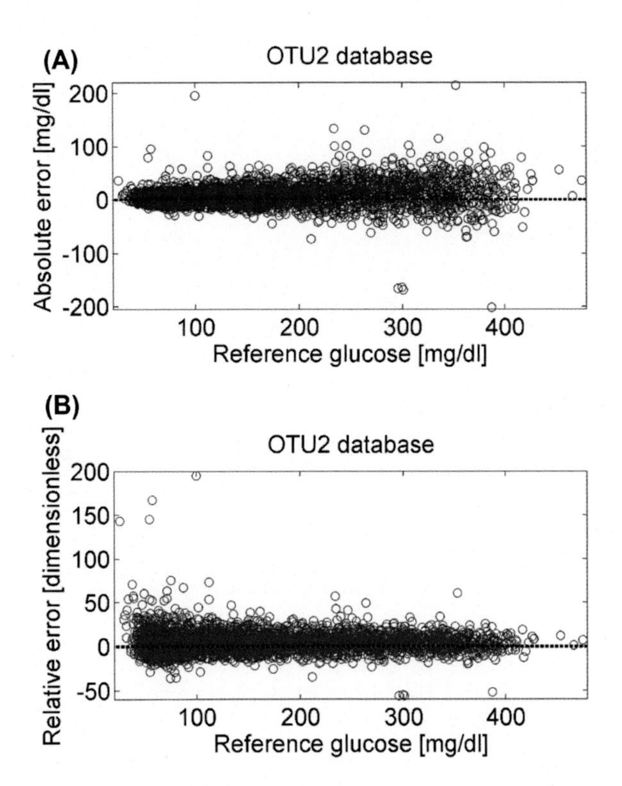

FIGURE 5.3

Scatter plots of absolute (panel A) and relative (panel B) error versus reference glucose for the training set of OTU2 dataset.

Adapted from Vettoretti M, Facchinetti A, Sparacino G, Cobelli C. A model of self-monitoring blood glucose measurement error. Journal of Diabetes Science and Technology 2017;11(4):724–735.

and a test set with cardinality $n_{test} = 1/3 \cdot n_{tot}$. The scatter plots of Fig. 5.3 report absolute (panel A) and relative (panel B) errors in the training set versus reference glucose. The scatter plots evidence that both absolute and relative errors are positively biased and their dispersion is not uniform over the entire glucose range.

In Fig. 5.4, the SD of absolute (panel A) and relative (panel B) error in the training set is displayed for increasing glucose values. In particular, the sample SD of absolute and relative error in the training set was computed on glucose intervals of half-width L = 15 mg/dL centered at the points of a glucose grid with uniform step S = 5 mg/dL. This representation allows to analyze how the SD of absolute and relative error varies across the glucose range. From the plot of Fig. 5.4, we can note that the SD of the absolute error (panel A) is approximately constant for glucose values lower than 75 mg/dL, while an increasing trend is observed for glucose values greater than 75 mg/dL; conversely, the SD of the relative error (panel B) presents a decreasing trend for glucose values lower than 75 mg/dL, while appearing approximately constant for glucose values greater than 75 mg/dL.

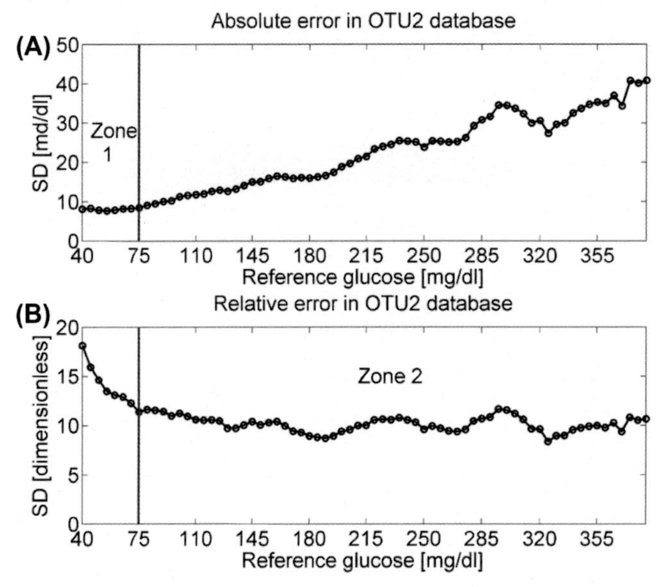

FIGURE 5.4

Absolute (panel A) and relative (panel B) error SD versus reference glucose for the training set of the OTU2 dataset. The threshold to separate zone 1 from zone 2 (75 mg/dL) is evidenced by a *vertical red line.*

Adapted from Vettoretti M, Facchinetti A, Sparacino G, Cobelli C. A model of self-monitoring blood glucose measurement error. Journal of Diabetes Science and Technology 2017;11(4):724–735.

Therefore two constant-SD zones were identified: zone 1, that is, BG $\leq$ 75 mg/dL, with constant-SD absolute error; zone 2, that is, BG > 75 mg/dL, with constant-SD relative error.

The histograms of Fig. 5.5 represent the relative frequency of absolute error in zone 1 (panel A) and relative error in zone 2 (panel B), calculated in the training set. For both error zones, we applied the Lilliefors test to test the null hypothesis that the absolute error in zone 1 and the relative error in zone 2 were normally distributed. With a 5% significance level, the test rejected the normality hypothesis for both absolute and relative error (*P*-value = .001). This result suggests that the Gaussian PDF is not a proper model to accurately describe the observed SMBG error distribution. As the error distributions in zone 1 and 2 did not show significant outliers, a simple skew-normal PDF model (black line in Fig. 5.5) was fitted by ML both in zone 1 (panel A) and 2 (panel B). Indeed, in this dataset, the skew-normal model alone was sufficient to well fit both the central part and the tails of the SMBG error distributions (as visible in Fig. 5.5), while the use of the exponential distributions did not improve the data fit (not shown). The values of location, scale, and skewness parameters identified in zone 1 and 2, as well as mean and SD of the identified models, are reported in Table 5.1. In particular, both error models present a significant positive mean (2.01 mg/dL in zone 1, 4.73% in zone 2) and positive skewness ($\alpha = 2.72$ in zone 1 and $\alpha = 1.41$ in zone 2).

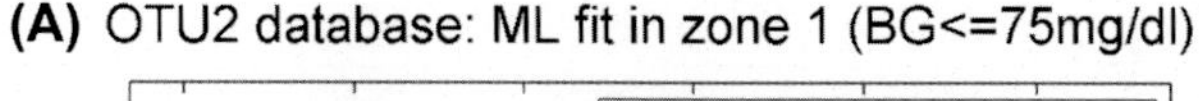

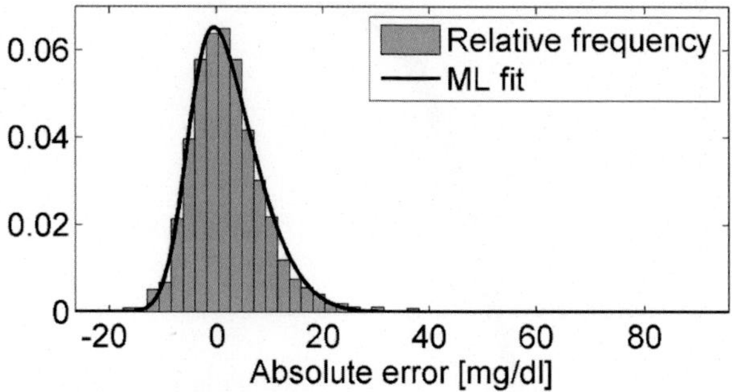

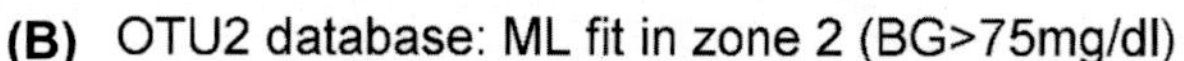

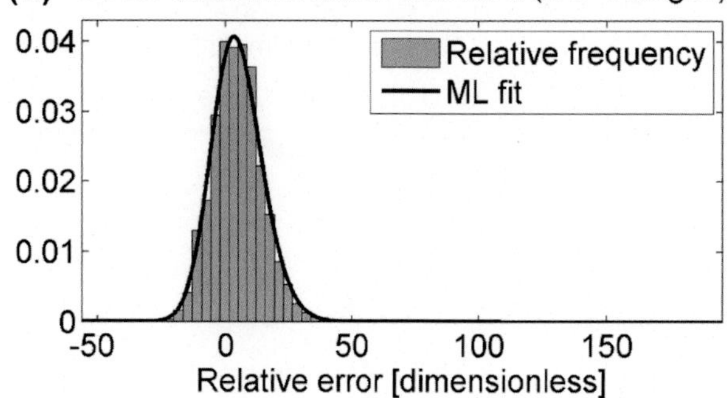

FIGURE 5.5

ML fit of the skew-normal PDF (*black line*) against histograms of the absolute error in zone 1 (panel A) and relative error in zone 2 (panel B) for the training set of the OTU2 dataset.

Adapted from Vettoretti M, Facchinetti A, Sparacino G, Cobelli C. A model of self-monitoring blood glucose measurement error. Journal of Diabetes Science and Technology 2017;11(4):724–735.

Table 5.1 OUT2 model parameters and second-order statistical description.

	Parameters of skew-normal PDF model			Second-order statistical description	
Zone	ξ	ω	α	**Mean**	**SD**
1	−5.37	9.86	2.72	2.01	6.54
2	−3.83	13.17	1.41	4.73%	10.00%

Data from Vettoretti M, Facchinetti A, Sparacino G, Cobelli C. A model of self-monitoring blood glucose measurement error. Journal of Diabetes Science and Technology 2017;11(4):724–735.

Model validation

For validation purposes, the identified PDF model was used to generate, for each zone, $N = 100$ replicates of $M = 500$ random samples having cardinality equal to the number of error data available in the test set for the same glucose zone. As visible in Fig. 5.6, the EDF of random samples simulated by the identified PDF model (blue solid lines) is very similar to the EDF calculated for test set data (red solid line) both in zone 1 (panel A) and 2 (panel B).

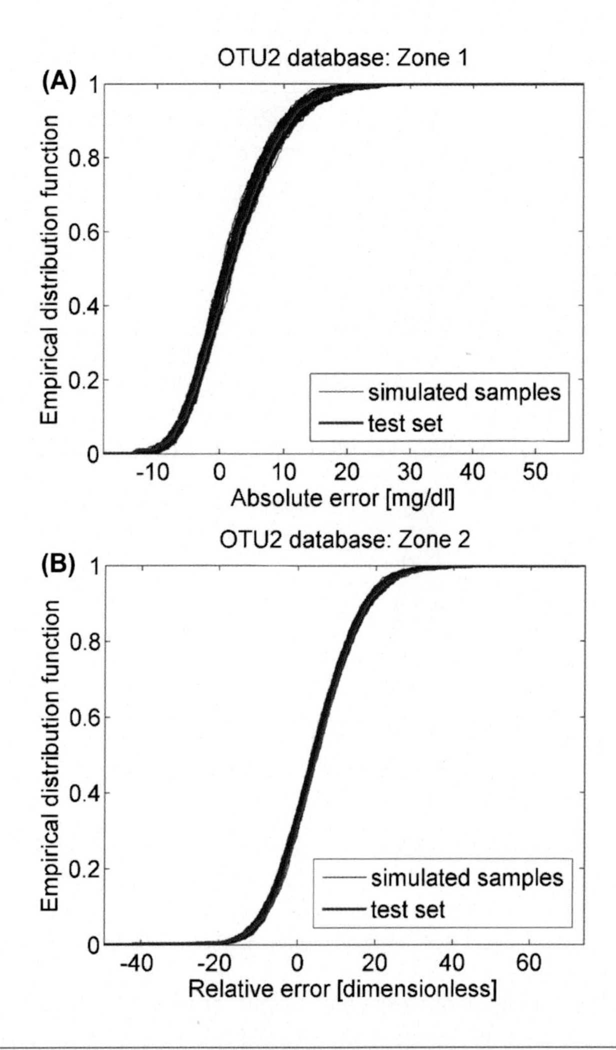

FIGURE 5.6

The EDF of the simulated random samples (*blue line*) and test set data (*red line*) is reported for zone 1 and 2 of the OTU2 dataset (panel A and panel B, respectively).

Adapted from Vettoretti M, Facchinetti A, Sparacino G, Cobelli C. A model of self-monitoring blood glucose measurement error. Journal of Diabetes Science and Technology 2017;11(4):724–735.

The mean, minimum, and maximum values of the average MAD between the EDF of the simulated datasets and the EDF of the test set data, obtained across the 100 groups of 500 simulated random samples, are reported in Table 5.2. Two-sample KS and CvM tests were also performed with significance level $\beta = 5\%$ on the N groups of M simulated random samples. On average (minimum–maximum), the KS test rejected H_0 for 0.53% (0.00%–1.40%) of zone 1 simulated samples and 1.94% (0.40%–4.6%) of zone 2 simulated samples, while the CvM test

Table 5.2 Validation of the OTU2 model. Mean (minimum–maximum) across 100 replicates of the average MAD between 500 simulated EDFs and test set EDF.

Two-zone skew-normal model		Two-zone Gaussian model		Single-zone Gaussian model
Zone 1	**Zone 2**	**Zone 1**	**Zone 2**	
0.0064 (0.0062 –0.0065)	0.0049 (0.0048 –0.0050)	0.0172 (0.0170 –0.0175)	0.0079 (0.0077 –0.0079)	0.0115 (0.0114 –0.0116)

Data from Vettoretti M, Facchinetti A, Sparacino G, Cobelli C. A model of self-monitoring blood glucose measurement error. Journal of Diabetes Science and Technology 2017;11(4):724–735.

rejected H_0 for 0.94% (0.00%–2.20%) of zone 1 simulated samples and 4.15% (1.40%–6.60%) of zone 2 simulated samples. As, for almost all the simulated random samples, H_0 cannot be rejected and the MAD between the EDFs of simulated samples and test set data results small (Table 5.2), we can conclude that the two-zone skew-normal model derived for OTU2 accurately reproduces the SMBG error distribution observed in the test set.

MAD and two-sample KS and CvM tests also demonstrate that the identified two-zone skew-normal model outperforms simpler Gaussian models like the single-zone Gaussian model, previously used in the literature to describe SMBG relative error distributions, and the two-zone Gaussian model in which a Gaussian (instead of skew-normal) PDF is used to describe absolute error in zone 1 and relative error in zone 2. Indeed, when the single-zone Gaussian model was used, the mean of average MAD (Table 5.2) was about twice the mean of average MAD obtained with the two-zone skew-normal model. In addition, the two goodness-of-fit tests rejected H_0 for almost 100% of the simulated samples (on average, the KS and CvM test rejected H_0 for 99.44% and 99.85% of the simulated samples). When the two-zone Gaussian model was used, the mean of average MAD was about twice the value obtained by the two-zone skew-normal model, and H_0 was rejected for more than 50% of zone 1 simulated samples (on average: 49.80%, for KS, and 70.00%, for CvM) and about 25% of zone 2 simulated samples (on average: 27.06%, for KS, and 24.03%, for CvM).

Case study 2: modeling the Bayer Contour Next measurement error

Dataset

The Bayer Contour Next (BCN) dataset was extracted from a larger dataset collected as part of a multicenter study conducted in 2014 (the original-specific aim was to assess the accuracy of a CGM sensor) [47]. The study involved 51 subjects (44 with T1D, 7 with T2D) who participated in a 12-h clinical session in which BG concentration was monitored every 30 min by an SMBG device, the BCN, and every 15 min by YSI. A total of 1410 SMBG measurements were collected (on average 27 per subject). Similarly, to the OTU2 dataset, diet, and insulin therapy were arranged to make BG vary in a wide range. Fig. 5.7 shows the SMBG and YSI data (blue triangles and red circles, respectively) collected during a clinical session for a representative subject.

Preprocessing of YSI and SMBG-YSI matching

As for the OTU2 dataset, YSI data were preprocessed to remove possible outliers (full red circles in Fig. 5.7) and derive a quasicontinuous smoothed profile (continuous red line in Fig. 5.7) by nonparametric Bayesian smoothing [45]. Then, SMBG measurements were matched to the nearest (in time) sample of the YSI smoothed profile, which was used as a reference sample. As done for the OTU2 dataset, when one or more YSI samples were missing, and thus two consecutive YSI measurements were 30 min or more far from each other, the reconstruction of the BG profile over the gaps was considered not reliable, and the SMBG measurements falling inside these gaps were excluded from the analysis (full blue triangles in Fig. 5.7). Specifically, 3.05% of available SMBG measurements were excluded from the analysis.

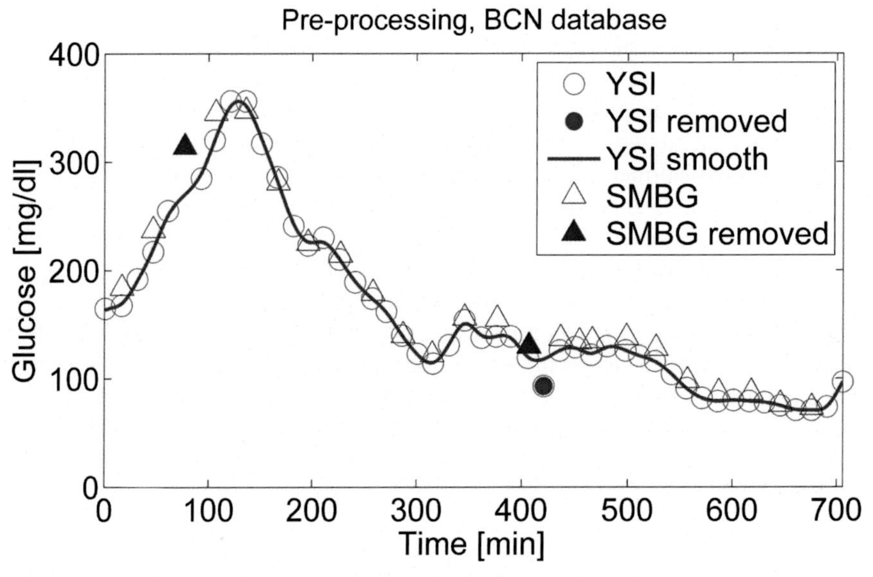

FIGURE 5.7

Preprocessing performed on a 12-h YSI sequence of the BCN dataset. The continuous line represents the quasicontinuous-time smoothed profile obtained from YSI measurements (*empty circles*) after removing inaccurate samples (*full circles*). *Triangles* represent all available SMBG measurements. SMBG samples within YSI gaps are removed from the analysis (*full triangles*).

Adapted from Vettoretti M, Facchinetti A, Sparacino G, Cobelli C. A model of self-monitoring blood glucose measurement error. Journal of Diabetes Science and Technology 2017;11(4):724–735.

After preprocessing the data, the SMBG samples selected for the analysis resulted well distributed in the glycemic range with 123 samples below 50 mg/dL, 159 between 50 and 70 mg/dL, 463 between 70 and 180 mg/dL, 399 between 180 and 250 mg/dL, and 222 above 250 mg/dL.

Model development

The obtained SMBG-BG matched pairs ($n_{tot} = 1366$) were used to calculate SMBG absolute and relative errors, which were divided into a training set ($n_{training} = 2/3n_{tot}$) and a test set ($n_{test} = 1/3n_{tot}$). As for the OTU2 dataset, both absolute and relative error of the training set did not present a uniform dispersion over the entire glucose range, as demonstrated by the scatter plots in Fig. 5.8.

Then, the sample SD of absolute and relative error in the training set was plotted for glucose intervals of half-width $L = 15$ mg/dL centered at the points of a glucose grid with uniform step $S = 5$ mg/dL (Fig. 5.9). Similar to the OTU2 dataset, also, in

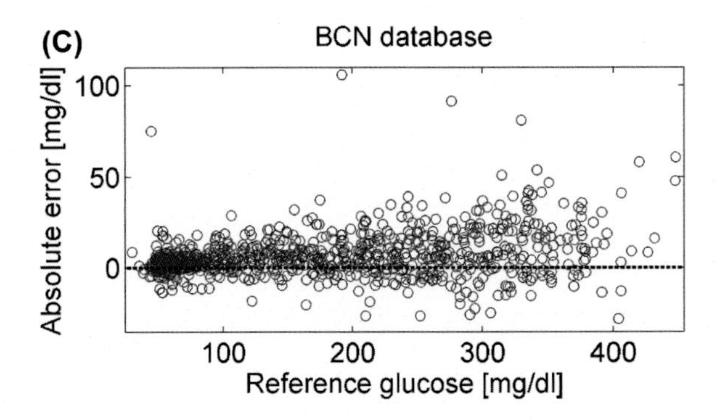

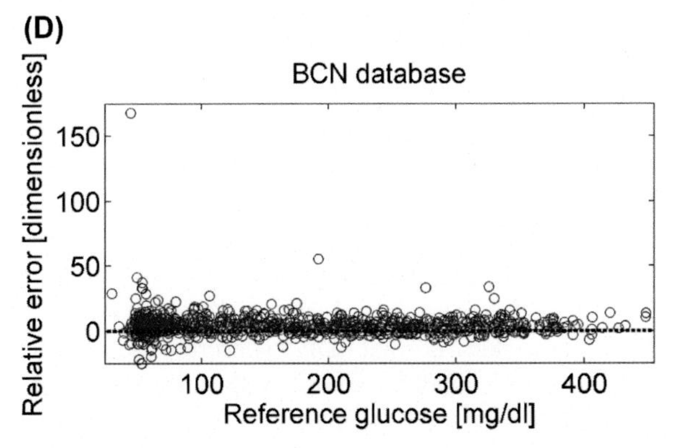

FIGURE 5.8

Scatter plots of absolute (panel C) and relative (panel D) error versus reference glucose for the training set of the BCN dataset.

Adapted from Vettoretti M, Facchinetti A, Sparacino G, Cobelli C. A model of self-monitoring blood glucose measurement error. Journal of Diabetes Science and Technology 2017;11(4):724–735.

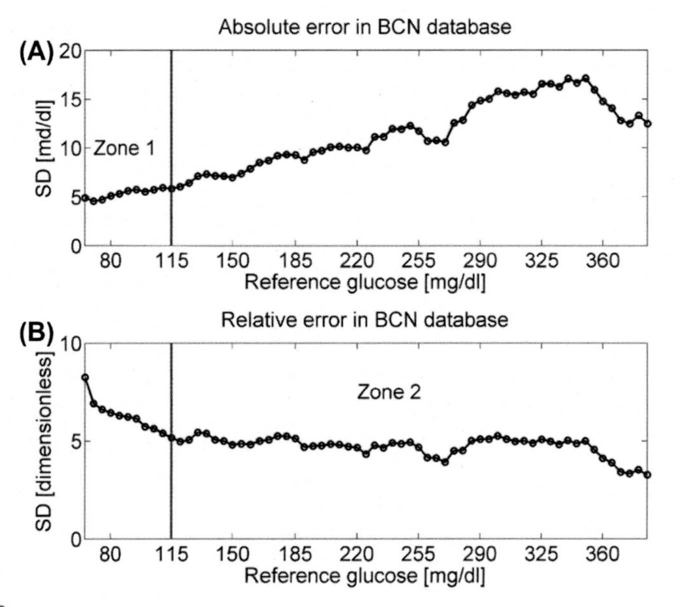

FIGURE 5.9

Absolute (panel C) and relative (panel D) error SD versus reference glucose for the training set of the BCN dataset. The threshold to separate zone 1 from zone 2 (115 mg/dL) is evidenced by a *vertical red line.*

Adapted from Vettoretti M, Facchinetti A, Sparacino G, Cobelli C. A model of self-monitoring blood glucose measurement error. Journal of Diabetes Science and Technology 2017;11(4):724–735.

this case, two zones can be identified: zone 1, that is, BG $\leq$ 115 mg/dL, with constant-SD absolute error; zone 2, that is, BG $>$ 115 mg/dL, with constant-SD relative error.

The histograms of the relative frequency of absolute error in zone 1 (panel C) and relative error in zone 2 (panel D) are reported in Fig. 5.10. For zones 1 and 2, the Lilliefors test for normality rejected the null hypothesis that the error (absolute for zone 1 and relative for zone 2) is normally distributed, therefore also in this case the Gaussian PDF model is not suitable to represent the error data distribution. The histograms of Fig. 5.10 evidence that some outliers are present in the error distribution both in zone 1 and in zone 2; thus, in this dataset, we fitted the error PDF model combining the skew-normal PDF, to describe the central part of the error distribution, and two exponential distributions, to describe the left and right outliers' distribution. The thresholds T_1 and T_2 to identify left and right outliers, respectively, were defined as follows: $T_1 = Q_1 - 1.5 \cdot (Q_3 - Q_1)$ and $T_2 = Q_3 + 1.5 \cdot (Q_3 - Q_1)$, where Q_1 and Q_3 represent the first and the third quartiles, respectively. According to this definition, in zone 1, values lower than -7.5 mg/dL were considered left outliers and values greater than 15.5 mg/dL were considered right outliers, while, in zone 2, we considered left outliers values under -11% and right outliers values over 17.5%.

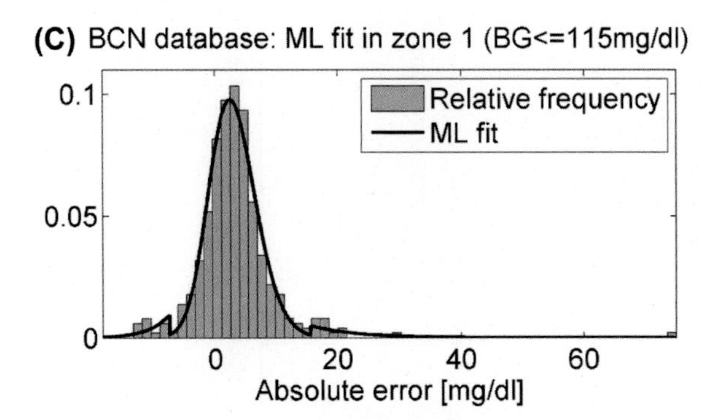

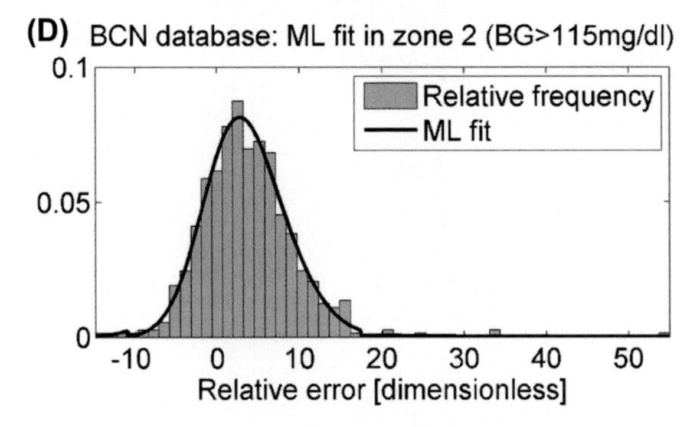

FIGURE 5.10

ML fit of the model obtained as a combination of the skew-normal PDF and the exponential PDF (*black line*) against histograms of the absolute error in zone 1 (panel C) and relative error in zone 2 (panel D) for the training set of the BCN dataset.

Adapted from Vettoretti M, Facchinetti A, Sparacino G, Cobelli C. A model of self-monitoring blood glucose measurement error. Journal of Diabetes Science and Technology 2017;11(4):724–735.

Left and right outliers' distributions were fitted by the exponential PDF models of Eqs. (5.5) and (5.6), while the nonoutliers' distribution was fitted by the skew-normal PDF of Eq. (5.3). The parameters of the identified models are reported in Table 5.3. Fig. 5.10 shows how the composite PDF obtained by Eq. (5.7) (black line) well approximates the error distribution of the training set data both in zone 1 (panel C) and in zone 2 (panel D).

Model validation

The identified PDF model was used to generate, for each zone, $N = 100$ groups of $M = 500$ random samples having cardinality equal to the number of error data available in the test set for the same glucose zone. The mean, minimum, and maximum values of average MAD between the EDF of simulated datasets (blue solid line in Fig. 5.11) and the EDF of test set data (red solid line in Fig. 5.11), obtained across the 100 groups of 500 simulated random samples, are reported in Table 5.4. The low average MAD values suggest that the two-zone skew-normal-exponential model is

Table 5.3 BCN model parameters and second-order statistical description.

	Nonoutliers, skew-normal model parameters			Left outliers, exponential model parameters			Right outliers, exponential model parameters			Second-order statistical description	
Zone	ξ	ω	α	λ_1	T_1	$p_V(1)$	λ_2	T_2	$p_V(2)$	**Mean**	**SD**
1	−0.23	5.05	1.36	0.30	−7.5	0.032	0.13	15.5	0.038	2.01	6.54
2	−0.71	6.54	1.46	0.46	−11	0.005	0.06	17.5	0.011	4.73%	10.00%

Data from Vettoretti M, Facchinetti A, Sparacino G, Cobelli C. A model of self-monitoring blood glucose measurement error. Journal of Diabetes Science and Technology 2017;11(4):724–735.

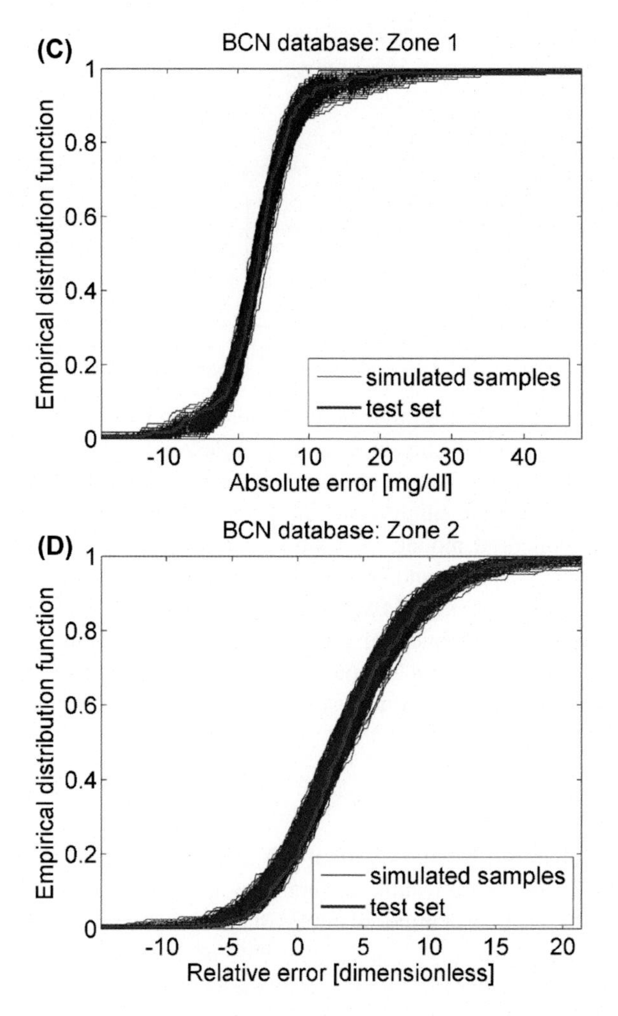

FIGURE 5.11

The EDF of the simulated random samples (*blue line*) and test set data (*red line*) is reported for zones 1 and 2 of the BCN dataset (panel C and panel D, respectively).

Adapted from Vettoretti M, Facchinetti A, Sparacino G, Cobelli C. A model of self-monitoring blood glucose measurement error. Journal of Diabetes Science and Technology 2017;11(4):724–735.

able to well reproduce the SMBG error distribution observed in the test set. This is confirmed by the results of the two-sample KS and CvM tests. Indeed, on average, the KS test rejected H_0 for 0.75% (0.00%–1.80%) and 1.43% (0.40%–2.60%) of zone 1 and 2 simulated samples, respectively, while the CvM test rejected H_0 for 0.95% (0.20%–2.20%) and 2.56% (1.40%–4.00%) of zone 1 and 2 simulated samples, respectively (5% significance level).

Table 5.4 Validation of the BCN model. Mean (minimum–maximum) across 100 replicates of the average MAD between 500 simulated EDFs and test set EDF.

Two-zone skew-normal model		Two-zone Gaussian model		Single-zone Gaussian model
Zone 1	Zone 2	Zone 1	Zone 2	
0.0131 (0.0128 −0.0134)	0.0156 (0.0150 −0.0161)	0.0296 (0.0292 −0.0301)	0.0270 (0.0264 −0.0275)	0.0393 (0.0391 −0.0397)

Data from Vettoretti M, Facchinetti A, Sparacino G, Cobelli C. A model of self-monitoring blood glucose measurement error. Journal of Diabetes Science and Technology 2017;11(4):724–735.

In addition, for this dataset, the validation results show that the two-zone skew-normal-exponential model outperforms the single-zone Gaussian model and the two-zone Gaussian model. Indeed, the mean of average MAD (Table 5.4) was about three times larger with the single-zone Gaussian model compared to the two-zone skew-normal-exponential model. In addition, the two goodness-of-fit tests always rejected H_0 both in zone 1 and in zone 2. When the two-zone Gaussian model was used, the mean of average MAD was about twice the value obtained with the two-zone skew-normal-exponential model, and H_0 was rejected for more than 50% of zone 1 simulated samples (on average: 59.21%, for KS, and 81.63%, for CvM) and more than 10% of zone 2 simulated samples (on average: 14.20%, for KS, and 20.67%, for CvM).

Remark

The threshold dividing the two constant-SD zones was 75 mg/dL for OTU2 and 115 mg/dL for BCN. These results are, not surprisingly, coherent with the requirements (in terms of accuracy) imposed on SMBG devices by the standard ISO 15197. Indeed, the 2003 standard [15] requires that 95% of the SMBG values should have an absolute error lower than 15 mg/dL for glucose concentration lower than 75 mg/dL and a relative error lower than 20% in the rest of the range, while the 2013 standard [16] requires that 95% of the SMBG values should have an absolute error lower than 15 mg/dL for glucose concentration lower than 100 mg/dL and a relative error lower than 15% in the rest of the range. In particular, the 75 mg/dL threshold we found for OTU2, which was approved by FDA in 2006, reflects the same partition defined by the standard ISO 15197:2003, while the 115 mg/dL threshold we found for BCN, which was approved by FDA in 2012, is similar to the one defined by the standard ISO 15197:2013.

Applications of the SMBG measurement error models

Models of SMBG measurement error can be used to generate synthetic SMBG data in many interesting applications. For example, a reliable model of SMBG accuracy can be used in in silico experiments for developing, testing, and optimizing insulin therapies to understand how much errors in SMBG readings can affect the calculation of insulin doses, and thus the quality of glycemic control [37,42]. Other possible applications include assessing the influence of SMBG accuracy and precision on calibration algorithms for CGM sensors and algorithms that optimize SMBG-based insulin dosing [48] or the frequency of SMBG testing [49].

An interesting application of the model of the BCN measurement error derived by Vettoretti et al. [40], and presented in this chapter, is its incorporation in the T1D patient decision simulator, a mathematical model of T1D patients making treatment decisions based on glucose monitoring devices, which has been recently developed by our research group at the University of Padova [50]. The T1D patient decision simulator, schematized in Fig. 5.12, receives input meal data (I) and patient-specific parameters describing both the patient's physiology (P1) and therapy (P2). The simulator output is the patient's BG concentration (O), which is simulated every minute, from which glycemic outcomes can be calculated. The T1D patient decision simulator includes four main components:

A. the UVA/Padova T1D simulator that describes glucose–insulin–glucagon dynamics in T1D subjects [26,27];
B. a model of the device used for glucose monitoring that simulates SMBG [40] and/or CGM measurements [31];
C. a model of the patient's behavior in making treatment decisions, for example, insulin dosing and hypoglycemia treatments;
D. a model of insulin delivery, for example, by insulin pump or multiple daily injections.

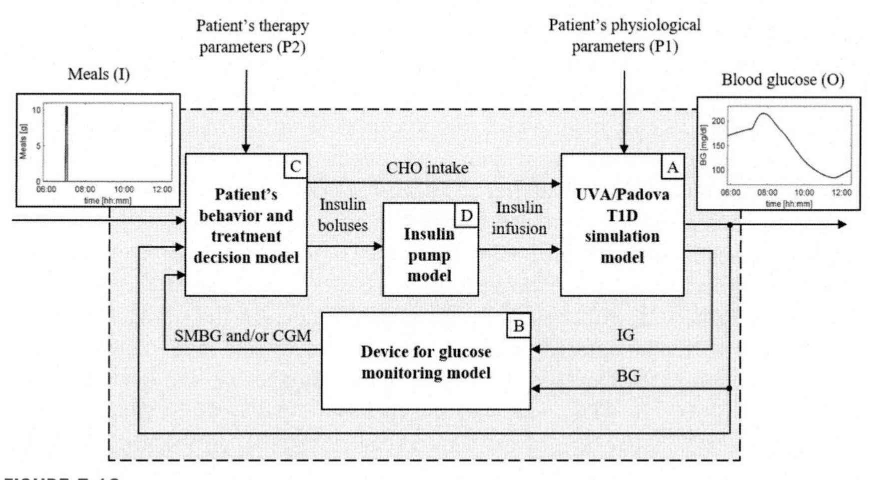

FIGURE 5.12

Schematic representation of the T1D patient decision simulator.

From Vettoretti M, Facchinetti A, Sparacino G, Cobelli C. Type-1 diabetes patient decision simulator for in silico testing safety and effectiveness of insulin treatments. IEEE Transactions on Biomedical Engineering 2018; 65(6):1281–1290; with permission.

The BCN error model is implemented within the *Device for glucose monitoring model* (block B in Fig. 5.12) to simulate SMBG measurements starting from the BG value returned by the UVA/Padova T1D simulator, whenever the patient's behavior and treatment decisions model (block C in Fig. 5.12) requests a BG check by SMBG. Specifically, if the simulated BG value, BG_{sim}, is in the zone 1 of the SMBG error PDF model (i.e., below 115 mg/dL), the SMBG absolute error, err_{abs}, is sampled from the PDF model identified in zone 1. Then, the simulated SMBG measurements, $SMBG_{sim}$, is obtained as follows:

$$SMBG_{sim} = err_{abs} + BG_{sim} \tag{5.11}$$

Conversely, if BG_{sim} is in zone 2, the SMBG relative error, err_{rel}, is sampled from the PDF model identified in zone 2. Then, $SMBG_{sim}$ is obtained as follows:

$$SMBG_{sim} = err_{rel} \cdot \frac{BG_{sim}}{100} + BG_{sim} \tag{5.12}$$

The T1D patient decision simulator was preliminary used to compare nonadjunctive CGM use, that is, the use of CGM to make treatment decisions, versus conventional therapy based on SMBG in 20 virtual adults [51]. Results of standard outcome metrics, for example, time in hypoglycemia, time in the target range, time in hyperglycemia, and the number of hypo/hyperglycemic events, calculated for both CGM and SMBG scenarios, supported the noninferiority of CGM nonadjunctive use versus SMBG use [51].

In 2016, the T1D patient decision simulator was used in collaboration with Dexcom Inc. (San Diego, CA) to run simulations to assess the safety and effectiveness of the nonadjunctive use of the Dexcom G5 Mobile sensor compared to conventional SMBG therapy [52]. In particular, a 2-week ISCT was performed on 40,000 virtual unique adults and pediatric patients, each defined by a different combination of physiology and therapy parameters, showing that, in both the adult and pediatric populations, the risk of hypo and hyperglycemia using CGM for insulin dosing was equivalent to, or even lower than, that obtained using SMBG. These results were included among all the material (clinical study data, analysis of human factors, opinion of clinicians and experts, and the testimony of CGM sensor users) that Dexcom Inc. presented to the Clinical Chemistry and Clinical Toxicology Devices Panel of the US Food and Drug Administration (FDA)'s Medical Device Advisory Committee on July 21, 2016 to ask for a change of label for the G5 Mobile device from adjunctive to nonadjunctive use [53,54]. The panel voted in favor of the safety and effectiveness of the Dexcom G5 Mobile nonadjunctive use, and 6 months later, in December 2016, the FDA approved the Dexcom G5 Mobile as the first CGM device for nonadjunctive use in the United States. This approval had a positive impact also on CGM reimbursement criteria, as in March 2017 Medicare announced covering the costs of therapeutic CGM devices, that is, CGM sensors approved for nonadjunctive use, for all patients with T1D and T2D on intensive insulin therapy [55]. These changes potentially will contribute to extend CGM use and stimulate the development of new CGM-based applications for personalized diabetes treatment and prevention [56].

Conclusion

In conclusion, models of SMBG measurement error are needed to generate synthetic SMBG data in ISCT, which can complement in vivo experiments, with a large saving of resources. Recently, our research group proposed a methodology to develop and validate SMBG measurement error models, which take into account the variability of error characteristics over the glucose range and the possible asymmetric distribution of the error. This methodology is general and, in principle, can be applied to any dataset containing SMBG measurements and BG references, like the OTU2 and BCN datasets presented in this chapter. In silico experiments based on SMBG measurement error models can play an important role in the regulatory approval of medical devices for diabetes therapy. Notably, the BCN model presented in this chapter was used, as part of the T1D patient decision simulator, to demonstrate the safety and effectiveness of CGM nonadjunctive use, in the regulatory process that brought to the FDA approval of the first nonadjunctive CGM system in the United States.

References

[1] Freckmann G, Baumstark A, Jendrike N, Zschornack E, Kocher S, Tshiananga J, Heister F, Haug C. System accuracy evaluation of 27 blood glucose monitoring systems according to DIN EN ISO 15197. Diabetes Technology and Therapeutics 2010;12(3): 221−31.

[2] Freckmann G, Schmid C, Baumstark A, Pleus S, Link M, Haug C. System evaluation of 43 blood glucose monitoring systems for self-monitoring of blood glucose according to DIN EN ISO 15197. Journal of Diabetes Science and Technology 2012;6(5):1060−75.

[3] Freckmann G, Baumstark A, Schmid C, Pleus S, Link M, Haug C. Evaluation of 12 glucose monitoring systems for self-testing: system accuracy and measurement reproducibility. Diabetes Technology and Therapeutics 2014;16(2):113−22.

[4] Bedini JL, Wallace JF, Pardo S, Petruschke T. Performance evaluation of three blood glucose monitoring systems using ISO 15197: 2013 accuracy criteria, consensus and surveillance error grid analyses, and insulin dosing error modeling in a hospital setting. Journal of Diabetes Science and Technology 2016;10(1):85−92.

[5] Zijlstra E, Heinemann L, Fischer A, Kapitza C. A comprehensive performance evaluation of five blood glucose systems in the hypo-, eu-, and hyperglycemic range. Journal of Diabetes Science and Technology 2016;10(6):1316−23.

[6] Baumstark A, Jendrike N, Pleus S, Haug C, Freckmann G. Evaluation of accuracy of six blood glucose monitoring systems and modeling of possibly related insulin dosing errors. Diabetes Technology and Therapeutics 2017;19(10):580−8.

[7] Jendrike N, Baumstark A, Pleus S, Liebing C, Beer A, Flacke F, Haug C, Freckmann G. Evaluation of four blood glucose monitoring systems for self-testing with built-in insulin dose advisor based on ISO 15197:2013: system accuracy and hematocrit influence. Diabetes Technology and Therapeutics 2018;20(4):303−13.

[8] Klonoff DC, Parkes JL, Kovatchev BP, Kerr D, Bevier WC, Brazg RL, Christiansen M, Bailey TS, Nichols JH, Kohn MA. Investigation of the accuracy of 18 marketed blood glucose monitors. Diabetes Care 2018;41(8):1681–8.
[9] Ramljak S, Lock JP, Schipper C, Musholt PB, Forst T, Lyon M, Pfützner A. Hematocrit interference of blood meters for patient self-monitoring. Journal of Diabetes Science and Technology 2013;7(1):179–89.
[10] Schmid C, Haug C, Heinemann L, Freckmann G. System accuracy of blood glucose monitoring systems: impact of use by patients and ambient conditions. Diabetes Technology and Therapeutics 2013;15(10):889–96.
[11] Lam M, Louie RF, Curtis CM, Ferguson WJ, Vy JH, Truong AT, Sumner SL, Kost GJ. Short-term thermal-humidity shock affects point-of-care glucose testing: implications for health professionals and patients. Journal of Diabetes Science and Technology 2014;8(1):83–8.
[12] Baumstark A, Pleus S, Schmid C, Link M, Haug C, Freckmann G. Lot-to-lot variability of test strips and accuracy assessment of systems for self-monitoring of blood glucose according to ISO 15197. Journal of Diabetes Science and Technology 2012;6(5):1076–86.
[13] Rodbard D. Characterizing accuracy and precision of glucose sensors and meters. Journal of Diabetes Science and Technology 2014;8(5):980–5.
[14] Klaff LJ, Brazg R, Hughes K, Tideman AM, Schachner HC, Stenger P, Pardo S, Dunne N, Parkes JL. Accuracy evaluation of Contour Next compared with five blood glucose monitoring systems across a wide range of blood glucose concentrations occurring in a clinical research setting. Diabetes Technology and Therapeutics 2015;17(1):1–8.
[15] International Organization for Standardization. In vitro diagnostic test systems – requirements for blood-glucose monitoring systems for self-testing in managing diabetes mellitus. ISO 15197:2003.
[16] International Organization for Standardization. In vitro diagnostic test systems – requirements for blood-glucose monitoring systems for self-testing in managing diabetes mellitus. ISO 15197:2013.
[17] U.S. Department of Health and Human Services, U.S. Food and Drug Administration. Self-monitoring blood glucose test systems for over-the-counter use: guidance for industry and food and drug administration staff. 2016. Available from: https://www.fda.gov/downloads/ucm380327. [Accessed 5 January 2019].
[18] Brazg RL, Klaff LJ, Parkin CG. Performance variability of seven commonly used self-monitoring of blood glucose systems: clinical considerations for patients and providers. Journal of Diabetes Science and Technology 2013;7(1):144–52.
[19] Heinemann L, Zijlstra E, Pleus S, Freckmann G. Performance of blood glucose meters in the low-glucose range: current evaluations indicate that it is not sufficient from a clinical point of view. Diabetes Care 2015;38:e139–40.
[20] Freckmann G, Pleus S, Link M, Baumstark A, Schmid C, Högel J, Haug C. Accuracy evaluation of four blood glucose monitoring systems in unaltered blood samples in the low glycemic range and blood samples in the concentration range defined by ISO 15197. Diabetes Technology and Therapeutics 2015;17(9):625–34.
[21] Viceconti M, Cobelli C, Haddad T, Himes A, Kovatchev B, Palmer M. In silico assessment of biomedical products: the conundrum of rare but not so rare events in two case studies. Proceedings – Institution of Mechanical Engineers, Part H 2017;231(5):455–66.

[22] Cobelli C, Dalla Man C, Sparacino G, Magni L, De Nicolao G, Kovatchev BP. Diabetes: models, signals, and control. IEEE Reviews in Biomedical Engineering 2009;2:54–96.
[23] Cobelli C, Dalla Man C, Pedersen MG, Bertoldo A, Toffolo G. Advancing our understanding of the glucose system via modeling: a perspective. IEEE Transactions on Biomedical Engineering 2014;61(5):1577–92.
[24] Fritzen K, Heinemann L, Schnell O. Modeling of diabetes and its clinical impact. Journal of Diabetes Science and Technology 2018;12(5):976–84.
[25] Kovatchev BP, Breton M, Dalla Man C, Cobelli C. In silico preclinical trials: a proof of concept in closed-loop control of type 1 diabetes. Journal of Diabetes Science and Technology 2009;3(1):44–55.
[26] Dalla Man C, Micheletto F, Lv D, Breton M, Kovatchev BP, Cobelli C. The UVA/PADOVA type 1 diabetes simulator: new features. Journal of Diabetes Science and Technology 2014;8(1):26–34.
[27] Visentin R, Campos-Náñez E, Schiavon M, Lv D, Vettoretti M, Breton M, Kovatchev BP, Dalla Man C, Cobelli C. The UVA/Padova type 1 diabetes simulator goes from single meal to single day. Journal of Diabetes Science and Technology 2018;12(2):273–81.
[28] Breton M, Kovatchev BP. Analysis, modeling, and simulation of the accuracy of continuous glucose sensors. Journal of Diabetes Science and Technology 2008;2(5):853–62.
[29] Lunn DJ, Wei C, Hovorka R. Fitting dynamic models with forcing functions: application to continuous glucose monitoring in insulin therapy. Statistics in Medicine 2011;30(18):2234–50.
[30] Facchinetti A, Del Favero S, Sparacino G, Castle JR, Ward WK, Cobelli C. Modeling the glucose sensor error. IEEE Transactions on Biomedical Engineering 2014;61(3):620–9.
[31] Facchinetti A, Del Favero S, Sparacino G, Cobelli C. Model of glucose sensor error components: identification and assessment for new Dexcom G4 generation devices. Medical and Biological Engineering and Computing 2015;53(12):1259–69.
[32] Boyd JC, Bruns DE. Quality specifications for glucose meters: assessment by simulation modelling of errors in insulin dose. Clinical Chemistry 2001;47(2):209–14.
[33] Boyd JC, Bruns DE. Monte Carlo simulation in establishing analytical quality requirements for clinical laboratory tests: meeting clinical needs. Methods in Enzymology 2009;467:411–33.
[34] Karon BS, Boyd JC, Klee GG. Glucose meter performance criteria for tight glycemic control estimated by simulation modelling. Clinical Chemistry 2010;56(7):1091–7.
[35] Karon BS, Boyd JC, Klee GG. Empiric validation of simulation models for estimating glucose meter performance criteria for moderate levels of glycemic control. Diabetes Technology and Therapeutics 2013;15(12):996–1003.
[36] Virdi NS, Mahoney JJ. Importance of blood glucose meter and carbohydrate estimation accuracy. Journal of Diabetes Science and Technology 2012;4(4):921–6.
[37] Breton MD, Kovatchev BP. Impact of blood glucose self-monitoring errors on glucose variability, risk for hypoglycaemia, and average glucose control in type 1 diabetes: an in silico study. Journal of Diabetes Science and Technology 2010;4(3):562–70.
[38] Pretty CG, Signal M, Fisk L, Penning S, Le Compte A, Shaw GM, Desaive T, Chase JG. Impact of sensor and measurement timing errors on model-based insulin sensitivity. Computer Methods and Programs in Biomedicine 2014;114(3):e79–86.
[39] Chan JCN, Wong RY, Cheung CK, Lam P, Chow CC, Yeung VT, Kan EC, Loo KM, Mong MY, Cockram CS. Accuracy, precision and user-acceptability of self blood glucose monitoring machines. Diabetes Research and Clinical Practice 1997;36(2):91–104.

[40] Vettoretti M, Facchinetti A, Sparacino G, Cobelli C. A model of self-monitoring blood glucose measurement error. Journal of Diabetes Science and Technology 2017;11(4): 724–35.
[41] Vettoretti M, Facchinetti A, Sparacino G, Cobelli C. Accuracy of devices for self-monitoring of blood glucose: a stochastic error model. Conference Proceedings IEEE Engineering in Medicine and Biology Society 2015;2015:2359–62.
[42] Campos-Náñez E, Fortwaengler K, Breton MD. Clinical impact of blood glucose monitoring accuracy: an in-silico study. Journal of Diabetes Science and Technology 2017; 11(6):1187–95.
[43] Azzalini A. The skew-normal distribution and related multivariate families. Scandinavian Journal of Statistics 2005;32(2):159–88.
[44] Christiansen M, Bailey T, Watkins E, Liljenquist D, Price D, Nakamura K, et al. A new-generation continuous glucose monitoring system: improved accuracy and reliability compared with a previous-generation system. Diabetes Technology and Therapeutics 2013;15(10):1–8.
[45] De Nicolao G, Sparacino G, Cobelli C. Nonparametric input estimation in physiological systems: problems, methods and case studies. Automatica 1997;33(5):851–70.
[46] YSI life sciences. User's manual YSI 2300 STAT PLUS. 2009. Available from: https://www.ysi.com/File%20Library/Documents/Manuals%20for%20Discontinued%20Products/YSI-2300-Stat-Plus-manual-j.pdf. [Accessed 9 January 2019].
[47] Bailey TS, Chang A, Christiansen M. Clinical accuracy of a continuous glucose monitoring system with an advanced algorithm. Journal of Diabetes Science and Technology 2015;9(2):209–14.
[48] García-Jaramillo M, Calm R, Bondia J, Tarín C, Vehí J. Insulin dosage optimization based on prediction of postprandial glucose excursions under uncertain parameters and food intake. Computer Methods and Programs in Biomedicine 2012;105(1):61–9.
[49] Wang Z, Paranjape R. A signal processing application for evaluating self-monitoring blood glucose strategies in a software agent model. Computer Methods and Programs in Biomedicine 2015;120(2):77–87.
[50] Vettoretti M, Facchinetti A, Sparacino G, Cobelli C. Type-1 diabetes patient decision simulator for in silico testing safety and effectiveness of insulin treatments. IEEE Transactions on Biomedical Engineering 2018;65(6):1281–90.
[51] Vettoretti M, Facchinetti A, Sparacino G, Cobelli C. Patient decision-making of CGM sensor driven insulin therapies in type 1 diabetes: in silico assessment. Conference Proceedings IEEE Engineering in Medicine and Biology Society 2015;2015:2363–6.
[52] Facchinetti A. Continuous glucose monitoring sensors: past, present and future algorithmic challenges. Sensors 2016;16(12):E2093.
[53] FDA advisory panel votes to recommend non-adjunctive use of Dexcom G5 Mobile CGM. Diabetes Technology and Therapeutics 2016;18:512–6.
[54] Edelman SV. Regulation catches up to reality: nonadjunctive use of continuous glucose monitoring data. Journal of Diabetes Science and Technology 2017;11(1):160–4.
[55] CGM rulings. Ruling No. [CMS-1682-R]. Centers for medicare & medicaid services. 2017. Available from: www.cms.gov/Regulations-and-Guidance/Guidance/Rulings/Downloads/CMS1682R.pdf. [Accessed 7 January 2019].
[56] Vettoretti M, Cappon G, Acciaroli G, Facchinetti A, Sparacino G. Continuous glucose monitoring: current use in diabetes management and possible future applications. Journal of Diabetes Science and Technology 2018;12(5):1064–71.

SECTION

Continuous glucose monitoring (CGM) devices

2

CHAPTER

CGM sensor technology

6

Andrew DeHennis, PhD [1], Mark Mortellaro, PhD [2]

[1]*Sr. Director of Engineering, R&D, Product Development Senseonics Incorporated, Germantown, MD, United States;* [2]*Director of Chemistry, Senseonics Incorporated, Germantown, MD, United States*

Introduction

Although self-monitoring of blood glucose has been an important tool for managing diabetes for several decades, this method has some significant limitations. It requires patients to prick their fingers using a lancing device to obtain a small blood sample, then apply the drop of blood onto a reagent strip, which is then read by a glucometer. Patients must do this multiple times a day, then analyze the readings to check their treatment or adjust their diet, insulin, antidiabetic medication, or exercise to keep their glucose levels within the target. Patients need to maintain a regular regimen of self-testing for self-management to be effective. Additionally, because the self-glucose checks are performed at only a few times during the day and capture only the blood glucose reading at that specific time, high and low spikes in blood glucose that can happen at any time of the day or night can be missed, as shown in Fig. 6.1. The identification and analysis of trends in blood glucose data difficult.

To improve the management of diabetes, Continuous Glucose Monitoring (CGM) systems have been developed, which incorporate advances in microelectronics, implantable materials, and wireless technology to provide an accurate long-term picture of glucose levels in both real time and over extended periods. This insight into glycemic variability enables users and physicians to make adjustments to their treatments and medications. Applications for the use of CGM systems are covered in [48].

This chapter describes the CGM technology and system components that are overviewed in Fig. 6.2 to enable CGM, focusing both on technology in systems that are currently available and systems being developed for future platforms. A discussion of performance factors will include accuracy metrics, sensor life/stability, and wireless connectivity of systems. A review of patient usability factors will include alerts and alarms, trends, plotting, and smartphone apps.

Glucose Monitoring Devices. https://doi.org/10.1016/B978-0-12-816714-4.00006-5

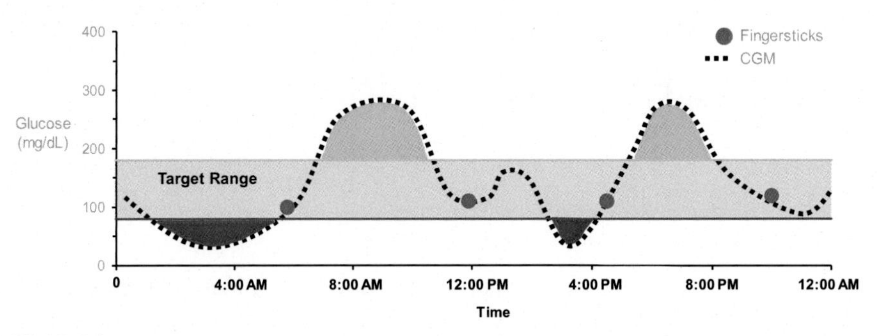

FIGURE 6.1

Intermittent self-monitoring blood glucose meter leads to unnoticed highs and lows.

Glucose transduction technologies

This section spans the various technologies that are used as transducers in converting glucose and glucose concentration to an electrical signal. As shown in Fig. 6.2, the technologies covered span those that are incorporated into current CGM as well as those that are still being developed for future systems.

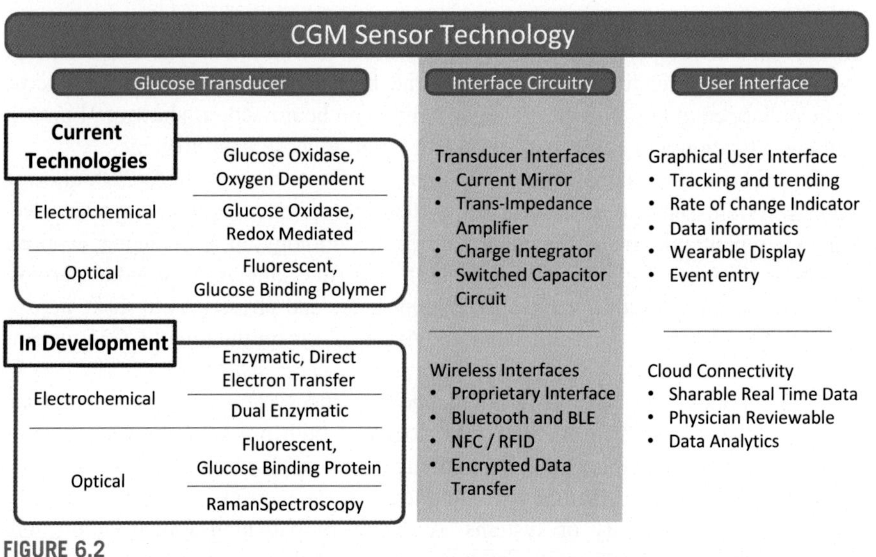

FIGURE 6.2

CGM sensor technology overview topics spanning glucose transducer technologies, transducer and wireless interface circuitry, as well as graphical user interface and cloud connectivity.

Current technologies

Transduction technologies used in commercially approved CGMs

Enzymatic, electrochemical-based sensors

The transduction technologies used in most commercial CGM systems (as well as most home blood glucose meters) rely on the measurement of an electrochemical signal generated from the reaction of an enzyme, glucose oxidase (GOx), with glucose [7,37]. Glucose oxidase-based sensing offers excellent selectivity for glucose over other compounds that are endogenous to biological fluids and is a well-established technology that has been in use for several decades. The first of such glucose sensors was developed in 1962 by Clark and Lyons from the Children's Hospital of Cincinnati. Their glucose sensor was composed of an oxygen electrode, an inner oxygen semipermeable membrane, a thin layer of GOx, and an outer dialysis membrane. Glucose concentrations were determined by measuring the decrease in oxygen concentration [43].

Two different approaches, referred to in the literature as "generations" of glucose oxidase sensors [7,37,44], are used in commercial CGM systems. In the first approach [4], glucose oxidase catalyzes the conversion of glucose and oxygen (O_2) to gluconic acid and hydrogen peroxide (H_2O_2). The net reaction can be shown as follows:

$$\text{Glucose} + O_2 \rightarrow \text{Gluconic Acid} + H_2O_2$$

Glucose concentration can be determined by monitoring either the consumption of O_2 or the generation of H_2O_2. A concentration-dependent current can be measured electrochemically via oxidation or reduction of those species at an electrode [39,65]. Systems from Dexcom (San Diego, CA) and Medtronic (Northridge, CA) measure the oxidation of hydrogen peroxide at the surface of a working electrode [49]. This approach requires the oxygen concentration to be in excess relative to the glucose concentration such that the reaction is glucose concentration limited [49]. As oxygen concentrations in the interstitial space fluid (ISF) are substantially less than glucose concentrations, the addition of glucose limiting membranes as components of the sensors is required to correct the oxygen-to-glucose imbalance and reduce or eliminate the oxygen dependency [49].

A second approach (i.e., generation) developed by Heller and colleagues [50] and used by Abbott Diabetes Care (Alameda, CA) employs a synthetic, polymeric redox-active mediator (Med) in place of oxygen, thus removing the dependency on sufficient oxygen concentration in the ISF and also the potential for local tissue irritation and sensor degradation that may arise from the overproduction of hydrogen peroxide. This reaction can be shown as follows:

$$\text{Glucose} + \text{Med} \rightarrow \text{Gluconic Acid} + \text{Med}_{(red)}$$

The polymeric mediator is covalently attached to the GOx such that it connects the enzyme reaction center to the surface of the electrode. The electrochemical oxidation of the reduced (Med_{red}) redox-active polymer (which regenerates the mediator) at the working electrode is measured, rather than the current generated by oxidation of hydrogen peroxide. Such CGM transduction systems are called as "wired enzymes."

The longevity of GOx-based CGM sensors may be limited by the in vivo stability of the enzyme. Loss of enzyme activity over time has been attributed to the presence of low molecular weight materials or build-up of hydrogen peroxide concentrations in the fluid surrounding the implanted sensor [51]. Incorporation of stabilizing additives into the sensor formulation is the most common means of increasing enzyme in vivo life, although molecular engineering of glucose oxidase to improve its stability has also been reported [51]. As all of the commercial GOx-based CGMs are based on needles that protrude through the skin during their usage lifetime, it is also possible that the irritation created by the movement of the needle with and prevention of wound closure does not allow for resolution of tissue inflammation that creates the enzyme inactivating species. In fact, an enzymatic and fully implantable CGM sensor under development by Glysens, Inc. (San Diego, CA) has been reported to function for up to 180 days in a human clinical trial, suggesting full implantation may be important to longevity [27].

Performance limitations frequently associated with enzymatic, electrochemical-based glucose sensors include sensitivity to fouling of the electrode surface by biomolecules (e.g., proteins, cells) over time resulting in loss of functionality, transient loss of glucose sensitivity upon physical compression at the sensor implant site (which typically occurs during sleeping), and chemical interferences from commonly used electroactive compounds such as ascorbic acid (vitamin C) and acetaminophen [65].

Nonenzymatic, optical-based sensors

Various CGM sensor technologies that do not require enzymes for glucose recognition or that use optical rather than electrochemical means of detection are reported in the literature [66,67]. However, the CGM system from Senseonics is the only such system commercially available in the United States and in Europe. Glucose concentration is measured by means of fluorescence from an abiotic (i.e., nonenzyme based), glucose-binding polymer that coats the surface of a sensor that is fully implanted into the dermal subcutaneous tissue (Fig. 6.3). Intensity of light emitted by the polymer changes in response to the glucose concentration in the ISF that surrounds the sensor. Fluorescence is measured by an optical system contained within the implanted sensor comprises a light-emitting diode, which serves as the excitation source for the fluorescent polymer, and spectrally filtered photodiodes that measure the glucose-dependent fluorescence intensity. As the sensor is fully implanted, it contains an internal antenna that wirelessly receives power from and communicates with an externally worn transmitter.

Glucose reversibly binds to the indicator boronic acid groups (which act as glucose receptors) in an equilibrium-binding reaction as shown in Fig. 6.3 [45].

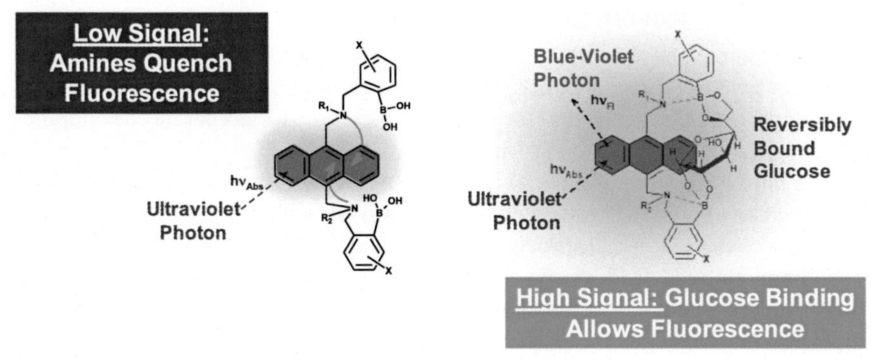

FIGURE 6.3

Equilibrium binding of glucose to the fluorescent, abiotic glucose indicating polymer component of the Senseonics CGM sensors (R_2 denotes connectivity to the polymer backbone).

When glucose is not present, anthracene fluorescence is quenched by intramolecular electron transfer from the unpaired electrons on the indicator tertiary amines. When glucose is bound to the boronic acids, the Lewis acidity of boron is increased, and weak boron—nitrogen bonds are formed. This weak bonding prevents electron transfer from the amines and consequently prevents fluorescence quenching.

As the fluorescent signal emanates from throughout the entire bulk of the fluorescent polymer, not just at the surface as with electrochemical sensors, it is not subject to the same degree of fouling as sensors that employ enzymatic, electrochemical transduction methods. Although substances frequently reported to interfere with enzymatic, electrochemical-based transcutaneous CGM systems, such as acetaminophen and ascorbic acid, have been shown to have no interfering effect, other substances such as tetracycline (which absorbs light over the spectral operating range of the sensor) and high concentrations of mannitol (known to occur when used as an irrigation solution during therapeutic transurethral procedures) have been shown to interfere with the sensor readings [65]. The fluorescent-based sensor is fully inserted into the interstitial tissue by a healthcare provider in an office-based procedure. As the sensor does not protrude through the skin, the insertion site is able to fully close and allow the resolution of the acute inflammatory response that may limit sensor life.

For this fluorophore, hydrogen peroxide has shown to be the reason for sensors to lose signal on the first day in CGM systems [70]. The Eversense uses a catalysis made out of platinum that sputter-coats the sensor core to prevent decomposition from hydrogen peroxide, which helps make it stable mechanically and chemically for longer periods of time [69].

Transduction technologies in development

Transduction technologies and CGM systems under development have been extensively reviewed elsewhere [46,52,66,67,68]. This section highlights some of the technologies that have progressed from academic to corporate development programs.

The third generation of glucose oxidase-based CGM sensors is being developed by DirectSens GmbH (Klosterneuburg, Austria), in which a genetically modified enzyme is covalently and electrochemically linked directly to the working electrode, thus eliminating the need for oxygen or redox-active mediators [7].

A glucose oxidase, electrochemical CGM sensor that is designed to be fully implanted into subcutaneous tissue is under development by researchers at the University of Connecticut in collaboration with Biorais, Inc. (Storrs, CT, USA) [53]. In contrast to the current, commercial CGM sensors that use an electrochemical glucose oxidase sensor that protrudes through the skin during use, the Biorais sensor is fully implanted into the subcutaneous tissue and wirelessly sends data and power through the skin via light. The sensor incudes a drug-eluting coating to suppress the foreign body response and thereby enable long (at least 3 months) implant life.

A fully implantable sensor that uses a combination of two immobilized enzymes, glucose oxidase, and catalase, immobilized in contact with an electrochemical oxygen sensor, is under development by Glysens, Inc (San Diego, CA) using technology developed by David Gough of the University of California, San Diego [54]. The combination of enzymes catalyzes the reaction:

$$\text{Glucose} + \tfrac{1}{2}\,O_2 = \text{gluconic acid}$$

The sensor measures glucose based on differential electrochemical oxygen detection. Oxygen remaining (i.e., not consumed by the enzymatic process) is measured by one enzyme-coated oxygen sensor and compared to a similar oxygen sensor that does not contain enzymes; the differential signal is used to determine glucose concentration. The sensor is designed to allow for a sufficient supply of oxygen to the enzyme region to avoid a stoichiometric oxygen deficit and to account for variations in oxygen concentration and local perfusion [54]. Another potential advantage of the dual enzyme configuration is that the immobilized catalase may prolong the in vivo lifetime of the glucose oxidase by preventing build of up of hydrogen peroxide, a potential source of enzyme inactivation and tissue inflammation [55]. This also allows for use of an excess of glucose oxidase to extend sensor life, as the overproduction of hydrogen peroxide is minimized by the catalase.

Several groups have applied enzymatic, electrochemical sensor technology onto microneedle-array sensors that are designed to penetrate and reside in the skin layers (dermis or epidermis) rather than in the subcutaneous tissue [56]. Two different approaches to microneedle sensor designs have been reported. In one approach, under development by Cass and coworkers (Imperial College of London, UK) and independently under development by Sano Intelligence (San Francisco, CA), the microneedles are designed to function as electrodes that measure current from the reaction of glucose oxidase (coated onto the sensor tips) with glucose. In a clinical study by Cass and coworkers, microneedles were shown to consistently penetrate into the epidermis when applied using manual pressure and sensor readings tracked with blood glucose readings [35]. In a second approach previously under development by Arkal Medical (Fremont, CA), the microneedles are hollow and designed to draw dermal interstitial fluid to the proximal, glucose oxidase-coated electrode. A clinical study of a hollow-needle microsensor array by Arkal demonstrated a good correlation with blood glucose readings over a 72-h wear period [21]. Biolinq (San Diego, CA) is developing a patch sensor the size of nickel using GOx coated onto microneedle arrays.

Profusa, Inc. (South San Francisco, CA) is developing tissue-integrating sensors using fluorescent hydrogels that can be injected under the skin and fluorescence intensity measured through the skin using an externally worn device. The small size, soft (flexible) nature, and unique porous structure minimizes the foreign body response, thus enabling long-term in vivo use time; sensors have been reported to function 4 years after implant [31]. A sensor for oxygen (Lumee™) is currently counter electrode (CE) marked and available in the EU and a glucose sensor has been reported to be under development. The glucose sensor hydrogel technology can conceivably be designed via either incorporation of glucose-sensing enzymes (such as glucose oxidase) that consume oxygen, or through the incorporation of a glucose-sensing dye in place of an oxygen-sensing dye.

Fluorescent glucose sensors that make use of the equilibrium binding between glucose and glucose/galactose-binding protein (GGBP) have been clinically evaluated by researchers at BD Technologies (Research Triangle Park, NC) [57]. The GGBP was modified by site-specific mutagenesis to improve stability and binding properties and was labeled with the fluorescent dye acrylodan. Reversible glucose binding to the protein changes the protein conformation that results in a change in the local environment around the dye and a corresponding shift in the peak fluorescent wavelength. The acrylodan-modified GGBP was attached to the tip of a fiber optic that was then placed either intradermally (<1 mm) or subcutaneously (>3 mm) through the skin. Fluorescent signals traveled through the fiber optics to the external reader and signals were shown to correlate with blood glucose measurements during the 12-h wear period.

Tissue interface for transcutaneous and subcutaneous transduction

All CGMs approved for use in the United States and the European Union are implanted through the skin and reside within the dermal subcutaneous tissue. Consequently, their functionality is dependent upon the extent by which the surrounding human tissue reacts with the sensor. A foreign body reaction (FBR) is initiated upon the creation of the wound during sensor insertion and may persist for the duration of the implant [1]. Short duration (acute) responses that may affect sensor performance include fouling of the sensor surface by proteins, reduction in glucose and/or oxygen in the tissue surrounding the sensor, localized decrease in pH, and chemical attack from cellular-generated reactive oxygen species (ROS); additional chronic effects include collagen encapsulation of the sensor that reduces access to glucose from the surrounding tissue and results in an increase in lag time between blood and sensor glucose responses [47,67,69]. Common strategies to reduce these effects include a selection of sensor materials that induce a minimal FBR, the addition of coatings that reduce protein absorption or elicit a favorable (healing) tissue response, and minimizing the size of the sensor. Nonetheless, the FBR is typically the predominant factor that limits the in vivo lifetime of implanted CGMs.

To reduce the chronic foreign body response and enable sensor longevity extending beyond a few weeks, several different medicinal agents have been incorporated into CGM sensors. The Senseonics implantable sensor contains a silicone polymer that releases micrograms of dexamethasone per day. The dexamethasone reduces inflammation in the local tissue surrounding the sensor thus reducing the generation of reactive oxygen species (ROS) that can oxidize the glucose-binding polymer. Similarly, Biorasis has shown that dexamethasone release from an erodible (i.e., dissolving) coating on their fully implantable CGM sensor reduces local inflammation over several months of implant time [25]. Vascular endothelial growth factor (VEGF) has been used to increase the density of blood vessels (and thus increase access to glucose) around an implantation site [38]. Klueh and coworkers demonstrate that VEGF enhances glucose sensor performance via increased vascularization while also having an antiinflammatory effect [23]. Nitric oxide has also been used to improve the tissue interface around an implanted glucose sensor. Nitric oxide serves as an inhibitor of bacterial cell proliferation and biofilm formation and prevents platelet activation and adhesion, infection, and subsequent thrombosis [6]. Nitric-oxide-releasing polymers developed by Mark Schoenfisch of the University of North Carolina at Chapel Hill have been demonstrated to enhance the performance of implanted glucose sensors [29] and that technology is in use by Clinical Sensors, Inc. (Research Triangle Park, NC).

Noninvasive technologies

Noninvasive (i.e., without implantation of a device or a chemical) CGMs are being developed using devices that clip onto the ear, that shine light through the skin, and that measure glucose sensing "smart" tattoos, and with glucose sensing contact lenses.

The ability of infrared light to penetrate the skin allows for the noninvasive measurement of glucose via Ramen spectroscopy [42]. One such CGM was developed by C8 Medisensors (San Jose, CA). A sensor worn on the skin shined near infrared light (NIR) light through the skin and measured the glucose Ramen signal that was reflected to the sensor. The sensor received CE mark approval but manufacturing and user to user variability reportedly prevented commercialization [34].

Dyes that change color or fluorescence intensity with changes in glucose concentration can be tattooed onto the skin and glucose measured using an optical measurement device applied over the tattoo [32]. Researchers at Harvard University and MIT have developed such glucose-responsive inks and demonstrated feasibility on an ex vivo pig skin model; a change in ink color from brown to blue was visible as the glucose concentration increased [30].

A noninvasive and removable glucose monitoring product that attaches to the ear (GlucoTrack) is available from Integrity Applications (Ashdod, Israel) but has not been approved by the FDA for use in the United States. This CGM technology is reported to measure the physiological effects of glucose in tissue via ultrasonic, electromagnetic, and thermal measurements, but certain factors such as body mass, gender, age, and ear piercings may affect measurement results [2].

Lakowicz and coworkers of the University of Maryland School of Medicine (Baltimore, MD) have reported on development of a contact lens that optically measures glucose in tear fluid [58]. Phenylboronic acid containing fluorescent dyes whose intensity changes with glucose concentrations are polymerized into hydrogels. One potential method for measuring the change in the fluorescent signal would use a handheld device that flashes light into the eye and measures the fluorescence intensity emanating from the contact lens. Similarly, color or fluorescent lifetime changes could be measured in response to glucose binding at the boronic acid sites using the appropriate dyes [8]. In one proposed type of contact lens, a change of color is monitored by a user by looking at a mirror and comparing glucose levels by a color strip [10]. Platform that utilize contact lenses with integrated, on-site electronics and wireless telemetry have also been developed [59]. These platforms could use either enzymatic or optical transduction, but the systems would need to compensate for the differences in the physiological differences between blood and tear glucose concentrations [60].

Sensor interface and system connectivity

A CGM System is composed of subcomponents and systems that span transduction, embedded systems, Apps, and PC/Cloud-based systems. An overview of the functional interaction of these components is shown in Fig. 6.4.

Stepping through this sequence begins with the acquisition of signal through the glucose transducer; the measurement, either analog or digital at that point, is then sent to a wearable transmitter using a short-range interface via wired or wireless connections. The commercial transcutaneous CGM sensors are wired and the subcutaneous sensors utilize near-field communication (NFC) wireless communication with a read range of <2 cm. This signal is then processed in the signal conversion electronics that is within the wearable transmitter. This wearable transmitter also enables the conversation of that signal into the glucose value. At this point, the signal is sent via a mid-range wireless to a unit that is implemented either using a custom RF protocol, or Bluetooth low energy, or using NFC direct to a hand-held device. With the measurement information in the handheld device, which is either a custom receiver or mobile phone, various informatics about the glycemic status and history

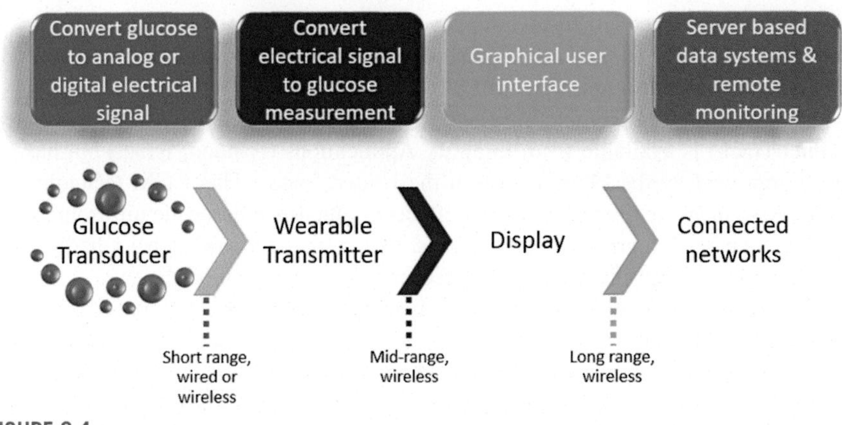

FIGURE 6.4

CGM connectivity block diagram for system components and wireless communication.

are displayed to the user through the graphical user interface. The information can then be sent using long-range wireless, which is typically WiFi-based, to then enable a review of the data using server-based data systems or even to transmit that information further to enable remote monitoring. Each of these is discussed in the following sections.

Sensor front end electronics

The initial processing of the signal from the glucose transducer goes through converting the electrical signal, which is typically current from the enzymatic and optical transducers, into a digitally encoded value. This is done through an analog interface circuit that is connected to an analog-to-digital converter. The analog signal processing is done as close to the sensor area as possible that can be either in the transmitter or as part of the subcutaneous implant. The initial circuit that processes the transduced signal is set up as a current sensor to sense either the current produced from the enzyme for an enzymatic sensor or to sense the current through a photodiode for an optical sensor. These types of interface circuitry is broadly described in Shenoy [62].

For enzymatic sensors, Shenoy describes the use of a configuring transducer interface in a three-electrode system to enable better control of the voltage across the sensing electrode by adding a CE into the electrochemical cell along with a working electrode (WE) and a reference electrode (RE) configured along with a potentiostat. This configuration has the RE connected to give negative feedback to the potentiostat in stabilizing the voltage between the RE and CE. This configuration can also be modeled along with the integrated circuit using a lumped circuit model RC circuit [62]. For optical sensors, a photodiode-based detectors serve as a robust basis for CGM systems. They also provide a path for integration because they can be formed through the same fabrication process as the analog and digital circuitry.

Further integration of the system-level functionality also enables scaling of the systems overall footprint as well as the detector size itself. This integration can build in circuitry to digitize the signal on-site to minimize the opportunity of introducing other noise sources into the system. Analog-to-digital converters can be implemented in the same transistor-based, integrated circuits as the potentiostat.

Fig. 6.5 shows system developed by Croce et al. that has the interface electronics are built into the subcutaneous implant along with the glucose transducer. The system shown in Fig. 6.5A shows the integrated system that has a current to frequency converter connected along with the potentiostat that uses a microcontroller to send and encode the frequency via Bluetooth. An optical integrated interface circuit is shown in Fig. 6.5B, which build in a Trans-Impedance Amplifier (TIA) as well as a near-field communication (NFC) interface to telemeter the data from the photodiodes.

Transmitter software and sensor calibration

The software used in a CGM sensor is designed to convert the electrical signal received from the transducer measurement into a glucose reading. Conversion

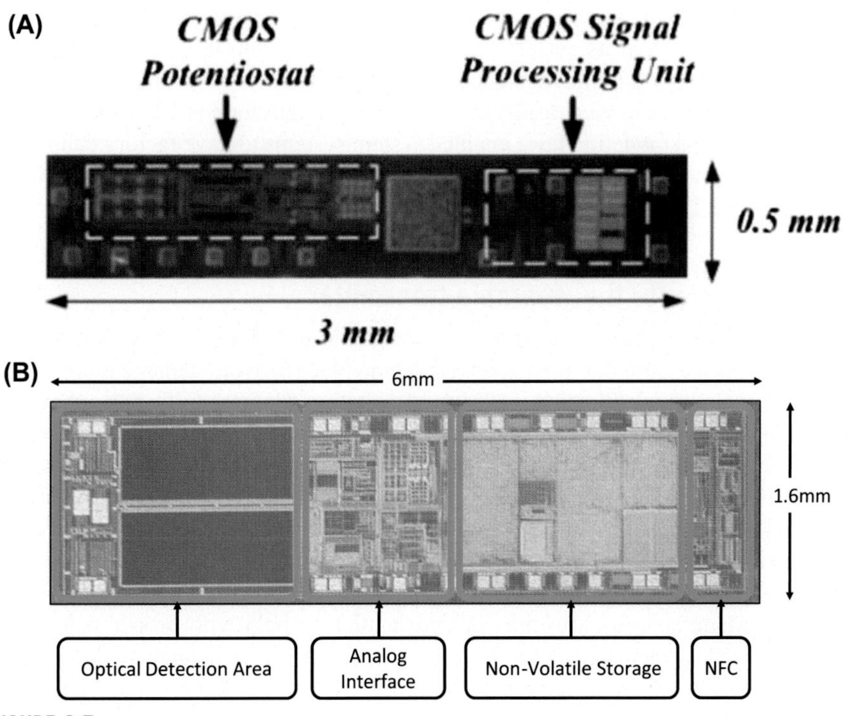

FIGURE 6.5

CGM systems with interface circuitry integrated with the transducers: (A) enzymatic [63] and (B) optical glucose transduction [61].

functions have been developed that convert raw measurements into glucose values. A sequence of the glucose values can then be utilized for the calculation of glucose trends. These compensation techniques also look to overcome any differences in glucose concentrations that may exist between blood glucose levels and interstitial glucose levels.

The time delay between a blood glucose reading and the value displayed by a continuous glucose monitor consists of the sum of the time lag between ISF and plasma glucose, in addition to the inherent electrical/chemical sensor delay due to the reaction process and any front-end signal-processing delays required to produce smooth traces [22]. Lag time can be caused by both physiologic and technologic issues. Physiologic lag results from the difference in glucose concentration in blood and ISF, which can increase or decrease before the blood glucose changes. Other physiologic interference can come from inside the body, such as blood clots or other biofouling, or be introduced from outside the body, such as medications or dietary factors. Technologic lags result from the time required for the sensor to analyze the sample and from the necessity of applying signal processing algorithms to filter noise or to average a series of readings to create a weighted glucose level over time [9].

Glucose sensor calibration methodologies are utilized to ensure that the system maintains its performance for the full duration of use. These methodologies range from systems that are in place to characterize individual sensors or sensor lots during manufacturing as well as systems that utilize a calibration update after insertion. Systems look to build in the transducer's fundamental glucose response characteristics [3,40] as well as in vivo distortion and noise compensation [14]. These advances in calibration and modeling have enabled system systems to have factory calibration [19]and accuracy that enables diabetes treatment decisions based on the CGM measurements [24].

Skin interface for the wearable transmitter

Continuous skin contact is critical to the successful use of a CGM system, so manufacturers must ensure that their products stay in place to avoid damage to the sensor or disruption of data collection. CGM manufacturers continue to face the challenge of finding an adhesive to use that will be strong enough to keep the transmitter attached to the skin during normal wear while also not irritating the wearer's skin. This can be particularly difficult in young patients whose smaller body surface area and active lifestyles can present a unique set of issues to overcome [13].

Dexcom describes their adhesive as "a pressure-sensitive acrylic adhesive coated on top of a polyester spunlace fabric" [41], while Senseonics uses a silicon-based adhesive. Manufacturers also provide information to help with proper wearing of a device, such as Medtronic's guide, "Tape Tips and Site Management." [28].

Some commonly used adhesives, such as acrylates, colophony, and Mastisol, have been found to cause allergy symptoms in wearers [36]. In a study of CGM users, Jadviscokova et al. found that minor local adverse effects, including hypersensitivity, itching, pain, redness, and burning, were found in less than 55% of subjects, and the effects were not severe enough to cause wearers to stop using the sensor [20].

System user interface and connectivity

For a person with diabetes, a continuous glucose monitoring system provides numerous advantages by making important information available both in real time and for historical review. This allows the user to both react to changes in blood glucose levels based on alerts from the system, and to identify trends and patterns over time that will help their diabetes healthcare team make adjustments to their treatment plan to aid in adjusting therapy to optimize diabetes management.

In a review of CGM systems, Rodbard noted that usability and system interface play a critical role in the adoption and success of any medical device. He suggested that usability issues to consider for CGM systems include the amount of time and training required for both clinicians and patients to learn the system and be able to interpret the data, reliability of results and ability to apply the information, and usefulness of the information to improve glycemic control and overall quality of life [33].

The latest generation of CGM systems has made great strides in developing user interfaces that provide information in a user-friendly way. The interface can be a stand-alone device provided by the manufacturer, such as the Freestyle Libre Reader, or an app that the user can download to a mobile device, such as the Guardian Connect.

All CGM systems provide users the ability to share information with a clinician by uploading it to cloud-based storage or having the information downloaded directly during an office visit to a data management application. These data management applications provide various graphical representations of historical data to help identify trends and patterns that can guide decisions about therapy adjustments.

One of the most beneficial features of most CGM systems is the ability to present real time and predictive alerts to notify the user of a hypoglycemic or hyperglycemic event when the measured glucose level reaches a predefined value. Alerts can be an audible sound or physical alert such as a vibration from the transmitter the user is wearing or a notification from the app on the user's phone. Alerts can be triggered when glucose values are either above or below the limits set by the user, are rising or falling above a defined rate, or are predicted to go above or below a defined limit. System-related alerts will also be presented, for example, to let the user know when to calibrate the device or that a system error has occurred. A user who has received an alert can then check the app to find out more and decide whether any action is required. Some products also provide the ability to share data to a remote monitoring app used by caregivers, such as family and friends. This means, for example, that a parent can receive an alert when their child's glucose is high or low.

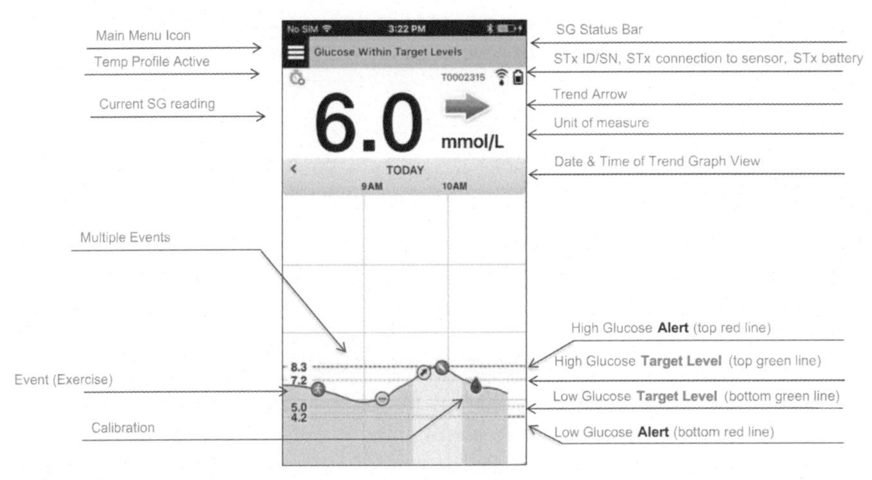

FIGURE 6.6

Sample mobile app home screen for a CGM system.

The user interface provided by the manufacturer is an important component of the user experience. It provides a variety of information to track glucose levels in the moment and over time. Fig. 6.6 is an example of a home screen for a CGM system. These display a variety of information in an easy-to-read format for quick reference by the user.

Information that is commonly displayed on CGM home screens includes:

- Current sensor glucose reading
- Trend arrow
- Trend graph
- Plotting of events, such as alerts, calibration, etc., over time
- System status

CGM interfaces also provide a means for the user to enter additional relevant information such as meals, exercise, or insulin injections. These user-entered events may also be displayed on the trend lines. When data collected by the system is combined with user-entered information, patterns and trends over time can be evaluated as inputs into changes in the treatment plan.

Connected systems

Continuous subcutaneous insulin infusion, commonly known as insulin pumps, provide a way for people with diabetes to receive a continuous supply of insulin 24/7, eliminating the need for multiple daily injections. Insulin, typically rapid-acting, is delivered from the pump two ways. Small microdoses, called basal insulin, are preprogrammed by the user to be delivered automatically as often as every 5 min throughout the day, and basal delivery profiles can be temporarily adjusted by the user to compensate for exercise, illness, etc. The user can deliver and program on-demand larger, bolus doses, to cover meals and to correct for high glucose. The user fills a cartridge with insulin and loads it into the pump. An infusion set is connected to the insulin cartridge on one end and connected to the user with a catheter placed subcutaneously and held in place with an adhesive patch. Infusion sets are typically changed about every 3 days. When inulin pumps were first introduced, patients had to "do the math" to calculate the amount of insulin needed to correct for high glucose or compensate for meals. Insulin pumps today typically include sophisticated bolus calculators that track insulin on board to avoid stacking of insulin, as well as offer advanced bolus features such as the ability to extend delivery of a bolus dose over a period of time to compensate for certain types of food and conditions such as gastroparesis. The latest generation of insulin pumps incorporates wireless communication technology to connect with CGM systems to allow users to collect more information and to automatically adjust basal insulin to avoid extreme high and low glucose [12].

Artificial pancreas

A CGM system can be integrated with an insulin pump to automatically adjust insulin delivery based on glucose levels and individual settings such as correction factor and insulin to carb ratios, with no user intervention required. This type of closed-loop, automated system is called an artificial pancreas, or AP, because it stimulates the function of a healthy pancreas with these automatic insulin adjustments. This closed-loop system works by communicating with both the sensor and the pump. Although no artificial pancreas system has been approved for use, the Medtronic Minimed 670G was introduced as a hybrid closed loop system in 2017.

Research at the University of Virginia and the University of Padova are developing "advanced algorithms that use CGM data to automatically adjust insulin delivery" [5]. These groups have developed the type 1 diabetes metabolic simulator (T1DMS), an in silico computer simulator that has been accepted by the FDA as a substitute for animal trials in certain cases. Researchers and manufacturers are utilizing these types of simulations to further develop autonomous systems for the administration of insulin. Fig. 6.7 shows the components of an artificial pancreas system.

As the figure shows, the artificial pancreas system uses CGM to measure changes in the wearer's glucose levels. Based on this information, the algorithm calculates adjustments to the amount of insulin being delivered by the pump.

Connected pens

Several manufacturers are working to develop a connected pen as an easier to use, less expensive alternative to insulin pumps, enabling an artificial pancreas system while also addressing the significant barrier to insulin pump adoption by users who are reluctant to wear a device 24/7. Although insulin pens have been widely used as they were introduced by Novo Nordisk in 1985, they did not provide data

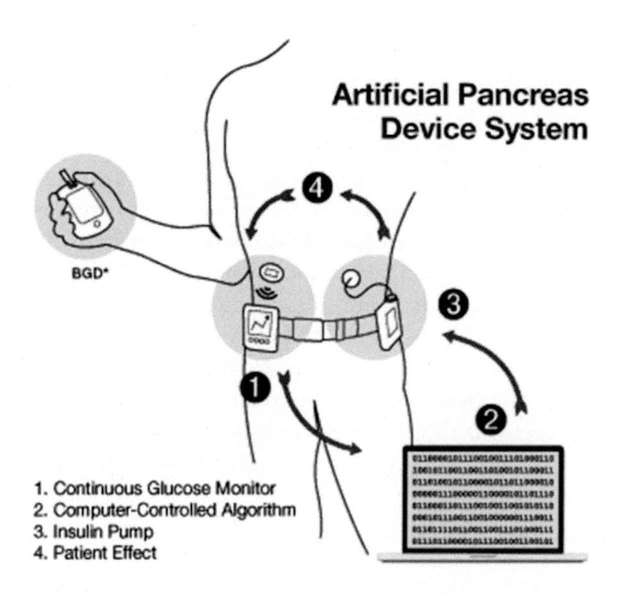

FIGURE 6.7

Artificial pancreas device system [18].

that a user or medical provider could use to identify trends and patterns for use in adjusting treatment plans. Several companies, including Companion Medical, Patients Pending, Innovation Zed, and Novo Nordisk, are attempting to change that by developing insulin pens that can provide data to be displayed by an app and used to track information like dosage amount, injection time, and missed injections.

Commercial systems

Overview

Currently, the four systems that are approved have unique differences in their implementation that enable a variety of choices for customer options in product features. One of the more prominent differences is that both transcutaneous and subcutaneous systems are currently available CGM systems. Transcutaneous sensors based systems consist of a very thin filament that is inserted through the skin and have a transmitter attached to the portion of the sensor above the skin. Users of transcutaneous CGM systems self-insert the sensor after being appropriately trained. Depending upon the manufacturer, currently available transcutaneous sensors last for 1 or 2 weeks, at which time a new sensor is inserted.

Implanted sensors are placed completely under the skin and provide CGM readings for months. Implantable sensors are inserted by a healthcare provider during an in-office procedure using local anesthetic and a small incision. Currently available implanted sensors communicate wirelessly with a transmitter worn over the sensor, and due to the nature of the sensor being placed completely under the skin, the transmitter can be removed and replaced without disturbing the sensor.

Four companies are currently making commercially available continuous glucose monitors. Dexcom, Abbott, Medtronic, and Senseonics all have products available in both the United States and Europe. The next sections further detail the features of their available CGM systems.

Abbott Freestyle Libre flash glucose monitor

The Freestyle Libre flash glucose monitor is a self-inserted transcutaneous CGM that is indicated for persons 18 years and older in the United States, and for ages 4 and older outside the United States. The Abbott Freestyle sensor uses an adhesive that is acrylic-based and worn for the life of the sensor, up to 14 days. Flash glucose monitoring is slightly different from CGM, as it requires the user to "flash" or scan the reader over the sensor to wirelessly collect and display a glucose reading and trend arrow. The system requires a scan every 8 h and there is a 1 h warm-up period at the start of each 14-day sensor session. Fingerstick blood glucose monitoring is not required except for situation labeled in the IFU such as during the first 12 h of wear, if symptoms do not match values, or other reasons discussed in the labeling. Depending upon the Libre product (e.g., a product marketed in the United States vs. other geographies), the system may or may not offer optional high and low glucose alerts.

The system is inserted by the user with the aid of a disposable, spring-loaded injection insertion device. The sensor and transmitter are combined in a single unit, which is worn at the back of the user's arm. Because the sensor and transmitter are combined, when removed, the user must replace it with a new unit. The connected app on the reader or smartphone provides glucose information.

Dexcom G6 CGM

The Dexcom G6 is a self-inserted transcutaneous CGM for ages 2 and older. Similar to the Freestyle Libre, the Dexcom G6 sensor is affixed to the body using a patch adhesive that is acrylic-based, and worn for the life of the sensor, up to 10 days. It is inserted using an injector device that deploys the sensor transcutaneous and enables connections to the attachable transmitter. The device needs to be unlocked and the sensor is loaded into the insertion device. Open the needle by removing the needle guard. Put the insertion tool on the patient's skin while ensuring that the skin is flat before pressing a button to insert the sensor [35]. Fingerstick blood glucose monitoring is not required unless CGM values do not match symptoms and other reasons contained in the labeling. The Dexcom G6 does not require calibration unless the sensor code is not entered, which the user can enter optionally. The warm-up period is 2 h for each 10-day sensor session. The G6 provides high, low, rate of change and predictive urgent low alerts.

Similar to the Freestyle Libre, the Dexcom G6 is indicated for nonadjunctive use, allowing the user to determine what treatment decisions to make without a fingerstick blood glucose value. Trend arrows on the home screenplay an important role in the user's treatment decisions. If an individual misses or miscalculates a dose, it is recommended not to base the next dosage on the arrows but calculate the value of glucose from the mealtime and use their insulin-to-carbohydrate ratio [26].

The Dexcom G6's transmitter life is 3 months, and the user receives two transmitters with each purchase. Similar to the Freestyle Libre, if the transmitter is removed or replaced, the sensor is also replaced. The Dexcom G6 wirelessly communicates with a commercially available insulin pump to automatically adjust some insulin doses.

Medtronic Guardian Connect CGM

The Medtronic Guardian Connect is a self-inserted transcutaneous CGM for people 14–75 years old. The sensor is inserted with the aid of a spring-loaded injector. Similar to other transcutaneous sensors, the adhesive patch is worn for the life of the 7-day sensor. The sensor is indicated for placement either on the abdomen or the arm. Fingerstick blood glucose monitoring is required for making treatment decisions, and calibration is required twice a day or as indicated by the system. The warm-up period is 2 h for each 7-day sensor session. The Medtronic Guardian's transmitter life is1 year.

Similar to other CGM systems, the glucose data are wirelessly sent to a reader or mobile app and displayed for the user. The Guardian Connect provides high, low, rate of change, and predictive high and low alerts.

The Medtronic CGM is part of a hybrid closed-loop system that automatically adjusts insulin delivery from select Medtronic insulin pumps.

Eversense CGM system

The Eversense CGM System is currently the only CGM system with an implantable sensor. The sensor is inserted by a health care provider in the upper arm and is intended to be used by people 18 and older. The adhesive patch is silicone-based and designed to be changed daily. Fingerstick blood glucose monitoring is required for making treatment decisions, and calibration is required twice a day. The warm-up period is 1 day in each 90- or 180-day sensor session. The Eversense Smart Transmitter warranty is one year.

The transmitter, worn over the sensor site, wirelessly powers the sensor and collects the glucose data. That data are sent wirelessly to a mobile app on the user's phone. The Eversense CGM System provides high, low, rate of change and predictive high and low alerts. The Eversense smart transmitter can be removed and replaced without ending the sensor session, and the transmitter provides uniquely patterned on-body vibratory alerts for high and low glucose. The area should be able to easily apply an adhesive patch. A 5 mm incision is made to then create a subcutaneous pocket below the skin surface where the sensor would be deployed with the insertion tools.

Put the insertion tool into the incision at about 45-degrees and lower to 5–10 degrees to match the marks of the blunt dissector. Insert the sensor into the pocket where the incision was made with the blunt dissector, which will lock into place. Confirm the sensor is in place by lightly touch the insertion area before removing the insertion tool from the incision. Close and correctly dress the incision with a skin adhesive, such as Steri-Strip.

Summary

This chapter has presented an array of glucose transduction techniques and system implementations that define the currently available commercial systems as well as proved the building blocks for future development. The glucose oxidase and fluorescence-based technologies have been built into systems that have been CE marked FDA approved for sale in globally. As part of the regulatory approval process, each of the systems went through pivotal studies to prove performance. Fig. 6.8 has a table that shows various features of each system.

For each of the systems, the duration of use varies; spanning from 7 to 90 days or more per sensor. Performance agreement has become a standard in assessing accuracy for CGM. Agreement is defined by looking the number of CGM values that are taken along with reference blood draw measurements and were within 20% for the reference values above 80 mg/dL or within 20 mg/dL for reference values at or

	Eversense CGM System	Abbott Freestyle Libre Flash Glucose Monitoring	Dexcom G6 CGM System	Medtronic Guardian Connect CGM System
Sensor wear	Up to 3 months	Up to 14 days	Up to 10 days	Up to 7 days
Performance 20/20% Agreement	94.3%	90.9%	91.7%	85.7%
Finger stick replacement	No	Yes	Yes	No
Interferents	Tetracyclines	Vitamin C, salicylic acid	Acetaminophen	Acetaminophen
Active, automatic alerts	High, low, rate of change, predictive high and low	U.S.- No; CE-Optional high, low, signal loss	High, low, rate of change, predictive urgent low	High, low, rate of change, predictive high and low
Calibrations/day	2/day	0; scan needed every 8 hrs	0; can optionally enter	2/day
Insertion	Professionally placed by HCP during in-office procedure	Self-insertion every 14 days	Self-insertion every 10 days	Self-insertion every 7 days
Remote monitoring	Yes	Yes	Yes	No
On-body, vibratory alerts	Yes	No	No	No

FIGURE 6.8

Feature comparison chart for the approved commercial CGM systems [11,15–17].

below 80 mg/dL. Thus the higher the percentage, the more agreement there is between the CGM and reference blood draw measurements. The systems also vary on whether they have been approved yet for treatment decisions such as dosing insulin. The Dexcom and Libre systems are approved for making treatment decisions without a confirmatory fingerstick blood glucose reading. The Eversense and Guardian system need two calibration updates per day, and Dexcom and Libre systems do not require calibration. As discussed in the transducer technology section, the method used in the conversion of glucose can leave it susceptible to interferents. Each of the interferents that they systems have been labeled against is listed in the figure as well. The capabilities of each of the systems with respect to active or automatic alerts, remote monitoring, and the capability for on-body vibration are also shown in Fig. 6.8.

These products and sensor technology provide the current basis for CGM systems for people with diabetes.

References

[1] Anderson JM, Rodriguez A, Chang DT. Foreign body reaction to biomaterials. Seminars in Immunology April 2008;20(2):86–100.

[2] Bahartan K, Horman K, Gal A, Drexler A, Mayzel Y, Lin T. Assessing the performance of a noninvasive glucose monitor in people with type 2 diabetes with different demographic profiles. Journal of Diabetes Research 2017;2017:1–8. https://doi.org/10.1155/2017/4393497.

[3] Breton M, Kovatchev B. Analysis, modeling, and simulation of the accuracy of continuous glucose sensors. Journal of Diabetes Science and Technology 2008;2(5):853–62. https://doi.org/10.1177/193229680800200517.

[4] Bruen D, Delaney C, Florea L, Diamond D. Glucose sensing for diabetes monitoring: recent developments. Sensors 2017;17(8):1866. https://doi.org/10.3390/s17081866.

[5] Center for Diabetes Technology – University of Virginia School of Medicine. n.d.

[6] Cha KH. Advances in glucose sensing techniques: novel non-invasive and continuous electrochemical glucose monitoring systems. 2018.

[7] Chen C, Zhao X-L, Li Z-H, Zhu Z-G, Qian S-H, Flewitt A. Current and emerging technology for continuous glucose monitoring. Sensors 2017;17(12):182. https://doi.org/10.3390/s17010182.

[8] Choi H. Recent developments in minimally and truly non-invasive blood glucose monitoring techniques. In: 2017 IEEE sensors. IEEE; 2017. p. 1–3. https://doi.org/10.1109/ICSENS.2017.8234291.

[9] Cobelli C, Man CD, Sparacino G, Magni L, Nicolao G De, Kovatchev BP. Diabetes: models, signals, and control. IEEE Reviews in Biomedical Engineering 2009;2:54–96. https://doi.org/10.1109/RBME.2009.2036073.

[10] Coyle S, Curto VF, Benito-Lopez F, Florea L, Diamond D. Wearable bio and chemical sensors. In: Wearable sensors. Elsevier; 2014. p. 65–83. https://doi.org/10.1016/B978-0-12-418662-0.00002-7.

[11] Dexcom Inc. Dexcom G6 users guide. 2018. Retrieved from: https://www.dexcom.com/guides.

[12] Ebmeier M. The artificial pancreas: what is it and when's it coming? — taking control of your diabetes. 2018.
[13] Englert K, Ruedy K, Coffey J, Caswell K, Steffen A, Levandoski L, Diabetes Research in Children (DirecNet) Study Group, for the D. R. in C. (DirecNet) S. Skin and adhesive issues with continuous glucose monitors: a sticky situation. Journal of Diabetes Science and Technology 2014;8(4):745—51. https://doi.org/10.1177/1932296814529893.
[14] Facchinetti A, Sparacino G, Cobelli C. An online self-tunable method to denoise CGM sensor data. IEEE Transactions on Biomedical Engineering 2010;57(3):634—41. https://doi.org/10.1109/TBME.2009.2033264.
[15] FDA. Summary of safety and effectiveness data (SSED), MiniMed 670G system. 2016. Retrieved from: https://www.accessdata.fda.gov/cdrh_docs/pdf16/P160017b.pdf.
[16] FDA. Summary of safety and effectiveness data (SSED), freeStyle libre flash glucose monitoring system. 2017. Retrieved from: https://www.accessdata.fda.gov/cdrh_docs/pdf15/P150021B.pdf.
[17] FDA. Summary of safety and effectiveness data (SSED), eversense continuous glucose monitoring system. 2018. Retrieved from: https://www.accessdata.fda.gov/%20cdrh_docs/pdf16/P160048B.pdf. https://www.accessdata.fda.gov/cdrh_docs/pdf16/P160048B.pdf. https://www.accessdata.fda.gov/cdrh_docs/pdf16/P160048B.pdf.
[18] FDA. What is the pancreas? What is an artificial pancreas device system? 2018. Retrieved from: https://www.fda.gov/MedicalDevices/ProductsandMedicalProcedures/HomeHealthandConsumer/ConsumerProducts/ArtificialPancreas/ucm259548.htm.
[19] Garg SK, Akturk HK. A new era in continuous glucose monitoring: food and drug administration creates a new category of factory-calibrated nonadjunctive, interoperable Class II medical devices. Diabetes Technology and Therapeutics June 2018. https://doi.org/10.1089/dia.2018.0142. United States.
[20] Jadviscokova T, Fajkusova Z, Pallayova M, Luza J, Kuzmina G. Occurrence of adverse events due to continuous glucose monitoring. Biomedical Papers of the Medical Faculty of the University Palacky, Olomouc, Czechoslovakia 2007;151(2):263—6.
[21] Jina A, Tierney MJ, Tamada JA, McGill S, Desai S, Chua B, et al. Design, development, and evaluation of a novel microneedle array-based continuous glucose monitor. Journal of Diabetes Science and Technology 2014;8(3):483—7. https://doi.org/10.1177/1932296814526191.
[22] Keenan DB, Mastrototaro JJ, Voskanyan G, Steil GM. Delays in minimally invasive continuous glucose monitoring devices: a review of current technology. Journal of Diabetes Science and Technology 2009;3(5):1207—14. https://doi.org/10.1177/193229680900300528.
[23] Klueh U, Antar O, Qiao Y, Kreutzer DL. Role of vascular networks in extending glucose sensor function: impact of angiogenesis and lymphangiogenesis on continuous glucose monitoring in vivo. Journal of Biomedical Materials Research Part A 2014;102(10): 3512—22. https://doi.org/10.1002/jbm.a.35031.
[24] Kovatchev BP, Patek SD, Ortiz EA, Breton MD. Assessing sensor accuracy for non-adjunct use of continuous glucose monitoring. Diabetes Technology and Therapeutics 2015;17(3):177—86. https://doi.org/10.1089/dia.2014.0272.
[25] Kropff J, Bruttomesso D, Doll W, Farret A, Galasso S, Luijf YM, et al. Accuracy of two continuous glucose monitoring systems: a head-to-head comparison under clinical research centre and daily life conditions. Diabetes, Obesity and Metabolism 2015; 17(4):343—9. https://doi.org/10.1111/dom.12378.

[26] Laffel LM, Aleppo G, Buckingham BA, Forlenza GP, Rasbach LE, Tsalikian E, et al. A practical approach to using trend arrows on the Dexcom G5 CGM system to manage children and adolescents with diabetes. Journal of the Endocrine Society 2017;1(12): 1461–76. https://doi.org/10.1210/js.2017-00389.

[27] Lucisano JY, Routh TL, Lin JT, Gough DA. Glucose monitoring in individuals with diabetes using a long-term implanted sensor/telemetry system and model. IEEE Transactions on Biomedical Engineering 2017;64(9):1982–93. https://doi.org/10.1109/TBME.2016.2619333.

[28] Medtronic. Getting started with continuous glucose monitoring. 2018. Retrieved from: www.medtronicdiabetes.com/sites/default/files/library/download-library/workbooks/Tape_Tips_and_Site_Management.pdf.

[29] Paul HS, Schoenfisch MH. Nitric oxide-releasing subcutaneous glucose sensors. In: In vivo glucose sensing. Hoboken, NJ, USA: John Wiley & Sons, Inc; 2009. p. 243–67. https://doi.org/10.1002/9780470567319.ch9.

[30] Powell A. Harvard researchers help develop 'smart' tattoos – Harvard Gazette. 2017.

[31] Profusa's Tiny implantable sensors keep working in patients even after four years | Medgadget. 2018.

[32] Rao PV, Gan SH. Recent advances in nanotechnology-based diagnosis and treatments of diabetes. Current Drug Metabolism 2015;16(5):371–5.

[33] Rodbard D. Continuous glucose monitoring: a review of successes, challenges, and opportunities. Diabetes Technology and Therapeutics 2016;18(Suppl. 2):S3–13. https://doi.org/10.1089/dia.2015.0417.

[34] Rohajn SY. Blood sugar crash. 2014. Retrieved from: https://www.technologyreview.com/s/529026/blood-sugar-crash.

[35] Sharma S, El-Laboudi A, Reddy M, Jugnee N, Sivasubramaniyam S, El Sharkawy M, et al. A pilot study in humans of microneedle sensor arrays for continuous glucose monitoring. Analytical Methods 2018;10(18):2088–95. https://doi.org/10.1039/C8AY00264A.

[36] Tsai A. Tips for tackling adhesive irritation and allergies: diabetes Forecast®, 2017.

[37] Vaddiraju S, Burgess DJ, Tomazos I, Jain FC, Papadimitrakopoulos F. Technologies for continuous glucose monitoring: current problems and future promises. Journal of Diabetes Science and Technology 2010;4(6):1540–62. https://doi.org/10.1177/193229681000400632.

[38] Vaddiraju S, Kastellorizios M, Legassey A, Burgess D, Jain F, Papadimitrakopoulos F. Needle-implantable, wireless biosensor for continuous glucose monitoring. In: 2015 IEEE 12th international conference on wearable and implantable body sensor networks (BSN). IEEE; 2015. p. 1–5. https://doi.org/10.1109/BSN.2015.7299421.

[39] Wang G, He X, Wang L, Gu A, Huang Y, Fang B, et al. Non-enzymatic electrochemical sensing of glucose. Microchimica Acta 2013;180(3–4):161–86. https://doi.org/10.1007/s00604-012-0923-1.

[40] Wang X, Mdingi C, DeHennis A, Colvin AE. Algorithm for an implantable fluorescence based glucose sensor. In: 2012 annual international conference of the IEEE engineering in medicine and biology society; 2012. p. 3492–5. https://doi.org/10.1109/EMBC.2012.6346718.

[41] What is the sensor adhesive made of? | Dexcom. n.d.

[42] Wróbel MS. Non-invasive blood glucose monitoring with Raman spectroscopy: prospects for device miniaturization. IOP Conference Series: Materials Science and Engineering 2016;104(1):012036. https://doi.org/10.1088/1757-899X/104/1/012036.

[43] Yoo E-H, Lee S-Y. Glucose biosensors: an overview of use in clinical practice. Sensors 2010;10(5):4558–76. https://doi.org/10.3390/s100504558.

[44] Heller A, Feldman B. Electrochemical glucose sensors and their applications in diabetes management. Chemical Reviews 2008;108(7):2482–505.

[45] James TD, Phillips MD, Shinkai S. Boronic acids in saccharide recognition. Royal Society of Chemistry; 2006.

[46] Kim J, Campbell AS, Wang J. Wearable non-invasive epidermal glucose sensors: a review. Talanta 2018;177:163–70.

[47] Nichols SP, Koh A, Storm WL, Shin JH, Schoenfisch MH. Biocompatible materials for continuous glucose monitoring devices. Chemical Reviews 2013;113(4):2528–49.

[48] Jia W. Continuous Glucose Monitoring. 2018. https://doi.org/10.1007/978-981-10-7074-7.

[49] Rossetti P, Bondia J, Veh J, Fanelli CG. Estimating plasma glucose from interstitial glucose: the issue of calibration algorithms in commercial continuous glucose monitoring devices. sensors 2010;10:10936–52.

[50] Schuhmann W, Ohara TJ, Schmidt HL, Heller A. Electron transfer between glucose oxidase and electrodes via redox mediators bound with flexible chains to the enzyme surface. Journal of the American Chemical Society 1991;113(4):1394–7.

[51] Harris JM, Reyes C, Lopez GP. Common causes of glucose oxidase instability in in vivo biosensing: a brief review. Journal of Diabetes Science and Technology 2013;7(4): 1030–8.

[52] Lee H, Hong YJ, Baik S, Hyeon T, Kim DH. Enzyme based glucose sensor: from invasive to wearable device. Adv. Healthcare Mater. 2018;7:1701150.

[53] Croce R, Vaddiraju S, Kondo J, Wang Y, Zuo L, Zhu K, Islam SK, Burgess D, Jain FC, Papadimitrakopoulos F. A miniaturized transcutaneous system for continuous glucose monitoring. Biomedical Microdevices 2013;15:151–60.

[54] Gough DA, Lucisano JY, Tse PH. Two-dimensional enzyme electrode sensor for glucose. Anal Chem 1985;57(12):2351–7.

[55] Tse PHS, Leypoldt JK, Gough DA. Determination of the intrinsic kinetic constants of immobilized glucose oxidase and catalase. Biotech Bioengin 1987;29:696–704.

[56] Chinnadayyala SR, Park KD, Cho S. Editors' choice—review—in vivo and in vitro microneedle based enzymatic and non-enzymatic continuous glucose monitoring biosensors. ECS Journal of Solid State Science and Technology 2018;7(7): Q3159–71. https://doi.org/10.1149/2.0241807jss.

[57] udge K, Morrow L, Lastovich AG, Kurisko D, Keith S, Hartsell J, Roberts B, McVey E, Weidemaier K, Win K, Hompesch M. Continuous glucose monitoring using a novel glucose/galactose binding protein: results of a 12-hour feasibility study with the becton dickinson glucose/galactose binding protein sensor. Diabetes Technology & Therapeutics 2011;13:309–17. https://doi.org/10.1089/dia.2010.0130.

[58] Badugu R, Lakowicz JR, Geddes CD. A glucose sensing contact lens: A non-invasive technique for continuous physiological glucose monitoring. J Fluoresc 2003;13(5): 371–4.

[59] Liao Y, Yao H, Parviz B, Otis B. A 3μW wirelessly powered CMOS glucose sensor for an active contact lens. San Francisco, CA: 2011 IEEE International Solid-State Circuits Conference; 2011. p. 38–40. https://doi.org/10.1109/ISSCC.2011.5746209.

[60] Sen DK, Sarin GS. Tear glucose levels in normal people and in diabetic patients. Br J Ophthalmol 1980;64(9):693–5 [PMC free article] [PubMed].

[61] DeHennis A, Getzlaff S, Grice D, Mailand M. An NFC-enabled CMOS IC for a wireless fully implantable glucose sensor. IEEE J Biomed Health Inform 2016;20(1):18–28. https://doi.org/10.1109/JBHI.2015.2475236.

[62] Shenoy V. CMOS Analog Correlator Based Glucose Sensor Readout Circuit. University at Texas Arlington; 2013. Dissertation.

[63] Croce RA, et al. A low power miniaturized CMOS-based continuous glucose monitoring system. Cambridge, MA, USA: 2013 IEEE International Conference on Body Sensor Networks; 2013. p. 1–4. https://doi.org/10.1109/BSN.2013.6575469.

[64] Wang H-C, Lee A-R. Recent developments in blood glucose sensors. Journal of Food and Drug Analysis 2015;23(2):191–200. https://doi.org/10.1016/J.JFDA.2014.12.001.

[65] Lorenz C, Sandoval W, Mortellaro M. Interference assessment of various endogenous and exogenous substances on the performance of the eversense long-term implantable continuous glucose monitoring system. Diabetes Technology & Therapeutics 2018; 20(5):344–52. https://doi.org/10.1089/dia.2018.0028.

[66] Chen C, Zhao X-L, Li Z-H, Zhu Z-G, Qian S-H, Flewitt A. Current and emerging technology for continuous glucose monitoring. Sensors 2017;17(12):182. https://doi.org/10.3390/s17010182.

[67] Vaddiraju S, Burgess DJ, Tomazos I, Jain FC, Papadimitrakopoulos F. Technologies for continuous glucose monitoring: current problems and future promises. Journal of Diabetes Science and Technology 2010;4(6):1540–62. https://doi.org/10.1177/193229681000400632.

[68] Rahman G. Recent trends in the development of electrochemical glucose biosensors. International Journal of Biosensors & Bioelectronics 2017;3(1). https://doi.org/10.15406/ijbsbe.2017.03.00051.

[69] Colvin AE, Jiang H. Increased in vivo stability and functional lifetime of an implantable glucose sensor through platinum catalysis. Journal of Biomedical Materials Research Part A 2013;101A(5):1274–82. https://doi.org/10.1002/jbm.a.34424.

[70] Mortellaro M, DeHennis A. Performance characterization of an abiotic and fluorescent-based continuous glucose monitoring system in patients with type 1 diabetes. Biosensors and Bioelectronics 2014;61:227–31. https://doi.org/10.1016/j.bios.2014.05.022.

CHAPTER

Clinical impact of CGM use

7

Chukwuma Uduku, MBBS, BSc, MRCP [1], Monika Reddy, MBChB, MRCP (UK), PhD [2], Nick Oliver, FRCP [3]

[1]*Clinical Research Fellow and Specialist Registrar in Endocrinology, Diabetes and Internal Medicine, Imperial College London, St. Mary's Hospital Medical School Building, London, United Kingdom;* [2]*Honorary Senior Clinical Lecturer, Consultant in Endocrinology, Diabetes and Internal Medicine, Imperial College London, St. Mary's Hospital Medical School Building, London, United Kingdom;* [3]*Wynn Professor of Human Metabolism, Consultant in Endocrinology, Diabetes and Internal Medicine, Imperial College London, St. Mary's Hospital Medical School Building, London, United Kingdom*

Introduction

History and general rationale for glucose monitoring

Diabetes mellitus is a chronic metabolic condition characterized by hyperglycemia and impaired glucose homeostasis. The rising global prevalence of diabetes places greater importance on preventing the immediate and long-term complications associated with suboptimal glucose management. Glycated hemoglobin (HbA1c) is the most commonly referenced metric for assessing long-term glycemic control and complication risk association. However, its inability to identify rapid changes in glycemia and detect glycemic extremes such as hypoglycemia highlights the need for dynamic, real-time glucose monitoring. Self-monitoring blood glucose (SMBG) technology allows clinicians and people with diabetes the ability to make timely therapeutic decisions based on their blood glucose levels and is a cornerstone in diabetes management. Blood glucose monitoring is particularly integral in delivering safe diabetes care in those at risk of iatrogenic hypoglycemia from intensified insulin therapy.

Normal glucose homeostasis is maintained by insulin and counterregulatory hormones responding to blood glucose levels in real time. Similar to most diabetes care initiatives, regular glucose monitoring attempts to safely achieve physiological glycemic control and modify complication-associated risk factors without compromising the patient experience. Early SMBG meters utilized light photometry to derive blood glucose levels from solid dry-reagent test strips. However, they were confined to use only in clinical areas due to expensive costs and their bulky profile. It was not until the late 1970s and 1980s when smaller meters with fewer operator-dependent steps made self-testing feasible. Current glucose meters apply enzyme electrode technology to estimate blood glucose levels and are a staple fixture across all levels of diabetes care.

Glucose Monitoring Devices. https://doi.org/10.1016/B978-0-12-816714-4.00007-7

The turn of the 21st century saw the first clinically approved continuous glucose monitoring (CGM) device using a subcutaneously sited sensor to measure changes in interstitial fluid glucose concentrations. Glucose levels are derived from proportionally produced electric currents and are automatically measured and accessible to users at 5-min intervals. The evolution of CGM over the last decade has seen devices become smaller with fewer required calibrations and longer sensor change intervals. The addition of mobile Bluetooth technology has also allowed wireless data transfer and integration of smartphone technology to display results on purpose-designed applications. This chapter will address the current and future benefits, limitations, and clinical applications of CGM technology.

Parameters of glucose control and risk association

HbA1c

The association between prolonged exposure to hyperglycemia and increased risk of diabetes-associated micro- and macrovascular is well documented [1–4]. Significant outcomes include the fourfold increased risk of coronary heart disease [5] and cardiovascular disease being responsible for a quarter of diabetes deaths [6]. The Diabetes Control and Complications Trial (DCCT) and the United Kingdom Prospective Diabetes Study revealed a reduced risk of microvascular complications in participants with type 1 (T1DM) and type 2 (T2DM) diabetes following intensive glycemic control, respectively [1,2]. These benefits extended long beyond the duration of the trial "legacy effect" with lower rates of microvascular disease persisting despite HbA1c rebounding to levels seen in the standard care groups [7–10]. Furthermore, the curvilinear relationship between HbA1c and vascular risk suggests that the greatest risk reduction is achieved when glycemic control is successfully improved from hyper- to normoglycemia with minimal benefit to be gained from further glucose reduction [3,11]. HbA1c levels above 6%–7% have been shown to significantly increase microvascular and cardiovascular risk in diabetes populations. However, many trials including ACCORD and VADT have failed to identify significant cardiovascular risk reduction with intensive glucose control and, in fact, highlight the importance of balancing glucose reduction with the increased risk of hypoglycemia.

Hypoglycemia

Despite pharmaceutical advances in the preparation and delivery of insulin replacement, hypoglycemia remains a common complication among insulin-requiring individuals with diabetes. Errors in mismatching insulin dose to ingested carbohydrates and the nonphysiological pharmacokinetics and pharmacodynamics of exogenous insulin are among various factors responsible for iatrogenic hypoglycemia. As established in the DCCT, hypoglycemia risk is further increased in attempts to diminish vascular risk by implementing intensive insulin therapy.

The American Diabetes Association defines typical hypoglycemia symptoms accompanied by a measured plasma glucose concentration of $\leq$3.9 mmol/L ($\leq$70 mg/dL) as documented symptomatic hypoglycemia. Recurrence of hypoglycemia is known to increase the risk of hypoglycemia unawareness and severe hypoglycemia, both linked with increased morbidity and mortality [12,13]. Severe hypoglycemia is a medical emergency where individuals are dependent on treatment from a third party. Delayed treatment can often result in immediate morbidity from seizure, coma, and even death.

Time in range and glucose variability

Hyperglycemia and severe hypoglycemia are established complications linked to short-term (within day) glucose variation, longitudinal lability in fasting plasma glucose, and HbA1c [14,15]. Despite evidence associating glycemic variability with high levels of oxidative stress and endothelial dysfunction [16,17], studies have been conflicting in attempts to establish glucose variability as a modifiable vascular risk factor. Long-term glucose variability in T1DM and T2DM has been shown to be associated with increased microvascular risk [18–21], and although supported in the early DCCT studies, subsequent data analyses failed to affirm this association [22,23]. Other studies have linked longitudinal and short-term glucose variability with increased cardiovascular risk outcomes [24,25]. However, longer and more robust trials are needed to clarify the benefit of reduced glucose variability on clinical outcomes.

Glucose monitoring

Limitations of SMBG

An association between frequent glucose monitoring and improved glycemic control is well established in the literature and becomes increasingly evident when nontesting individuals self-monitor glucose at least 3–4 times a day [26–29]. The UK National Institute for Health and Care Excellence (NICE) and the American Diabetes Association (ADA) encourage individuals on multiple daily insulin injections (MDI) to monitor blood glucose levels at least four times a day to achieve target glycemic control [30,31]. Unfortunately, compliance using conventional glucose meters remains a challenge and is often hindered by the inconvenience and pain associated with multiple daily fingerprick tests [32]. Compliance rates as low as 40% have been reported in T1DM populations with up to one-third of individuals being unaware of the recommended SMBG guidelines [28,33]. Lack of time and forgetfulness have been cited as explanations for low SMBG compliance, particularly in younger, recently diagnosed, and individuals in full-time employment [33]. These findings highlight psychosocial barriers hindering SMBG compliance and remind us of the importance of structured education and appropriate psychological support to harness the full potential of glucose monitoring technology.

Benefits of CGM

Operational advantage

As an alternative point-of-care modality to SMBG, CGM conveniently provides instant access to real-time glucose levels, the trajectory of glucose levels, the rate of change of glycemia, and a visual indication of time spent in a target glycemic range. The operational process necessary to obtain a capillary glucose measurement using conventional dry-reagent strips can be time consuming and inconvenient, particularly in the working environment. Many CGM systems utilize a familiar and universally adopted technology such as the smartphone to immediately present data and overcome potential operational barriers and stigma attached to using capillary glucose testing kits. Automated CGM measurements uploaded directly to the user's smartphone or proprietary receiver make monitoring glucose levels intuitive, and aid individuals that find SMBG difficult to remember. The increasing life span of current sensors exceeding beyond 1 week, and one of the latest sensors in the market requiring no calibration, further contributes toward reducing the burden of skin pricking associated with conventional SMBG.

Direction, pattern and trends, investigative tool

Planning and undertaking events with a significant glycemic impact can be challenging when uncertain about the imminent direction or trajectory of glucose levels. Most CGM systems can inform users if glucose levels are rising, falling, or steady and, in turn, assist insulin dose decisions particularly when planning meals or physical activity.

Using CGM offers unprecedented insight into a full 24-h glucose profile and allows users and healthcare professionals the opportunity to contextualize glucose excursions and glycemic variability. The glycemic load and response to meals represent the biggest influence on blood glucose levels and often demonstrate significant variability within and between individuals. Applying CGM to individually characterize the glycemic impact of meal types helps with calculating accurate insulin-to-carbohydrate ratios and the consideration of using extended insulin boluses. Overnight, people often report significant variation in blood glucose levels at a time when performing frequent SMBG is limited by sleep. The dawn phenomenon best illustrates this and describes an early morning circadian elevation in hormones responsible for an increase in fasting glucose in a population of individuals with impaired β-cell function [34–37]. Different from the Somogyi effect, it is not preceded by iatrogenic hypoglycemia; however, both can be responsible for unpredictable overnight hyperglycemia. Overnight CGM usefully illustrates this glycemic pattern in individuals struggling with fasting hyperglycemia and assists in compensatory basal insulin dose adjustments. Extended use of CGM ultimately facilitates the fine-tuning of glycemic control by lifestyle and therapeutic changes in response to real-time and retrospectively identified temporal and behavioral trends.

Alerts

An additional benefit of CGM is the ability to alert users when glycemia veers outside target parameters. This function encourages timely intervention to avoid labile swings (variability) in blood glucose and severe hypo- and hyperglycemia. Hypoglycemia is normally heralded by autonomic and eventually neuroglycopenic symptoms prompting a capillary glucose measurement and appropriate treatment once confirmed. Impaired awareness of hypoglycemia describes the failure to identify a significant decline in blood glucose or the failure to elicit autonomic symptoms before neuroglycopenia. With a prevalence of just under 20% in adults with T1DM [38], the CGM alarm feature provides a safety mechanism to prevent forthcoming hypoglycemia in this large vulnerable group. Most global health organizations recommend physical activity as part of a diabetes care plan [30,39]. Unfortunately for many people with diabetes exercise is fraught with fear of hypoglycemia resulting in an avoidance of physical activity or additional calorie intake as a precautionary step. As an adjunct to capillary glucose testing, CGM can alert users of impending hypoglycemia during exercise and mitigate the risk by facilitating appropriate insulin dosing before and after exercise.

Clinical application of CGM

In clinical settings, CGM can be used as a permanent adjunct to facilitate daily glucose monitoring or as a short-term investigative tool to identify trends and optimize treatment. Incongruence between HbA1c and capillary glucose measurements is a common clinical scenario and often represents periods of hypo- and hyperglycemia missed by SMBG. CGM fills the gaps missed by SMBG and sheds light on blood glucose levels when users are asleep or unable to perform capillary glucose monitoring. Permanent CGM use is currently prioritized for vulnerable individuals with hypoglycemia unawareness and individuals struggling to improve glucose control despite optimizing all other avenues of treatment. The UK NICE guidelines reserve long-term CGM application for those willing to comply with use for at least 70% of the time, committed to calibrating when required, and struggling with any of the following: severe hypoglycemia, complete hypoglycemia unawareness, frequent asymptomatic hypoglycemia, extreme fear of hypoglycemia and a HbA1c >9% despite 10 times daily SMBG [40]. Although this list of indications may appear prescriptive their subjectivity allows clinical discretion to take precedent if necessary. The ADA recommends CGM be considered in both T1DM and T2DM where individuals are failing to achieve treatment goals despite being on intensive insulin therapy, with greater precedent reserved for those experiencing troublesome hypoglycemia [41].

CGM efficacy

Given the recent availability of CGM and limited long-term complication data, heterogeneous outcomes of glycemic control have been adopted as measures of clinical efficacy. These include HbA1c, time in range, time spent in hypo- or hyperglycemia, and various metrics of glycemic variability.

Retrospective CGM studies

Without immediate access to CGM data, retrospective (blinded) CGM users are unable to make real-time therapeutic changes and study outcomes derived from this population are largely based on interventions following retrospective data analysis.

Observational trials

Observational retrospective studies have reported significant HbA1c reductions across various age groups in individuals with T1DM, with some studies maintaining insulin dose neutrality [42–44]. Although critiqued for low participant numbers, not showing significant hypoglycemia reduction, and no comparative control arm, these studies reinforce the value of CGM in identifying glucose trends and facilitating effective therapeutic intervention.

Randomized controlled trials

Apart from a 3-month randomized controlled crossover study reporting a significant reduction in HbA1c during the retrospective CGM phase [45], randomized controlled trials (RCTs) have largely failed to establish superior glycemic control using retrospective CGM. A large 12-week trial of 128 randomized T1DM children found no significant difference in HbA1c change when comparing retrospective CGM against SMBG [46]. Similarly, a 3-month adult trial comparing adjunctive blinded CGM against SMBG failed to show a difference in the improvement of HbA1c despite identifying half the cases of asymptomatic hypoglycemia [47]. Interestingly, the control arms in both studies saw participants performing SMBG more than the clinically recommended 4 times a day, an achievement most individuals would find challenging under normal circumstances. These shortcomings are heavily biased toward SMBG when comparing against CGM and must be considered before transferring outcomes into real-world clinical settings.

Real-time CGM studies

Real-time CGM (RT-CGM) provides users with immediate access to glucose levels, rate and direction of change, and glycemic profile patterns.

Observational

Observational studies following adults and children using RT-CGM also report improvement in glycemic control and highlight the benefits of dynamic therapeutic modification. Reduced time spent in hypo- and hyperglycemia has been successfully used to demonstrate improved glucose control when using implantable and transcutaneous CGM [48,49], while studies measuring HbA1c have observed the greatest reductions in participants starting RT-CGM with poor baseline glycemic control (HbA1c >9%) [50]. These outcomes favor identifying groups that would benefit most from CGM and applying a targeted approach to clinical implementation. However, crediting CGM alone for the significant HbA1c improvements ignores the fact that individuals with the above target HbA1c may struggle to adhere to or may not have received optimal treatment before CGM. Therefore it is commonplace in clinical practice to ensure that those considered for CGM partake in a form of validated structured education and a period of optimal treatment before commencing CGM.

Randomized controlled trials

Continuous RT-CGM has been shown to have the greatest effect in reducing HbA1c in a head-to-head RCT comparing continuous CGM, biweekly real-time CGM and five-times daily SMBG [51]. This study population of adults and children with baseline HbA1c >8% despite intensive treatment further supports the targeted application of CGM in individuals struggling to achieve optimal control. Two large RCTs, the Juvenile Diabetes Research Foundation (JDRF) CGM study and the DIAMOND clinical trial, both showed real-time CGM to reduce HbA1c in pediatric and adult T1DM populations [52,53]. The JDRF multicenter RCT compared various real-time CGM devices against SMBG and explored the impact of age and frequency of CGM use on HbA1c outcomes. Following 26 weeks of CGM, the subgroup of participants older than 25 years old showed a significant reduction in HbA1c and increased time in target range (71–180 mg/dL) without increasing the risk of biochemical hypoglycemia (≤70 mg/dL). This group was shown to use CGM significantly more frequently than the younger participants with 83% using CGM at least six times a week. Although the CGM cohort of 8–14-year olds failed to achieve a statistically significant HbA1c reduction, they had more participants with HbA1c <7% and a ≥10% relative reduction in HbA1c. These findings further highlight the importance of selective CGM application in adults willing to engage, as reflected in the NICE guideline's requirement for at least 70% compliance. Although most individuals with T1DM deliver insulin by MDI most CGM users manage their diabetes using continuous subcutaneous insulin infusion (CSII). The DIAMOND clinical trial randomized adults using MDI to either real-time CGM or usual care for 24 weeks to establish the efficacy of CGM in MDI users [52]. The CGM group produced a 1% and 1.1% reduction in HbA1c at 12 and 24 weeks, respectively, with over 90% using CGM at least 6 days a week. Similarly, in a crossover RCT comprising T1DM adults with suboptimal baseline glucose control (HbA1c >7.8%), the GOLD trial demonstrated a −0.43% mean difference in HbA1c in the CGM arm compared to conventional therapy [54]. Both the JDRF and DIAMOND studies reported reduced glycemic variability with real-time CGM [52,55]. DIAMOND showed that real-time CGM significantly increased the average minutes per day spent in target range (3.9–10 mmol/L) and in turn reduced the median minutes per day spent above and below target. Whereas, the JDRF study successfully demonstrated a reduction in glycemic variability across various metrics except for the lability index and mean absolute glucose change per unit time [55].

The SWITCH study group randomized 124 adult and pediatric participants with T1DM to 6 months with and without real-time CGM (sensor on and off) and crossed over both arms following a 4-month washout period [56]. In addition to significant HbA1c reductions, participants in the CGM arm displayed a change in insulin administration behavior with a greater tendency to deliver more frequent insulin boluses [56]. The greater number of people using real-time CGM has seen a rise in users applying preemptive intervention (including insulin correction) to proactively address blood glucose deviations, a glucose management strategy coined as "sugar surfing" by Dr. Stephen Ponder MD.

The frequency of hypoglycemia as an index of glucose control and diabetes risk has become increasingly relevant as a greater impetus toward lowering HbA1c has seen more individuals receive intensive insulin therapy. The application of RT-CGM for 26 weeks in adults and children with T1DM has been shown to significantly lower the time spent in hypoglycemia (interstitial glucose <63 mg/dL) compared to a control arm with access to blinded CGM every second week for 5 days [57]. Individuals with a history of severe hypoglycemia and hypoglycemia unawareness are among a vulnerable subset and are likely to benefit most from the features offered by RT-CGM. Two multicenter RCTs (IN CONTROL and HypoDE), focused on successfully demonstrating the efficacy of RT-CGM in reducing hypoglycemia, severe hypoglycemia, and impaired awareness when applied in this high-risk population. The use of RT-CGM during the intervention arm of the IN CONTROL trial significantly improved the time spent in normoglycemia while, in turn, reducing the time spent in hypo- and hyperglycemia. Furthermore, RT-CGM significantly reduced the accounts of severe hypoglycemia during the 16 weeks intervention period versus the control SMBG arm (14 vs. 34 events, $P = .033$) [58]. Similarly, HypoDE showed RT-CGM to lower the frequency of hypoglycemic events from 10.8 (SD 10.0) to 3.5 (4.7) per 28 days, and ultimately reduce the incidence of hypoglycemia by 72% in high-risk individuals [59].

CGM pregnancy data

Well-controlled glycemia is strongly advocated before and during pregnancy to reduce the risk of complications including preeclampsia, macrosomia, and congenital malformations. Despite intensive antenatal care follow-up, the rate of severe hypoglycemia is five times greater in early pregnancy compared to nonpregnant women and 85% fail to achieve target HbA1c [43,60]. Although not routinely offered during pregnancy, CGM is often considered when challenged with severe hypoglycemia and labile glucose control. A prospective randomized control trial in pregnant women with T1DM and T2DM saw lower third-trimester average HbA1c levels, lower birth weights, and lower macrosomia risk when using retrospective CGM compared to usual antenatal care [61]. In a large multicenter RCT (CONCEPTT), pregnant women and women planning a pregnancy with T1DM were randomized to adjunctive real-time CGM with capillary glucose monitoring or capillary glucose monitoring alone [62]. The pregnant CGM cohort reported more time in the target range, reduced glycemic variability, and a minimal HbA1c reduction at 34 weeks gestation. In addition, neonatal outcomes including large for gestational age, neonatal hypoglycemia, and admission to neonatal intensive care >24 h were all significantly reduced [62]. Glycemic control outcomes revealed no HbA1c improvements in the women planning pregnancy; however, this cohort was not powered to identify significant changes seen in the pregnancy cohort [62].

CGM quality of life data

An important aspect of managing chronic disease is understanding the impact both the condition and the required long-term therapy have on an individual's quality of life. Self-monitoring the impact of daily activities on glycemia provides users with a contextual understanding of how to modify their management to suit their lifestyle and not vice versa. Applying the thematic analysis to semistructured interviews, RT-CGM has been shown to enhance a sense of safety and control, while reducing stress among a majority of T1DM participants at high risk of hypoglycemia [63]. However, a section of participants found CGM to be intrusive, frustrating, and serve to remind us of the heterogeneous expectations of what is considered an improved quality of life [63]. The subjective nature of these outcomes encourages further investigation across varying diabetes populations.

CGM limitations

Despite the significant increase in CGM users over the last decade, SMBG remains the universally accepted method for glucose monitoring for individuals with insulin-requiring diabetes. Later we have outlined the factors limiting CGM uptake as a user, physician, and device dependent.

User dependent

Similar to many chronic diseases, the daily experience of living with diabetes can be extremely personal, and expectations can dynamically vary between individuals. To successfully yield the proven benefits of CGM, healthcare professionals must be insightful in identifying individuals that would struggle to apply the technology and those that would be resistant change and what it entails. Abandoning well-served methods and adopting new practices can be frightening, and users require the support of robust education programs to help facilitate the transition. As established, forgetfulness is often cited as a reason behind SMBG noncompliance. Frequent SMBG and blood glucose data analysis can be exhausting with many users choosing not to engage as a conscious or subconscious avoidance strategy [64]. Challenges for users include changing sensors, wirelessly pairing transmitters, performing steady-state calibration, and analyzing large blood glucose datasets, all of which may reinforce avoidant behavior. Conversely, willingness to engage may not suffice in individuals where other factors such as impaired visual acuity, reduced manual dexterity or inability to overcome the learning curve would deem them unsuitable for CGM. The addition of nonnumerical variables such as glucose trend arrows has ambiguous outcomes with regard to the extent glucose levels will be affected. This, in turn, can introduce an element of uncertainty when the user is making insulin dose decisions.

Healthcare provider dependent

CGM application remains relatively niche among the general diabetes population and most physicians have limited experience interpreting CGM datasets. A standard guideline or reference algorithm does not exist to assist healthcare professionals when making CGM-driven treatment interventions resulting in varied proposed interventions. A prospective observational study saw 2 days of CGM data from 20 pregnant women with T1DM presented to four physicians to give daily treatment adjustment recommendations. Significant differences were observed when reviewing the proposed interventions between the CGM days [65]. Although limited by a small number of analyzing physicians, these results highlight the subjective approach to CGM interpretation and call for a unified thinking process when educating healthcare providers. The time requirements necessary to analyze CGM data in a clinical setting pose a barrier in healthcare systems restricted by limited trained professionals and consultation time constraints. Healthcare providers may deem the additional time and resources required to educate healthcare professionals to deliver CGM as economically nonviable and as a result limit CGM uptake.

Device dependent

CGM is designed to be used as an adjunctive tool alongside conventional SMBG and not a direct replacement. Unfortunately, this is a common misconception resulting in misplaced expectations among potential users. The application of CGM for therapeutic decisions is assumed on interstitial glucose being interchangeable with traditional capillary blood glucose measurements. Although glucose levels across both compartments are established by a process of diffusion, CGM calibration against corresponding steady-state capillary glucose is recommended to ensure accuracy. MARD (mean absolute relative difference) is the most used metric for assessing CGM sensor accuracy with smaller percentages reflecting measurements closer to reference glucose values. A value 10% is regarded as the accuracy threshold based on in silico simulation demonstrating insignificant hypoglycemia with lower percentages [66]. As a measure of accuracy, MARD is a variable depending on glucose concentrations and rate of change, thus creating significant inaccuracies when comparing data between trials. Sensor lag describes the delay in glucose diffusion across the capillary bed, into the interstitial space, and finally being registered by the CGM sensor. Technological advances have seen sensor lag reduce to a few minutes from the 15 min reported in early CGM systems. Nonetheless, the cumulative effect of this delay can result in sensor measurements drifting further away from corresponding capillary blood glucose values. Systems susceptible to sensor drift are required to undergo twice-daily CGM calibration during glycemic steady states to overcome the significant inaccuracy resulting from sensor drift. Significant improvements in CGMs accuracy has seen the UK Driver and Vehicle Licensing Agency update their guidelines to permit its use among car and motorcycle drivers on insulin therapy. However, individuals are obligated to measure their capillary glucose if they experience symptoms of hypoglycaemia or blood glucose levels are 4mmol/L or below. Alarm fatigue is well described in CGM where hypo- and hyperglycemia alarm notifications are intrinsic components in establishing clinical efficacy and in extreme cases cannot be disabled. Shivers et al. addressed the artificially higher thresholds set for low glucose alarms to increase detection sensitivity, however inversely reducing specificity and subjecting users to more false-positive alarms [67]. This, in turn, increases alarm fatigue, drives users to ignore alarms, and ultimately abandon CGM [67]. The signal loss to receiver/smartphone and false measurements due to physical sensor compression are additional hardware issues with significant implications when responsible for automated insulin delivery within a closed-loop AP system. Lastly, unless financially reimbursed by healthcare providers, permanent use of CGM is an expensive venture and economically out of reach for most individuals.

Available CGM systems

Positive clinical trial outcomes have contributed to a rise in the popularity and number of available CGM systems. Listed below are a summary of randomised controlled CGM trials and the currently available CGM systems in the United Kingdom (Tables 7.1 and 7.2).

Dexcom

The Dexcom G6 is a standalone CGM system with superior accuracy (MARD 9%) compared to its competitors. The sensor is easily inserted and requires a 2 h warm-up period before glucose levels are uploaded for user access. Once active, glucose measurements are transmitted via Bluetooth to either a proprietary Dexcom receiver, the Dexcom smartphone application, or a sensor-augmented insulin pump (e.g., Tandem T-slim). The G6 comes equipped with a simple to use one-touch sensor applicator, boasts an extended sensor life span of 10 days, and is not affected by acetaminophen. Of greatest significance to users is the

Table 7.1 Randomized controlled CGM trials.

Study	Method	Participants	Duration of CGM	Outcomes
Deiss et al. [51]	*Crossover RCT*	T1DM 81 children 81 adults	3-month crossover	Mean HbA1c % reduction from baseline versus control 1 month (0.6 vs. 0.2, $P = .008$) 3 months (1.0 vs. 0.4, $P = .003$)
Tamborlane et al. [53]	RCT	T1DM 98 age ≥25 years 110 age 15–24 years 114 age 8–14 years	26 weeks	Mean HbA1c % reduction from baseline ≥25 years (0.5 vs. 0.02, $P < .001$) 15–24 years (0.18 vs. 0.21, $P = .52$) 8–14 years (0.37 vs. 0.22, $P = .20$)
Beck et al. [52]	RCT	T1DM 158 adults	24 weeks	Mean HbA1c % reduction from baseline 12 weeks 1.1 versus 0.5, $P < .001$ 24 weeks 1.0 versus 0.4, $P < .001$ Median duration hypo <70 mg/dL 43 min/day versus 80 min/day
Battelino et al. [56]	Crossover RCT	T1DM 72 children 81 adults	6 months	Mean difference in HbA1c CGM versus control −0.46 (95% CI −0.26, −0.66; $P < .001$)
Lind et al. [54]	Crossover RCT	T1DM 142 adults	26 weeks	Mean difference in HbA1c % from control −0.43 (95% CI −0.57, −0.29; $P < .001$)
Van Beers et al. [58]	Crossover RCT	T1DM Gold score ≥4 56 adults	16 weeks	Mean difference in % time spent in normoglycemia (72–180 mg/dL) versus control 9.6% (95% CI 8.0, 11.2; $P < .0001$)
Batellino et al. [57]	Crossover RCT	T1DM 53 children 67 adults	26 weeks	Time spent in hypoglycemia (≤63 mg/dL) versus control Ratio of means 0.49 (95% CI 0.26, 0.76; $P = .03$)

(continued)

Table 7.1 Randomized controlled CGM trials.—*cont'd*

Study	Method	Participants	Duration of CGM	Outcomes
Heineman et al. [59]	Parallel RCT	T1DM MDI 149 adults	26 weeks	Mean number of hypoglycemia events ($\leq$54 mg/dL) per 28 days versus baseline CGM 3.5 (SD 4.7) versus 10.8 (10) Control 13.7 (11.6) versus 14.4 (12.4) CGM 72% reduced incidence of hypo The incidence rate ratio 0.28 (95% CI 0.20, 0.39; $P < .0001$)
Murphy et al. [61]	Prospective RCT	46 T1DM 25 T2DM Pregnant women	Up to 7 days supplementary CGM at 4–6-week intervals between 8- and 32-week gestation	Mean HbA1c % from 32 to 36 weeks' gestation versus standard antenatal care 5.8% (SD 0.6) versus 6.4% (0.7) Median birth weight centiles 69% versus 93% ($P = .02$) (exclusion of five twins $P = .04$)
Feig et al. [62]	RCT	T1DM 215 pregnant ($\leq$13 weeks and 6 days gestation) 110 planning pregnancy	Pregnancy—CGM from randomization to 34 weeks Planning pregnancy—CGM from randomization to 24 weeks	Percentage time in target pregnant CGM versus control (68% vs. 61%, $P = .0034$), time in hypoglycemia (27% vs. 32%, $P = .0279$) Pregnant CGM lower incidence of large for gestational age, odds ratio 0.51 (95% CI 0.28, 0.90; $P = .0210$) No significant CGM benefit in planning a pregnancy

Table 7.2 Available CGM and Flash glucose monitoring systems.

	CGM				Flash
	Dexcom G6	**Medtronic Guardian Connect**	**Medtrum S7/A6**	**Senseonics Eversense**	**Abbott Freestyle Libre**
Accuracy (MARD)	9%	10%	9%	8%	10%
Warm up	2 h	2 h	S7—2 h A6—2 h	24 h	1 h
Calibration	Not indicated	Twice daily	S7—twice on insertion, once-daily thereafter A6—twice daily	Twice daily	Not indicated (*Factory Calibrated*)
Sensor life cycle	10 days	Seven days	S7—14 days A6—14 days	Up to 180 days	14 days
Transmitter life cycle	3 months	Six-day rechargeable battery life Over 12 months of operational life	S7—3 months A6—charge during sensor change (14 days)	Charge daily 12 months of operational life	
High and low glucose alert	Yes	Yes	S7 and A6	Yes	No
Predicted glucose alert	Yes	Yes	Yes	Yes	No

elimination of compulsory capillary blood glucose calibration, previously required on sensor insertion and every 12 h with its predecessor. The transmitter can be reused when changing sensors, however, must be discarded after exceeding its 3-month life span. Dexcom CGMs allow the sharing of data reports with healthcare professionals and carers via the Dexcom Clarity software and various open-access data storage services.

Medtronic

The Guardian Connect is Medtronic's latest offering combining their Guardian 3 sensor and Guardian 3 Link transmitter to deliver an easily deployed slim profile CGMs. Housing a rechargeable battery, the Guardian 3 Link transmitter requires charging when replacing sensors after 7 days of use and can be reused for over 1 year. A 2-h warm-up period is required following insertion before glucose levels are updated at 5-min intervals. Data are transmitted using Bluetooth technology and are accessible via the Guardian Connect smartphone app, the MiniMed 640G, and the new 670G insulin pumps. Integration with the newly available sensor-augmented Minimed 670G utilizes Medtronic's SmartGuard self-adjusting insulin delivery system (see section Future of CGM). Although both systems promise almost similar levels of accuracy, Guardian Connect is dependent on twice daily CBG calibration to achieve a MARD of 10%. The inclusion of a 10–60 min predictive alert feature for anticipating hypo- and hyperglycemia events allows users to lower glycemic variability by preemptive intervention. Automated data uploads and data-sharing functions are managed via the Medtronic Carelink Software.

Medtrum

The S7 EasySense by Medtrum delivers an easy-to-use CGM system with an impressive 14-day sensor life span. Insertion of the S7 sensor is a simple process requiring calibration twice on initial application and once daily thereafter. A life span of 3 months allows the S7 transmitter to be reused during intervening sensor changes. Updating and storing real-time data on the EasySense app and locally within the CGM transmitter provide accessibility to a variety of smartphone platforms and ensure data redundancy. Medtrum's A6 Touchcare system integrates their A6 CGM device, P6 disposable patch pump, and proprietary handheld personal diabetes manager (PDM) to deliver a sensor-augmented pump with low glucose and predictive low glucose suspend features. The A6 sensor boasts a MARD of 9%, with a 7-day sensor life span and requirement for twice-daily calibration. Furthermore, the A6 Touchcare is reliant on the PDM device to view real-time data, deliver bolus inulin doses, change basal insulin rates, and apply the predicted low glucose suspend feature. Data storage is managed on the PDM and transmitter and can be paired to the easy touch smartphone app.

Senseonics

The Eversense is the first implantable CGM system promising a sensor life span of 180 days and an 8.8% MARD. The small long-term sensor requires professional insertion within the subcutaneous space and is wirelessly powered by an externally worn smart transmitter. Real-time glucose levels, trends, and predictive alerts are transmitted to the mobile app via Bluetooth technology. Nondirect contact communication between the implanted sensor and external transmitter permits the removal or changing of the smart transmitter without the need to reinsert a new sensor. Additional benefits of an implanted sensor include being able to partake in activities requiring submersion in water for long periods. The practical benefits extend to the smart transmitter that is rechargeable, water-resistant, submerged in 1 m for 30 min, and capable of delivering on-body vibration alerts to notify hypo and hyperglycemia even when separated from the smartphone receiver.

Flash glucose monitoring

Flash glucose monitoring, also called as intermittent CGM, is a subset of CGM where glucose levels are reported only after the user scans the device by closely passing the Freestyle Libre reader or a smartphone with the Freestyle Librelink App over the sensor. In addition to presenting the current glucose value to the smartphone or card device, sensor scanning also displays the last 8 h of glucose data and a trend arrow to represent the current glucose direction. Similar to CGM, both technologies use enzyme electrode technology to measure changes in interstitial fluid glucose and require the user to wear a subcutaneously sited sensor. Flash sensors have a life span of 14 days and are factory calibrated overcoming the need to perform twice-daily steady-state capillary glucose calibration. This is a welcomed feature for users looking to minimize fingerprick testing; however, it can potentially result in "sensor drift" disparity toward the end of the sensor life cycle. Different from CGM where 5-min interval glucose levels are continuously being uploaded and saved for detailed retrospective analysis, flash sensors store readings every 15 min, and retrospective analysis is limited to the last 8 h. The absence of a continuous glucose data stream precludes the application of flash glucose monitoring in the sensor-augmented pump and artificial pancreas systems. Alarm notification of hypo- and hyperglycemia is another significant CGM feature not present in flash glucose monitoring systems. The exclusion of this feature relies on the user to scan the sensor when hypoglycemia is suspected and could lead to missed events, particularly among individuals with diminished hypoglycemia awareness.

Randomized controlled trials in T1DM and T2DM have reported reduced incidence of hypoglycemia when flash glucose monitoring is applied as an adjunct alongside SMBG [68,69]. The IMPACT trial demonstrated a greater reduction in the incidence, magnitude, and time spent in hypoglycemia (<70 mg/dL, <3.9 mmol/L) after 6 months of using flash monitoring in well-controlled adults with T1DM [68].

Improved glycemic control outcomes extended to significantly lower measures of glucose variability and better Diabetes Treatment Satisfaction Questionnaire scores [68]. In T2DM populations, flash glucose monitoring has been shown to significantly reduce hypoglycemia (<70 mg/dL, <3.9 mmol/L) risk by 55% with reduced SMBG frequency [69]. Selective application of flash monitoring based on these outcomes would place individuals at high risk of hypoglycemia as the most likely to benefit.

The I HART CGM study compared the efficacy of flash glucose monitoring and CGM at reducing hypoglycemia in a prospective parallel-group study among T1DM adults at high risk [70]. Following a 2-week run with blinded CGM, participants were randomized to either real-time CGM or flash monitoring for 8 weeks with an option for all participants to continue real-time CGM for an additional 8 weeks. The group switching from flash monitoring to real-time CGM showed a significant reduction in time spent in hypoglycemia (<54 mg/dL, <3.0 mmol/L) and increased time in target range (<70–180 mg/dL, 3.9–10 mmol/L) [70]. In contrast, no significant changes were observed in the participants that remained on real-time CGM suggesting superiority when addressing risk in T1DM adults vulnerable to hypoglycemia.

The Abbot Freestyle Libre is currently the only available flash glucose monitoring system available. Both the sensor and transmitter are housed in a single disc-shaped unit sited in the upper arm. A MARD of 10% delivers accuracy nearly comparable to available CGMs; however, false-positive hypoglycemia readings suggest a decline in sensor accuracy in the low glucose range. The transmission of data is currently limited to the proprietary Freestyle Libre receiver and smartphones with the Freestyle Librelink App. The soon to be available next-generation FreeStyle Libre 2 system builds on the success of its predecessor by introducing an optional customizable alarm feature for hypoglycemia, hyperglycemia, and Bluetooth signal loss. This inclusion overcomes a significant limitation of flash glucose monitoring and makes strides toward bridging the gap with CGM.

Further utility of CGM

Combining technology

The sensor-augmented pump (SAP) combines CGM and a CSII pump to deliver real-time glucose readings to the pump interface. This provides convenience and continuous blood glucose feedback to support insulin dose adjustment. The ability to automatically suspend insulin pump delivery once glucose levels reach a preconfigured threshold “low glucose or threshold-suspend” is a feature adopted by SAP systems to help mitigate hypoglycemia risk and bypass user error when responding to CGM alarm notifications. Sensor-augmented pumps and low glucose suspend have been shown to successfully reduce HbA1c and overnight hypoglycemia. A 1-year multicenter randomized control trial comparing SAP to MDI demonstrated a weight neutral significant reduction in HbA1c, 7.5% versus 8.1% ($P < .001$) [71].

The ASPIRE trial showed the impact of the low glucose suspend feature in T1DM adults with documented nocturnal hypoglycemia by successfully demonstrating a 32% reduction in the frequency of nocturnal hypoglycemia and a 38% lower mean area under the curve when combining threshold suspend alongside SAP compared with SAP alone [72]. Successful integration of diabetes technologies is considered a crucial step toward developing a fully automated closed-loop insulin delivery system "artificial pancreas." Closed-loop studies have yielded significant improvements in measures of glucose control, especially overnight. These include, increased percentage time in range, and reduced time spent in hypo and hyperglycaemia [72a]. Hybrid closed-loop systems leave prandial insulin bolus delivery in the hands of the user and apply automated insulin delivery throughout the rest of the day. As part of the Medtronic ecosystem, combining the Guardian Connect CGMs and the MiniMed 670G pump takes advantage of Medtronic's self-adjusting insulin delivery technology (Medtronic SmartGuard). Functioning as a hybrid closed-loop artificial pancreas, SmartGuard is capable of 5-min dose titration against CGM-derived glucose levels.

Inpatient CGM

Worsening glucose control and hyperglycemia are commonly seen during intercurrent illness in diabetes. Diabetes care in the critically unwell was previously focused on achieving tighter glucose control and preventing possible effects of stress-induced hyperglycemia [73]. Studies have since produced conflicting evidence regarding the impact of critical illness-induced stress hyperglycemia with some studies conversely associating tight glycemic control with an increased risk of mortality [74]. As a result, critical care strategies have evolved to target moderate glycemic control while considering the clinical impact of premorbid HbA1c, relative hypoglycemia, and glucose variability. Despite many intensive care units (ICU) adopting intravenous insulin infusion and frequent capillary glucose monitoring to achieve predetermined glucose parameters, CGM use has not successfully crossed over to inpatient settings. A review of inpatient CGM clinical trials showed no additional glycemic control benefit when CGM is applied to ICU patients already undergoing frequent CBG for intravenous insulin infusion titration [75]. However, it must be stated that ICU CGM outcomes are limited by the small number of available studies and the significant heterogeneity with regards to the severity of illness, type of diabetes, and choice of CGM among study participants. Outside of ICU, CGM outcomes fall closer in line with outpatient results with superior detection of hypoglycemia compared to CBG [75]. Further work is required comparing CGM outcomes in T1DM and T2DM populations at varying levels of critical care dependency.

Summary

In an endeavor to reduce the morbidity and mortality associated with diabetes-related complications, CGM successfully facilitates frequent glucose monitoring with proven improvements in HbA1c, time in glycemic range, and hypoglycemia. Discrete access to automated glucose monitoring, digestible retrospective glucose analysis, and a reduction in fingerprick glucose testing are among the practical benefits offered to users. The inclusion of high and low glucose alert features helps address previously unidentified glucose fluctuations and supports a population of individuals vulnerable to the deleterious impact of hypoglycemia unawareness. As a research tool, access to continuous glucose data has proven to be insightful and remains pivotal in establishing the clinical significance of glycemic variability and other metrics of glucose control beyond HbA1c.

In acknowledging the many benefits of CGM, users and healthcare providers must not lose sight of the proven efficacy of CGM as an adjunct and not a replacement for capillary glucose testing. Furthermore, evidence suggests CGM to be more effective when applied as a real-time interventional tool among adults willing to comply with frequent use. Appreciation of these variables and the provision of appropriate educational support must be considered when selecting individuals that would reap the benefits of CGM. Ongoing dialogue to ensure that the limitations and operational requirements are understood is crucial to retain compliance and manage user expectations. Although lacking a glucose alert feature, the arrival of flash glucose monitoring provides an unobtrusive means of glucose monitoring in individuals unable to improve glycemia despite the intensive escalation of therapy.

In our aspiration to replicate physiological glucose control, CGM technology is currently the most effective means of providing frequent efferent glucose data to aid human or algorithmic insulin pump directed intervention. Sensor-augmented insulin pumps with integrated low glucose suspend capabilities have proven successful in improving glycemic control and reducing hypoglycemia. With the growing presence of hybrid closed-loop systems on the scene, CGM continues to maintain its position at the forefront of innovative diabetes care.

References

[1] Diabetes Control and Complications Trial Research Group, et al. The effect of intensive treatment of diabetes on the development and progression of long-term complications in insulin-dependent diabetes mellitus. The New England Journal of Medicine 1993;329:977–86.

[2] Stratton IM, et al. Association of glycaemia with macrovascular and microvascular complications of type 2 diabetes (UKPDS 35): prospective observational study. BMJ 2000;321:405–12.

[3] Zoungas S, et al. Association of HbA1c levels with vascular complications and death in patients with type 2 diabetes: evidence of glycaemic thresholds. Diabetologia 2012; 55:636–43.

[4] Duckworth W, et al. Glucose control and vascular complications in veterans with type 2 diabetes. The New England Journal of Medicine 2009;360:129–39.

[5] Kannel WB, McGee DL. Diabetes and cardiovascular disease. The Framingham study. Journal of the American Medical Association 1979;241:2035–8.

[6] Moss SE, Klein R, Klein BE. Cause-specific mortality in a population-based study of diabetes. American Journal of Public Health 1991;81:1158–62.

[7] Diabetes Control and Complications Trial/Epidemiology of Diabetes Interventions and Complications (DCCT/EDIC) Research Group *et al.* Modern-day clinical course of type 1 diabetes mellitus after 30 years' duration: the diabetes control and complications trial/epidemiology of diabetes interventions and complications and Pittsburgh epidemiology of diabetes complications experience (1983-2005). Archives of Internal Medicine 2009;169:1307–16.

[8] Holman RR, Paul SK, Bethel MA, Matthews DR, Neil HA. 10-year follow-up of intensive glucose control in type 2 diabetes. The New England Journal of Medicine 2008; 359:1577–89.

[9] Martin CL, et al. Neuropathy among the diabetes control and complications trial cohort 8 years after trial completion. Diabetes Care 2006;29:340–4.

[10] Diabetes Control and Complications Trial/Epidemiology of Diabetes Interventions and Complications Research Group, et al. Retinopathy and nephropathy in patients with type 1 diabetes four years after a trial of intensive therapy. The New England Journal of Medicine 2000;342:381–9.

[11] Zhong GC, Ye MX, Cheng JH, Zhao Y, Gong JP. HbA1c and risks of all-cause and cause-specific death in subjects without known diabetes: a dose-response meta-analysis of prospective cohort studies. Scientific Reports 2016;6:24071.

[12] McCoy RG, et al. Increased mortality of patients with diabetes reporting severe hypoglycemia. Diabetes Care 2012;35:1897–901.

[13] Dahlquist G, Kallen B. Mortality in childhood-onset type 1 diabetes: a population-based study. Diabetes Care 2005;28:2384–7.

[14] Monnier L, et al. Toward defining the threshold between low and high glucose variability in diabetes. Diabetes Care 2017;40:832–8.

[15] Ceriello A, Monnier L, Owens D. Glycaemic variability in diabetes: clinical and therapeutic implications. The Lancet Diabetes and Endocrinology 2019;7:221–30.

[16] El-Osta A, et al. Transient high glucose causes persistent epigenetic changes and altered gene expression during subsequent normoglycemia. Journal of Experimental Medicine 2008;205:2409–17.

[17] Ceriello A, et al. Oscillating glucose is more deleterious to endothelial function and oxidative stress than mean glucose in normal and type 2 diabetic patients. Diabetes 2008;57:1349–54.

[18] Picconi F, et al. Retinal neurodegeneration in patients with type 1 diabetes mellitus: the role of glycemic variability. Acta Diabetologica 2017;54:489–97.

[19] Yang YF, et al. Visit-to-Visit glucose variability predicts the development of end-stage renal disease in type 2 diabetes: 10-year follow-up of taiwan diabetes study. Medicine (Baltimore) 2015;94:e1804.

[20] Dorajoo SR, et al. HbA1c variability in type 2 diabetes is associated with the occurrence of new-onset albuminuria within three years. Diabetes Research and Clinical Practice 2017;128:32–9.

[21] Virk SA, et al. Association between HbA1c variability and risk of microvascular complications in adolescents with type 1 diabetes. The Journal of Cinical Endocrinology and Metabolism 2016;101:3257–63.

[22] Lachin JM, et al. Effect of glycemic exposure on the risk of microvascular complications in the diabetes control and complications trial–revisited. Diabetes 2008;57: 995–1001.
[23] Kilpatrick ES, Rigby AS, Atkin SL. Mean blood glucose compared with HbA1c in the prediction of cardiovascular disease in patients with type 1 diabetes. Diabetologia 2008;51:365–71.
[24] Forbes A, Murrells T, Mulnier H, Sinclair AJ. Mean HbA1c, HbA1c variability, and mortality in people with diabetes aged 70 years and older: a retrospective cohort study. The Lancet Diabetes and Endocrinology 2018;6:476–86.
[25] Tang X, et al. Glycemic variability evaluated by continuous glucose monitoring system is associated with the 10-y cardiovascular risk of diabetic patients with well-controlled HbA1c. Clinica Chimica Acta 2016;461:146–50.
[26] Schutt M, et al. Is the frequency of self-monitoring of blood glucose related to long-term metabolic control? Multicenter analysis including 24,500 patients from 191 centers in Germany and Austria. Experimental and Clinical Endocrinology and Diabetes 2006;114:384–8.
[27] Miller KM, et al. Evidence of a strong association between frequency of self-monitoring of blood glucose and hemoglobin A1c levels in T1D exchange clinic registry participants. Diabetes Care 2013;36:2009–14.
[28] Karter AJ, et al. Self-monitoring of blood glucose levels and glycemic control: the Northern California Kaiser Permanente Diabetes registry. The American Journal of Medicine 2001;111:1–9.
[29] Evans JM, et al. Frequency of blood glucose monitoring in relation to glycaemic control: observational study with diabetes database. BMJ 1999;319:83–6.
[30] National clinical guideline centre (UK). 2015.
[31] American Diabetes Association. 6. Glycemic targets: standards of medical care in diabetes-2018. Diabetes Care 2018;41:S55–64.
[32] Heinemann L. Finger pricking and pain: a never ending story. Journal of Diabetes Science and Technology 2008;2:919–21.
[33] Mostrom P, Ahlen E, Imberg H, Hansson PO, Lind M. Adherence of self-monitoring of blood glucose in persons with type 1 diabetes in Sweden. BMJ Open Diabetes Research and Care 2017;5:e000342.
[34] Schmidt MI, Hadji-Georgopoulos A, Rendell M, Margolis S, Kowarski A. The dawn phenomenon, an early morning glucose rise: implications for diabetic intraday blood glucose variation. Diabetes Care 1981;4:579–85.
[35] Schmidt MI, Lin QX, Gwynne JT, Jacobs S. Fasting early morning rise in peripheral insulin: evidence of the dawn phenomenon in nondiabetes. Diabetes Care 1984;7.
[36] Campbell IW. The somogyi phenomenon. A short review. Acta Diabetologia Latina 1976;13:68–73.
[37] Porcellati F, Lucidi P, Bolli GB, Fanelli CG. Thirty years of research on the dawn phenomenon: lessons to optimize blood glucose control in diabetes. Diabetes Care 2013; 36:3860–2.
[38] Geddes J, Schopman JE, Zammitt NN, Frier BM. Prevalence of impaired awareness of hypoglycaemia in adults with type 1 diabetes. Diabetic Medicine 2008;25:501–4.
[39] Colberg SR, et al. Physical activity/exercise and diabetes: a position statement of the American diabetes association. Diabetes Care 2016;39:2065–79.

[40] (UK), N. C. G. C. In: Type 1 diabetes in adults: diagnosis and management. UK: National Institute for Health and Care Excellence; 2015.
[41] Danne T, et al. International consensus on use of continuous glucose monitoring. Diabetes Care 2017;40:1631–40.
[42] Bode BW, Gross TM, Thornton KR, Mastrototaro JJ. Continuous glucose monitoring used to adjust diabetes therapy improves glycosylated hemoglobin: a pilot study. Diabetes Research and Clinical Practice 1999;46:183–90.
[43] Schaepelynck-Belicar P, Vague P, Simonin G, Lassmann-Vague V. Improved metabolic control in diabetic adolescents using the continuous glucose monitoring system (CGMS). Diabetes and Metabolism 2003;29:608–12.
[44] Kaufman FR, et al. A pilot study of the continuous glucose monitoring system: clinical decisions and glycemic control after its use in pediatric type 1 diabetic subjects. Diabetes Care 2001;24:2030–4.
[45] Ludvigsson J, Hanas R. Continuous subcutaneous glucose monitoring improved metabolic control in pediatric patients with type 1 diabetes: a controlled crossover study. Pediatrics 2003;111:933–8.
[46] Tanenberg R, et al. Use of the continuous glucose monitoring system to guide therapy in patients with insulin-treated diabetes: a randomized controlled trial. Mayo Clinic Proceedings 2004;79:1521–6.
[47] Chico A, Vidal-Rios P, Subira M, Novials A. The continuous glucose monitoring system is useful for detecting unrecognized hypoglycemias in patients with type 1 and type 2 diabetes but is not better than frequent capillary glucose measurements for improving metabolic control. Diabetes Care 2003;26:1153–7.
[48] Garg SK, Schwartz S, Edelman SV. Improved glucose excursions using an implantable real-time continuous glucose sensor in adults with type 1 diabetes. Diabetes Care 2004;27:734–8.
[49] Garg S, Jovanovic L. Relationship of fasting and hourly blood glucose levels to HbA1c values: safety, accuracy, and improvements in glucose profiles obtained using a 7-day continuous glucose sensor. Diabetes Care 2006;29:2644–9.
[50] Bailey TS, Zisser HC, Garg SK. Reduction in hemoglobin A1C with real-time continuous glucose monitoring: results from a 12-week observational study. Diabetes Technology and Therapeutics 2007;9:203–10.
[51] Deiss D, et al. Improved glycemic control in poorly controlled patients with type 1 diabetes using real-time continuous glucose monitoring. Diabetes Care 2006;29:2730–2.
[52] Beck RW, et al. Effect of continuous glucose monitoring on glycemic control in adults with type 1 diabetes using insulin injections. Journal of the American Medical Association 2017;317:371.
[53] Juvenile Diabetes Research Foundation Continuous Glucose Monitoring Study Group, et al. Continuous glucose monitoring and intensive treatment of type 1 diabetes. The New England Journal of Medicine 2008;359:1464–76.
[54] Lind M, et al. Continuous glucose monitoring vs conventional therapy for glycemic control in adults with type 1 diabetes treated with multiple daily insulin injections: the GOLD randomized clinical trial. Journal of the American Medical Association 2017;317:379–87.
[55] El-Laboudi AH, Godsland IF, Johnston DG, Oliver NS. Measures of glycemic variability in type 1 diabetes and the effect of real-time continuous glucose monitoring. Diabetes Technology and Therapeutics 2016;18:806–12.

[56] Battelino T, et al. The use and efficacy of continuous glucose monitoring in type 1 diabetes treated with insulin pump therapy: a randomised controlled trial. Diabetologia 2012;55:3155–62.

[57] Battelino T, et al. Effect of continuous glucose monitoring on hypoglycemia in type 1 diabetes. Diabetes Care 2011;34:795–800.

[58] van Beers CA, et al. Continuous glucose monitoring for patients with type 1 diabetes and impaired awareness of hypoglycaemia (IN CONTROL): a randomised, open-label, crossover trial. The Lancet Diabetes and Endocrinology 2016;4:893–902.

[59] Heinemann L, et al. Real-time continuous glucose monitoring in adults with type 1 diabetes and impaired hypoglycaemia awareness or severe hypoglycaemia treated with multiple daily insulin injections (HypoDE): a multicentre, randomised controlled trial. Lancet 2018;391:1367–77.

[60] Murphy HR, et al. Improved pregnancy outcomes in women with type 1 and type 2 diabetes but substantial clinic-to-clinic variations: a prospective nationwide study. Diabetologia 2017;60:1668–77.

[61] Murphy HR, et al. Effectiveness of continuous glucose monitoring in pregnant women with diabetes: randomised clinical trial. BMJ 2008;337:a1680.

[62] Feig DS, et al. Continuous glucose monitoring in pregnant women with type 1 diabetes (CONCEPTT): a multicentre international randomised controlled trial. Lancet 2017; 390:2347–59.

[63] Vloemans AF, et al. Keeping safe. Continuous glucose monitoring (CGM) in persons with type 1 diabetes and impaired awareness of hypoglycaemia: a qualitative study. Diabetic Medicine 2017;34:1470–6.

[64] Polonsky WH, Fisher L, Hessler D, Edelman SV. What is so tough about self-monitoring of blood glucose? Perceived obstacles among patients with type 2 diabetes. Diabetic Medicine 2014;31:40–6.

[65] Kerssen A, de Valk HW, Visser GH. Day-to-day glucose variability during pregnancy in women with type 1 diabetes mellitus: glucose profiles measured with the continuous glucose monitoring system. BJOG 2004;111:919–24.

[66] Kovatchev BP, Patek SD, Ortiz EA, Breton MD. Assessing sensor accuracy for non-adjunct use of continuous glucose monitoring. Diabetes Technology and Therapeutics 2015;17:177–86.

[67] Shivers JP, Mackowiak L, Anhalt H, Zisser H. "Turn it off!": diabetes device alarm fatigue considerations for the present and the future. Journal of Diabetes Science and Technology 2013;7:789–94.

[68] Bolinder J, Antuna R, Geelhoed-Duijvestijn P, Kroger J, Weitgasser R. Novel glucose-sensing technology and hypoglycaemia in type 1 diabetes: a multicentre, non-masked, randomised controlled trial. Lancet 2016;388:2254–63.

[69] Haak T, et al. Flash glucose-sensing technology as a replacement for blood glucose monitoring for the management of insulin-treated type 2 diabetes: a multicenter, open-label randomized controlled trial. Diabetes Therapy 2017;8:55–73.

[70] Reddy M, Jugnee N, Anantharaja S, Oliver N. Switching from flash glucose monitoring to continuous glucose monitoring on hypoglycemia in adults with type 1 diabetes at high hypoglycemia risk: the extension phase of the I HART CGM study. Diabetes Technology and Therapeutics 2018;20:751–7.

[71] Bergenstal RM, et al. Effectiveness of sensor-augmented insulin-pump therapy in type 1 diabetes. The New England Journal of Medicine 2010;363:311–20.
[72] Bergenstal RM, et al. Threshold-based insulin-pump interruption for reduction of hypoglycemia. The New England Journal of Medicine 2013;369:224–32.
[72a] Brown SA, Kovatchev BP, Raghinaru D, et al. Six-month randomized, multicenter trial of closed-loop control in type 1 diabetes. N. Engl. J. Med. 2019;381:1707–17.
[73] van den Berghe G, et al. Intensive insulin therapy in critically ill patients. The New England Journal of Medicine 2001;345:1359–67.
[74] NICE-SUGAR Study Investigators, et al. Intensive versus conventional glucose control in critically ill patients. The New England Journal of Medicine 2009;360:1283–97.
[75] Levitt DL, Silver KD, Spanakis EK. Inpatient continuous glucose monitoring and glycemic outcomes. Journal of Diabetes Science and Technology 2017;11:1028–35.

CHAPTER

Accuracy of CGM systems

8

William L. Clarke, MD [1], Boris Kovatchev, PhD [2]

[1] *Profesor, Emeritus of Pediatric Endocrinology, Department of Pediatrics, University of Virginia, Charlottesville, VA, United States;* [2] *Professor and Director, Center for Diabetes Technology, Department of Psychiatry and Neurobehavioral Sciences, University of Virginia, Charlottesville, Virginia, United States*

Introduction

Continuous glucose monitoring (CGM) was introduced into the management options available for assisting patients with diabetes in controlling their disease nearly 20 years ago [1–3]. The advent of this technology has led to remarkable changes in day-to-day living and the possibility of long-term reduction in the risk of serious complications. Since then, significant progress has been made toward versatile and reliable CGM devices that not only monitor the entire course of blood glucose (BG) day and night but also provide feedback to the patient, such as alarms when BG reaches preset low or high levels. A number of early studies have documented the benefits of CGM and charted guidelines for its clinical use [4–9]. Perhaps most importantly, CGM has provided a missing critical piece needed for the development of automated closed-loop control, known as the "artificial pancreas" [10–14]. With CGM systems, patients can immediately confirm their symptoms of elevated or low BG levels and determine the direction of their BG trends and the rate of BG change. With real-time CGM, alarms for low and high BG have made it possible for patients and their families to sleep throughout the night.

However, while CGM has the potential to revolutionize the control of diabetes, it also generates data streams that are both voluminous and complex. The utilization of such data requires an understanding of the physical, biochemical, and mathematical principles and properties involved in this new technology. It is important to know that CGM devices measure glucose concentration in a different compartment—the interstitium. Interstitial glucose (IG) fluctuations are related to BG presumably via the diffusion process [15–17]. Historically, to account for the gradient between BG and IG, CGM devices have been calibrated with capillary glucose; more recently, factory calibration is used to bring the typically lower IG concentration to BG levels. Successful calibration would adjust the amplitude of IG fluctuations with respect to BG but would not eliminate the possible time lag due to BG-to-IG glucose transport. Because such a time lag could greatly influence the accuracy of CGM, a number of studies have been dedicated to its investigation,

Glucose Monitoring Devices. https://doi.org/10.1016/B978-0-12-816714-4.00008-9

yielding various results [18–21]. For example, it was hypothesized that if a glucose fall is due to peripheral glucose consumption the physiologic time lag would be negative, that is, fall in IG would precede fall in BG [15,22]. In most studies, IG lagged behind BG (most of the time) by 4–10 min, regardless of the direction of BG change [17,18]. The formulation of the push–pull phenomenon offered reconciliation of these results and provided arguments for a more complex BG–IG relationship than a simple constant or directional time lag [21,23–25]. In addition, errors from calibration, loss of sensitivity, and random noise confound CGM data [26]. Nevertheless, the accuracy of CGM has been steadily increasing [27–31] and has reached a point where CGM could be used as a replacement for traditional BG measurement [32]. Increased overall accuracy and reduction of outliers were due to both improved sensor technology and advances in algorithmic signal processing used to convert the raw electrochemical sensor data into calibrated BG values [33–35].

In addition to presenting frequent data (e.g., every 5–10 min), CGM devices typically display directional trends and BG rate of change and are capable of alerting the patient of upcoming hypo- or hyperglycemia. These features are based on methods that predict blood glucose and generate alarms and warning messages. In the past 15 years, these methods have evolved from a concept [36] and early implementation in CGM devices [37,38], to elaborate predictive low glucose suspend [39–41], advisory [42], and artificial pancreas systems [43–46].

Because the clinical adoption of CGM increases [47] and contemporary systems use a CGM signal to discontinue insulin delivery (e.g., low glucose suspend), or modulate insulin delivery up and down (e.g., artificial pancreas), it is imperative that the information CGM provides be accurate and reliable. The methods for assessment of CGM accuracy are reviewed below.

Clinical accuracy

Accuracy of CGM systems includes both clinical and numerical (statistical) components and each must be assessed differently. Clinical accuracy always asks the question "*Will the treatment decision made on the basis of a particular BG monitoring system reading be correct?*" The information provided by CGM is much different from that provided by a single self-monitoring of blood glucose (SMBG) result. SMBG results are static. They represent a single point in time that may or may not bear a relationship to previous test results. CGM readings are part of a process in time, a moving dataset that includes speed and direction as well as a BG value [48]. A good analogy is a difference between a still photo of a cue ball on a billiard table and videos that show that cue ball rolling left or right, fast or slowly, toward the eight balls or in a direction that will miss contact with any other ball. In other words, BG fluctuations are a continuous process in time. Each point of that process is characterized by its location, speed, and direction of change. Thus, at any point in time, the BG value is not only a number but a vector with a specific bearing. Fig. 8.1 presents this concept depicted over the grid of the original error grid

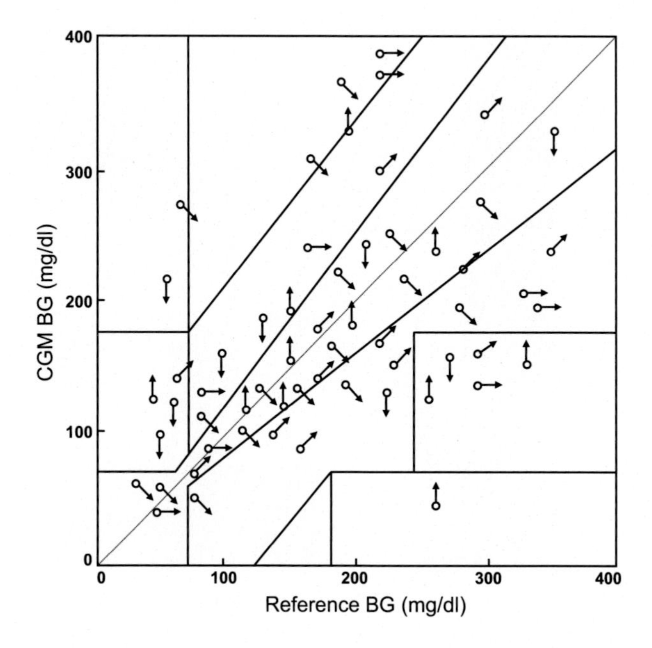

FIGURE 8.1

An ensemble of CGM data points plotted over the original error grid. Each CGM data point is represented by a vector that has a position and direction of change.

analysis, known as the Clarke EGA [49]—a two-dimensional plot of CGM BG levels (*y*-axis) plotted against reference BG measured at the same time (*x*-axis):

More information means a better chance of making a correct decision, but also means that accuracy must be assessed on more than one component. What is needed is an error grid that quantifies the clinical accuracy of CGM by both absolute BG value (point accuracy) and change in BG (rate accuracy), and takes into account the biological factor of the time lag between blood and interstitial tissue. This continuous glucose-error grid analysis (CG-EGA) also needs to be applicable to the original EGA [49], the consensus EGA (CEG, [50]), and the surveillance error grid (SEG, [51]). In addition, it is imperative that the distribution of hypoglycemic, euglycemic, and hyperglycemic BG levels evaluated be similar to that observed in individuals with diabetes because rates and direction of BG change may have different clinical interpretations in different ranges. For example, a low BG that is falling rapidly signals a different treatment decision than a low BG level that is rising rapidly. An error in rate or direction in the low BG change could lead to a very dangerous clinical decision.

The CG-EGA includes two different grids: one for point accuracy (P-EGA) and another for rate accuracy (R-EGA). When the reference rate of BG change is within −1 to 1 mg/dL/min, it is considered to not be clinically significant and the original EGA is used. Thus, in this case, the P-EGA is equivalent to the original EGA [49] depicted in Fig. 8.2 and represents a scatterplot of CGM versus Reference

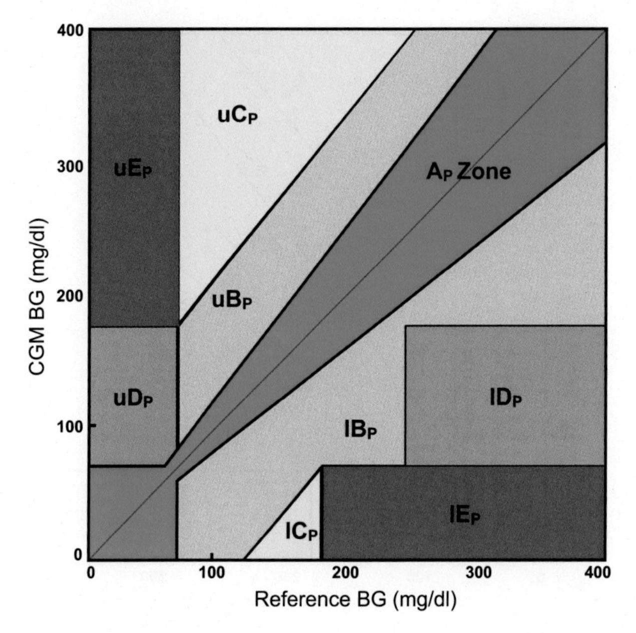

FIGURE 8.2

Point error grid (P-EGA) plot.

BG divided into zones A_P, B_P, C_P, D_P, and E_P (suffix P stands for point, prefix "u" or "l" stands for upper or lower). Chapter 3 in this volume discusses the original EGA and the subsequent consensus and surveillance error grids in detail.

However, when the reference BG rate of change increases beyond ±1 mg/dL/min, the zones of the P-EGA change—different P-EGAs are used for different rates of BG change. This is important because as stated previously, continuous BG values are a process in time and each point must be considered in relation to the previous point. These grids take into account an average time lag between blood and interstitial BG of 7 min. In addition to the original EGA, four different P-EGA grids are presented in Fig. 8.3:

1. If the rate of fall is between −2 and −1 mg/dL/min, the upper zones A and B are expanded by 10 mg/dL—red lines in Fig. 8.3.1. This allows for the CGM BG to reach the corresponding traditional EGA zone within 7 min (1.5 mg/dL/min X 7 min);
2. If the CGM BG is falling faster then −2 mg/dL/min, the upper limits of zones A, B, and D are expanded by 20 mg/dL (3 mg/dL/min X 7—red lines in Fig. 8.3.2);
3. If the CGM BG is rising at a rate of +1−2 mg/dL/min, the lower limits of zones A, B, and D are expanded by 10 mg/dL—red lines in Fig. 8.3.3;
4. If the CGM BG is rising faster than +2 mg/dL/min, the lower limits of zones A, B, and D are expanded by 20 mg/dL—red lines in Fig. 8.3.4.

During the analysis, these adjustments are made for every (reference, CGM BG) data pair and therefore require the use of appropriate software.

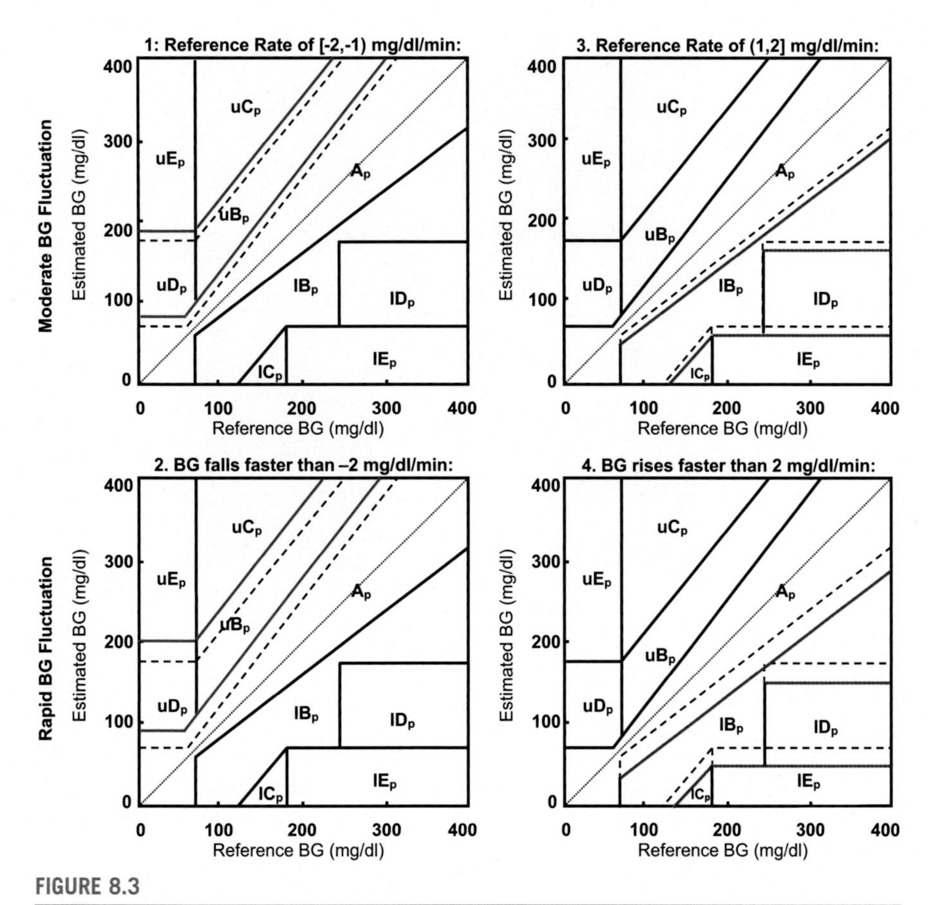

FIGURE 8.3

Expansion of the P-EGA zones compensating for interstitial time lag at different rates of blood glucose change. Upper zones are expanded to compensate for falling BG levels; lower zones are expanded to compensate for increasing BG levels.

The rate error grid (R-EGA) is derived by plotting CGM BG rates of change versus reference BG rates of change (Fig. 8.4), For this analysis, it is suggested that a sampling frequency of at least 10–15 min be used. This frequency should be sufficient to capture representative BG changes. Sampling should include blocks of time (~4 h) representative of changes in subjects' conditions (food, exercise, insulin administration). The boundaries of the R-EGA have been set at −4 to +4 mg/dL/min based on maximum and minimum BG fluctuations collected from numerous datasets. The R-EGA is divided into zones A through E, which have clinical treatment designations similar to the P-EGA:

1. In zone A, the CGM rate of change is within −1 to +1 mg/dL/min of reference rate and is therefore clinically accurate. The zone A boundaries are expanded to −2 and +2 mg/dL/min at extreme rates of −4 and +4 mg/dL/min because these rates of rise and fall are rare, and correctly recognizing their direction results in a clinically accurate decision;

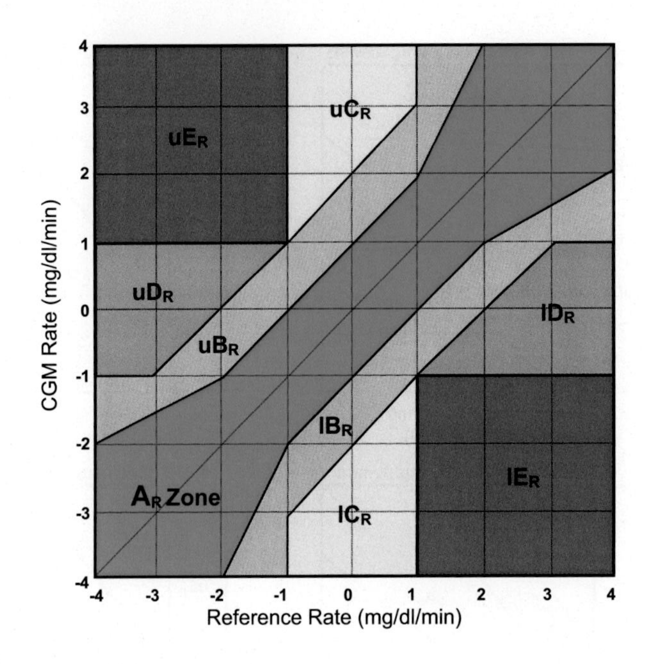

FIGURE 8.4

Rate error grid (R-EGA).

2. In zone C (overtreatment errors), the reference rate is -1 to $+1$ mg/dL/min but the CGM rate suggests treatment that might be overtreatment;
3. D zone data points represent failures to detect and treat rapid rises or falls in reference BG, and
4. E zone points are erroneous errors because the CGM rate is opposite of the reference BG rate, and treatment would be exactly opposite to that needed;
5. Zone B points are benign errors. The CGM rate errors would not lead to inaccurate treatments, or if they do, that treatment would not be likely to lead to a significant negative outcome.

The final step in CG-EGA is combining the P-EGA and R-EGA data in the three clinically relevant BG regions—hypoglycemia, euglycemia, and hyperglycemia [48,52,53]. For the purposes of this presentation, hypoglycemia is defined as BG $<$ 70 mg/dL, euglycemia as the target or treatment goal, BG 71$-$180 mg/dL, and hyperglycemia as BG $>$ 180 mg/dL. As with the original EGA used for describing clinical accuracy with SMBG systems [49], the target or treatment range in CG-EGA can be modified as needed. The CG-EGA error matrix shown in Fig. 8.5 displays the relevant (for the three clinical BG ranges) P-EGA results versus the R-EGA results. For each of the 88 resulting combinations, it is necessary to decide if a P-EGA zone combined with a particular R-EGA zone would result in a clinically accurate, benign, or erroneous treatment decision. CGM can be accurate in terms of P-EGA but inaccurate in terms of R-EGA or vice versa. In the

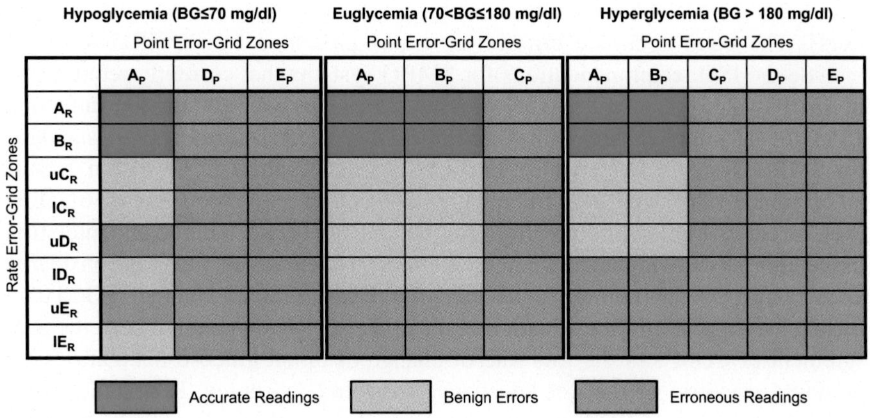

Hypoglycemia (BG≤70 mg/dl) — Point Error-Grid Zones: A_P, D_P, E_P

Euglycemia (70<BG≤180 mg/dl) — Point Error-Grid Zones: A_P, B_P, C_P

Hyperglycemia (BG > 180 mg/dl) — Point Error-Grid Zones: A_P, B_P, C_P, D_P, E_P

Rate Error-Grid Zones: A_R, B_R, uC_R, lC_R, uD_R, lD_R, uE_R, lE_R

Accurate Readings / Benign Errors / Erroneous Readings

FIGURE 8.5

Error matrix combining P-EGA and R-EGA.

matrix, clinical accuracy occurs when both P-EGA and R-EGA values are within zones A or B. Clinically, benign errors are those where P-EGA values fall within zones A or B and there are zone C, D, or E errors in the R-EGA that would be unlikely to result in clinical action or negative outcomes. Hypoglycemic BG values always require treatment regardless of rate data. Therefore, when the P-EGA value is in the A zone during hypoglycemia, only R-EGA values that fail to detect rapidly falling CGM rates (upper D zone) or signify rapidly rising CGM rates when the reference rate is actually rapidly falling (upper E zone) are erroneous readings. When the CGM accurately detects euglycemia, only R-EGA values in the E zone (rapidly rising or falling when reference BG is rapidly changing in the opposite direction) would suggest a clinically negative outcome. When the CGM accurately detects hyperglycemia, only failure to detect a rapid rise in BG (lower zone D error) or detecting a rapid rise or fall in BG when the opposite change is occurring (E zone erroneous rate errors) would lead to a clinically negative situation.

Numerical (statistical) accuracy

Numerical accuracy of CGM has been analyzed using familiar statistical techniques including average BG, standard deviation, linear regression, correlation coefficients, and coefficients of variation. Most data have been presented in terms of differences between CGM and reference BG values expressed as mean and median relative difference and/or mean absolute relative difference. The FDA has affirmed the Clinical and Laboratory Standards Institute (CLSI) guidelines on performance metrics for CGM POCT05-A [54] as the appropriate criteria for premarket regulatory evaluation of CGM. Those guidelines state that "there is no consensus on criteria for good technical point accuracy performance of continuous glucose monitors." The point accuracy criteria for CGM presented by POCT05-A do not include the ISO recommendations for SMBG systems that currently require 95% of SMBG values to be within 15% of reference values over 100 mg/dL and within 15 mg/dL of reference BG values under 100 mg/dL [55]. Current FDA standards for SMBG systems also require 95% of all SMBG values to be within 15% of all reference values and 99% to be within 20% of all reference values [56]. However, no compensation or corrections for the time lag between blood and interstitial tissue are routinely included when the numerical accuracy of CGM is considered. To account for the numerical accuracy of CGM rate of change, POCT05-A included a suggestion for evaluation of the numerical agreement between the measurement of CGM and the true rate of change of blood glucose fluctuations and proposed a metric of the rate of change (or trend) accuracy, R-deviation (RD) [54]. The CLSI guidelines recommend the inclusion of trend analysis using RD as part of the data submitted for approval of new CGM, as well as the absolute R-deviation (ARD) that can be used to assess the proximity between the underlying rate of blood glucose fluctuations and the rate displayed by CGM. RD and ARD are computed as presented in Fig. 8.6 using the following formulas: RD is defined as the difference between reference and sensor instantaneous rates of change: $RD = \frac{\Delta R - \Delta S}{\Delta t}$, while $ARD = \left|\frac{\Delta R - \Delta S}{\Delta t}\right|$ is defined as the absolute value of the RD. Thus, the mean RD corresponds to mean error in point accuracy; the mean ARD corresponds to mean absolute error in point accuracy.

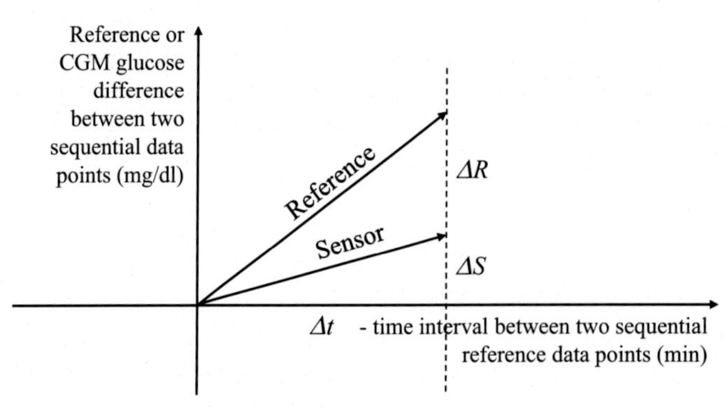

FIGURE 8.6

Computing R-deviation as the difference between CGM and references rates of change over a certain time interval. The pace of computing R-deviation is defined by the sampling rate of the reference data.

Conclusions

CGM continues to evolve, and improvements in clinical and numerical accuracy and performance can be anticipated. The criteria for market approval need to be reviewed and updated to assure that they are appropriate for the sophisticated intended use of these devices. Quantitative and qualitative measures for analyzing the data streams generated by CGM will also need to be refined and standardized to optimize their contribution to efficient and effective patient care. Although some statistical techniques provide appropriate measures of the point accuracy of CGM, they do not address the rate, or trend, components of the accuracy of CGM systems. Continuous monitoring results in a data stream, a process in time, which means that each data point is dependent on the prior data point and therefore not an independent value as typically assumed by standard statistics [27,57]. Thus the criteria for using standard statistical techniques to analyze these data are, generally, not met. The order of the data points is important: the sequence of BG readings of 74, 80, 85, 86 mg/dL is not clinically the same as the sequence of BG readings of 86, 85, 80, and 74 mg/dL, even though the mean and standard deviation of the two data series are identical. Thus the assessment of trend (or rate) accuracy is important for the evaluation of CGM devices, both in terms of clinical and numerical metrics.

Although CG-EGA is admittedly complex, the information that it provides is invaluable in deciding whether a particular CGM system would be effective in different clinical situations. For instance, patients with a history of reduced awareness of hypoglycemia might benefit from using a CGM system that is highly accurate in the hypoglycemic range. Comparing the clinical accuracy of CGM systems in the three clinically relevant ranges could help with the selection process. Systems that feature alarms to signal impending hypo- or hyperglycemia would need to be particularly accurate in measuring the rate and direction of change. CG-EGA has been and is being used to evaluate the clinical accuracy of CGM systems as part of the information required by regulatory agencies for marketing approval. To date, the original EGA remains the only point error grid modified to support the evaluation of the clinical accuracy of CGM systems.

Acknowledgments

Author disclosures

BPK reports patents related to diabetes technology managed by the UVA Licensing and Ventures Group, speaking engagements for Sanofi and Dexcom, consulting for Sanofi and Tandem, and material support (to the University of Virginia) from Tandem Diabetes Care, Roche Diagnostics, and Dexcom.

This work was supported in part by the UVA Strategic Investment in Precision Individualized Medicine for Diabetes (PrIMeD Project).

References

[1] Mastrototaro JJ. The MiniMed continuous glucose monitoring system. Diabetes Technology and Therapeutics 2000;2(Suppl. 1). S-13 – S-18.
[2] Bode BW. Clinical utility of the continuous glucose monitoring system. Diabetes Technology and Therapeutics 2000;2(Suppl. 1). S-35 – S-42.
[3] Feldman B, Brazg R, Schwartz S, Weinstein R. A continuous glucose sensor based on wired enzyme technology – results from a 3-day trial in patients with type 1 diabetes. Diabetes Technology and Therapeutics 2003;5:769–78.
[4] Deiss D, Bolinder J, Riveline J, Battelino T, Bosi E, Tubiana-Rufi N, Kerr D, Phillip M. Improved glycemic control in poorly controlled patients with type 1 diabetes using real-time continuous glucose monitoring. Diabetes Care 2006;29:2730–2.
[5] Garg K, Zisser H, Schwartz S, Bailey T, Kaplan R, Ellis S, Jovanovic L. Improvement in glycemic excursions with a transcutaneous, real-time continuous glucose sensor. Diabetes Care 2006;29:44–50.
[6] Kovatchev BP, Clarke WL. Continuous glucose monitoring reduces risks for hypo- and hyperglycemia and glucose variability in diabetes. Diabetes 2007;56(Suppl. 1):0086OR.
[7] The Juvenile Diabetes Research Foundation Continuous Glucose Monitoring Study Group. Continuous glucose monitoring and intensive treatment of type 1 diabetes. New England Journal of Medicine 2008;359:1464–76.
[8] Klonoff DC. Continuous glucose monitoring: roadmap for 21st century diabetes therapy. Diabetes Care 2005;28:1231–9.
[9] Hirsch IB, Armstrong D, Bergenstal RM, Buckingham B, Childs BP, Clarke WL, Peters A, Wolpert H. Clinical application of emerging sensor technologies in diabetes management: consensus guidelines for continuous glucose monitoring. Diabetes Technology and Therapeutics 2008;10:232–46.
[10] Hovorka R. Continuous glucose monitoring and closed-loop systems. Diabetic Medicine 2006;23:1–12.
[11] Klonoff DC. The artificial pancreas: how sweet engineering will solve bitter problems. Journal of Diabetes Science and Technology 2007;1:72–81.
[12] Clarke WL, Kovatchev BP. The artificial pancreas: how close we are to closing the loop? Pediatric Endocrinology Reviews 2007;4:314–6. PMID: 17643078.
[13] Cobelli C, Renard E, Kovatchev BP. Perspectives in diabetes: artificial pancreas: past, present, future. Diabetes 2011;60:2672–82. PMID: 22025773; PMCID: PMC3198099.
[14] Kovatchev BP. Diabetes technology: monitoring, analytics, and optimal control. Cold Spring Harbor Perspectives in Medicine 2018. https://doi.org/10.1101/cshperspect.a034389. PMID: 30126835.
[15] Rebrin K, Steil GM, van Antwerp WP, Mastrototaro JJ. Subcutaneous glucose predicts plasma glucose independent of insulin: implications for continuous monitoring. American Journal of Physiology Endocrinology and Metabolism 1999;277:E561–71.
[16] Rebrin K, Steil GM. Can interstitial glucose assessment replace blood glucose measurements? Diabetes Technology and Therapeutics 2000;2:461–72.
[17] Steil GM, Rebrin K, Hariri F, Jinagonda S, Tadros S, Darwin C, Saad MF. Interstitial fluid glucose dynamics during insulin-induced hypoglycaemia. Diabetologia 2005;48:1833–40.

[18] Boyne M, Silver D, Kaplan J, Saudek C. Timing of changes in interstitial and venous blood glucose measured with a continuous subcutaneous glucose sensor. Diabetes 2003;52:2790–4.
[19] Kulcu E, Tamada JA, Reach G, Potts RO, Lesho MJ. Physiological differences between interstitial glucose and blood glucose measured in human subjects. Diabetes Care 2003; 26:2405–9.
[20] Stout PJ, Racchini JR, Hilgers ME. A novel approach to mitigating the physiological lag between blood and interstitial fluid glucose measurements. Diabetes Technology and Therapeutics 2004;6:635–44.
[21] Wentholt IME, Hart AAM, Hoekstra JBL, DeVries JH. Relationship between interstitial and blood glucose in type 1 diabetes patients: delay and the push-pull phenomenon revisited. Diabetes Technology and Therapeutics 2004;9:169–75.
[22] Wientjes KJ, Schoonen AJ. Determination of time delay between blood and interstitial adipose tissue glucose concentration change by microdialysis in healthy volunteers. The International Journal of Artificial Organs 2001;24:884–9.
[23] Aussedat B, Dupire-Angel M, Gifford R, Klein JC, Wilson GS, Reach G. Interstitial glucose concentration and glycemia: implications for continuous subcutaneous glucose monitoring. American Journal of Physiology Endocrinology and Metabolism 2000;278:E716–28.
[24] Basu A, Dube S, Slama M, Errazuriz I, Amezcua JC, Kudva YC, Peyser T, Carter RE, Cobelli C, Basu R. Time lag of glucose from intravascular to interstitial compartment in humans. Diabetes 2013;62:4083–7.
[25] Basu A, Dube S, Veettil S, Slama M, Kudva YC, Peyser T, Carter RE, Cobelli C, Basu R. Time lag of glucose from intravascular to interstitial compartment in type 1 diabetes. Journal of Diabetes Science and Technology 2014. https://doi.org/10.1177/1932296814554797.
[26] Kovatchev BP, Clarke WL. Peculiarities of the continuous glucose monitoring data stream and their impact on developing closed-loop control technology. Journal of Diabetes Science and Technology 2008;2:158–63.
[27] Clarke WL, Kovatchev BP. Continuous glucose sensors: continuing questions about clinical accuracy. Journal of Diabetes Science and Technology 2007;1:669–75.
[28] The Diabetes Research in Children Network (DirecNet) Study Group. The accuracy of the Guardian® RT continuous glucose monitor in children with type 1 diabetes. Diabetes Technology and Therapeutics 2008;10:266–72.
[29] Kovatchev BP, Anderson SM, Heinemann L, Clarke WL. Comparison of the numerical and clinical accuracy of four continuous glucose monitors. Diabetes Care 2008;31: 1160–4.
[30] Garg SK, Smith J, Beatson C, Lopez-Baca B, Voelmle M, Gottlieb PA. Comparison of accuracy and safety of the SEVEN and the navigator continuous glucose monitoring systems. Diabetes Technology and Therapeutics 2009;11:65–72.
[31] Christiansen M, Bailey T, Watkins E, Liljenquist D, Price D, Nakamura K, Boock R, Peyser T. A new-generation continuous glucose monitoring system: improved accuracy and reliability compared with a previous-generation system. Diabetes Technology and Therapeutics 2013;15:881–8.

[32] Kovatchev BP, Patek SD, Ortiz EA, et al. Assessing sensor accuracy for non-adjunct use of continuous glucose monitoring. Diabetes Technology and Therapeutics 2015; 17:177–86.

[33] Facchinetti A, Sparacino G, Guerra S, et al. Real-time improvement of continuous glucose monitoring accuracy: the smart sensor concept. Diabetes Care 2013;36: 793–800.

[34] Peyser TA, Nakamura K, Price D, et al. Hypoglycemia accuracy and improved low glucose alerts of the latest Dexcom G4 Platinum continuous glucose monitoring system. Diabetes Technology and Therapeutics 2015;17. https://doi.org/10.1089/dia.2014.0415.

[35] Kovatchev BP. Hypoglycemia reduction and accuracy of continuous glucose monitoring. Diabetes Technology and Therapeutics 2015;17. https://doi.org/10.1089/dia.2015.0144.

[36] Heise T, Koschinsky T, Heinemann L, Lodwig V. Glucose monitoring study group. Hypoglycemia warning signal and glucose sensors: requirements and concepts. Diabetes Technology and Therapeutics 2003;5:563–71.

[37] Bode B, Gross K, Rikalo N, Schwartz S, Wahl T, Page C, Gross T, Mastrototaro J. Alarms based on real-time sensor glucose values alert patients to hypo- and hyperglycemia: the guardian continuous monitoring system. Diabetes Technology and Therapeutics 2004;6:105–13.

[38] McGarraugh G, Bergenstal R. Detection of hypoglycemia with continuous interstitial and traditional blood glucose monitoring using the FreeStyle Navigator continuous glucose monitoring system. Diabetes Technology and Therapeutics 2009;11:145–50.

[39] Buckingham B, Cobry E, Clinton P, Gage V, Caswell K, Kunselman E, Cameron F, Chase HP. Preventing hypoglycemia using predictive alarm algorithms and insulin pump suspension. Diabetes Technology and Therapeutics 2009;11:93–7. PMID: 19848575.

[40] Battelino T, Nimri R, Dovc K, Phillip M, Bratina N. Prevention of hypoglycemia with predictive low glucose insulin suspension in children with type 1 diabetes: a randomized controlled trial. Diabetes Care 2017;40:764–70. PMID: 28351897.

[41] Pinsker JE, Dassau E. Predictive low-glucose suspend to prevent hypoglycemia. Diabetes Technology and Therapeutics 2017;19:271–6. PMID: 28426238.

[42] Breton MD, Patek SD, Lv D, Schertz E, Robic J, Pinnata J, Kollar L, Barnett C, Wakeman C, Oiveri M, Fabris C, Chernavvsky D, Kovatchev BP, Anderson SM. Continuous glucose monitoring and insulin informed advisory system with automated titration and dosing of insulin reduces glucose variability in type 1 diabetes mellitus. Diabetes Technology and Therapeutics 2018;20:531–40. PMID: 29979618.

[43] Bergenstal RM, et al. Safety of a hybrid closed-loop insulin delivery system in patients with type 1 diabetes. Journal of the American Medical Association 2016;316:1407–8. PMID: 27629148.

[44] Kovatchev BP. The artificial pancreas in 2017: the year of transition from research to clinical practice. Nature Reviews Endocrinology 2018;14:74–6. PMID: 29286043.

[45] Brown S, Raghinaru D, Emory E, Kovatchev BP. First look at control-IQ: a new-generation automated insulin delivery system. Diabetes Care October 10, 2018: dc181249. PMID: 30305346.

[46] Kovatchev BP. Automated closed-loop control of diabetes: the artificial pancreas. Bioelectronic Medicine 2018;4:14. https://doi.org/10.1186/s42234-018-0015-6.

[47] Danne T, Nimri R, Battelino T, Bergenstal RM, Close K, Devries JH, Garg S, Heinemann L, Hirsch I, Amiel SA, Beck R, Bosi E, Buckingham B, Cobelli C, Dassau E, Doyle 3rd FJ, Heller S, Hovorka R, Jia W, Jones T, Kordonouri O, Kovatchev B, Kowalski A, Laffel L, Maahs D, Murphy HR, Nørgaard K, Parkin CG, Renard E, Saboo B, Scharf M, Tamborlane WV, Weinzimer SA, Phillip M. International consensus on use of continuous glucose monitoring. Diabetes Care 2017;40:1631–40. PMID: 29162583.

[48] Kovatchev B, Gonder-Frederick L, Cox D, Clarke W. Evaluating the accuracy of continuous glucose monitoring sensors. Diabetes Care 2004;27:1922–8.

[49] Clarke W, Cox D, Gonder-Frederick L, Carter W, Pohl S. Evaluating the clinical accuracy of self-blood glucose monitoring systems. Diabetes Care 1987;10:622–8.

[50] Parkes J, Slatin S, Pardo S, Ginsberg B. A new consensus error grid to evaluate the clinical significance of inaccuracies in the measurement of blood glucose. Diabetes Care 2000;23:1143–5.

[51] Klonoff D, Lias C, Vigersky R, Clarke W, Parkes J, Sacks D, Kirkman S, Kovatchev B, The Error Grid Panel. The surveillance error grid. Journal of Diabetes Science and Technology 2014;8:658–72.

[52] Clarke W, Anderson S, Farhy L, Breton M, Gonder-Frederick L, Cox D, Kovatchev B. Evaluating the clinical accuracy of two continuous glucose sensors using continuous glucose-error grid analysis. Diabetes Care 2005;28:2412–7.

[53] Clarke W, Anderson S, Kovatchev B. Evaluating clinical accuracy of continuous glucose monitoring systems: continuous glucose – error grid analysis (CG-EGA). Current Diabetes Reviews 2008;4:193–9.

[54] Clinical and Laboratory Standards Institute (CLSI). Performance metrics for continuous interstitial glucose monitoring; approved guideline. CLSI document POCT05-A. Wayne, PA: CLSI; 2008.

[55] International Organization for Standardization. In vitro diagnostic test systems – requirements for blood glucose monitoring systems for self-testing in managing diabetes mellitus. ISO International Standard 2013:15197.

[56] Self-monitoring blood glucose test systems for over-the-counter use: guidance for industry and food and drug administration staff, document issued on October 11, 2016. Available from: www.fda.gov/downloads/MedicalDevices/DeviceRegulationandGuidance/GuidanceDocuments/UCM380327.pdf.

[57] Clarke W, Kovatchev B. Statistical tools to analyze continuous glucose monitor data. Diabetes Technology and Therapeutics 2009;11(Suppl. 1):S1–10.

CHAPTER

Calibration of CGM systems

9

Giada Acciaroli, PhD, Martina Vettoretti, PhD, Andrea Facchinetti, PhD, Giovanni Sparacino, PhD

Department of Information Engineering, University of Padova, Padova, Italy

Minimally invasive continuous glucose monitoring (CGM) sensors measure a signal that reflects glucose concentration only indirectly. Indeed, the wired-based sensor placed through a needle in the subcutaneous tissue measures an electrical current signal derived from the glucose-oxidase electrochemical reaction [1,2]. The calibration process consists in the estimation of a mathematical model that converts the electrical current signal (given in fractions of ampere) into glucose concentration values (in mg/dL). The parameters of the calibration model are usually estimated by matching a few self-monitoring of blood glucose (SMBG) samples suitably collected by the patient as reference measurements.

Most commercialized minimally invasive CGM systems perform the first calibration a few hours (e.g., one or two) after sensor insertion when the sensor warm-up period has completed, and the subsequent ones every 12−24 h, usually employing a simple first-order time-independent linear model as the calibration function [3−6]. Given the complex nonlinear and time-dependent relationship between measured current and glucose concentration, the use of a simple linear function as an approximation of the more complex behavior is acceptable within time intervals of limited duration. Thus, frequent recalibrations are required to maintain sensor accuracy, as recommended by the manufacturers' instructions [7−9]. Patients' discomfort associated with the frequent calibration of the device, and the need to improve CGM sensors' accuracy and reliability called for the development of more sophisticated calibration techniques. In the last decade, several signal processing, modeling, and machine-learning methods have been proposed by the academic community to address the calibration issue, which led to improvements in CGM sensor accuracy and user acceptability. We refer the reader to Refs. [10,11] for an extended discussion of CGM technologies and current trends and to Ref. [12] for a comprehensive review of the calibration process.

In this chapter, after a formal description of the calibration problem, we will illustrate some calibration techniques proposed in the literature for minimally invasive CGM sensors. We will then present an example of the implementation of a recently proposed calibration technique based on Bayesian estimation.

Glucose Monitoring Devices. https://doi.org/10.1016/B978-0-12-816714-4.00009-0

Calibration of minimally invasive CGM sensors

Problem statement

The calibration process consists in the estimation of a mathematical model that converts the electrical current signal measured by the sensor (in fractions of ampere) into glucose concentration values (in mg/dL).

Letting $u(t)$ be the glucose concentration profile, $y(t)$ the electrical current signal, and $f(\cdot)$ the function of parameters $\boldsymbol{P} = [p_1, p_2, \ldots, p_n]$ that relates $u(t)$ and $y(t)$, the calibration process can be schematized in two steps (illustrated in panels (A) and (B) of Fig. 9.1). The first step consists in the identification of the calibration model parameters $\boldsymbol{P}$. In formal terms, the current signal $y(t)$ collected by the sensor and corrupted by measurement error $w(t)$, and the blood glucose (BG) measurements (samples of $u(t)$) acquired by the patient at correspondent time instants, are described by the model (see Fig. 9.1, panel (A)):

$$y(t) = f(\boldsymbol{P}, u(t)) + w(t) \tag{9.1}$$

from which a numerical value $\widehat{\boldsymbol{P}}$ of the calibration parameter vector can be provided using, for instance, parametric estimation techniques. This step can be repeated each time a new BG reference is available, with consequent updates of the calibration parameters $\widehat{\boldsymbol{P}}$ (e.g., every 12–24 h by acquiring SMBG samples). The second step leads to the estimation of the glucose concentration profile (see Fig. 9.1, panel (B)). Formally, from the vector of estimated parameters $\widehat{\boldsymbol{P}}$ and the measured current profile $y(t)$, the calibrated glycemic profile $\widehat{u}(t)$ is obtained in real time by inverting the calibration function $f(\cdot)$:

$$\widehat{u}(t) = f^{-1}\left(\widehat{P}, y(t)\right) \tag{9.2}$$

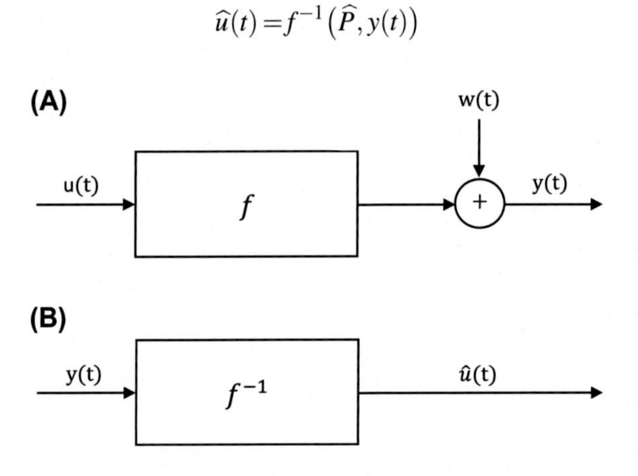

FIGURE 9.1

Schematic representation of the two steps characterizing the calibration process. Panel (A): first step, identification of the parameters of the calibration function f from the glucose concentration samples, $u(t)$ and the sensor measurements, $y(t)$, corrupted by measurement noise, $w(t)$. Panel (B): second step, estimation of the glucose concentration profile, $\widehat{u}(t)$, from the sensor measurements, $y(t)$, by inverting the calibration function.

The choice of the calibration function $f(\cdot)$ is critical. It has to be invertible, and it has to precisely describe the relationship between the electrical current signal and glucose concentration, which can be, in the most general case, nonlinear and time variant (in this case, time t would be, explicitly, an input of $f(\cdot)$). Moreover, the choice of using either the electrical current or the BG measurements as an independent variable in the calibration model may affect the calibration performance [13].

The most common and simplest calibration model adopted by manufacturers of CGM systems is a first-order time-independent linear function [14–17], with parameters $\boldsymbol{P} = [s, b]$, where s and b are called as sensor sensitivity and baseline (or offset), respectively. In this case, the model of the measurements reported in Eq. (9.1) in a general form turns into

$$y(t) = f(\boldsymbol{P}, u(t)) + w(t) = s \cdot u(t) + b + w(t) \tag{9.3}$$

The numerical determination of the model parameters $\widehat{s}$ and $\widehat{b}$ is, thus, required. For such a scope, if for instance two BG references $u(t_1)$ and $u(t_2)$ are available at times t_1 and t_2, knowing the electrical current values given by the sensor at the same time instants, $y(t_1)$ and $y(t_2)$, the so-called two-point calibration can be performed [18], which allows the estimation of sensitivity, $\widehat{s}$, and baseline, $\widehat{b}$, from the two measured pairs as follows:

$$\begin{cases} \widehat{s} = \dfrac{y(t_2) - y(t_1)}{u(t_2) - u(t_1)} \\[2ex] \widehat{b} = y(t_2) - \dfrac{y(t_2) - y(t_1)}{u(t_2) - u(t_1)} \cdot u(t_2) \end{cases} \tag{9.4}$$

In general, when multiple pairs of electrical current and BG samples are available at times t_i $(i = 1, 2, \ldots, N)$, as shown in Fig. 9.2, a linear regression is used to fit the sensitivity and baseline to the data. In particular, including the measurement noise $w(t_i)$, the model of the measurements becomes

$$y(t_i) = s \cdot u(t_i) + b + w(t_i) \tag{9.5}$$

and the numerical determination of model parameters is done by minimizing the residual sum of squares, that is, the differences between the measurements, $y(t_i)$, and the model predictions, $\widehat{y}(t_i) = s \cdot u(t_i) + b$:

$$\left[\widehat{s}, \widehat{b}\right] = \underset{s,b}{\operatorname{argmin}} \sum_{i=1}^{N} w(t_i)^2 = \underset{s,b}{\operatorname{argmin}} \sum_{i=1}^{N} (y(t_i) - \widehat{y}(t_i))^2 \tag{9.6}$$

Finally, the calibrated glucose profile $\widehat{u}(t)$ is obtained from the measured current signal $y(t)$ and the estimated calibration parameters $\widehat{s}$ and $\widehat{b}$ by inverting the calibration function:

$$\widehat{u}(t) = \frac{y(t) - \widehat{b}}{\widehat{s}} \tag{9.7}$$

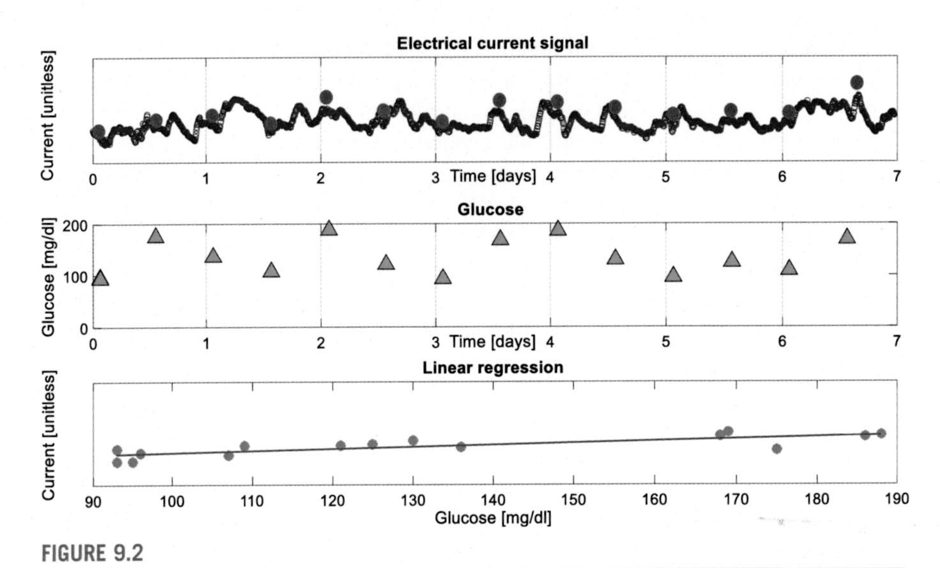

FIGURE 9.2

Top panel, electrical current profile $y(t)$ (in *blue circles*) with *red circles* indicating the electrical current samples corresponding to time instants at which the blood glucose samples $u(t)$, shown in the middle panel by *green triangles*, are acquired. Bottom panel, linear regression between $u(t)$ and $y(t)$.

The quality of the estimate of the calibration parameters is expected to increase with N, that is, the more electrical current-BG pairs that are available, the more accurately the calibration parameters are estimated. On the other hand, increasing N is difficult to satisfy, for practical reasons, for example, for the discomfort related to the acquisition of SMBG samples, and because CGM manufacturers push to minimize the calibration points to facilitate the ease of use of their devices. Moreover, in the presence of measurement uncertainty and/or when only a few data points are available, the standard two-point calibration of Eq. (9.4) could be simplified to a one-point calibration by considering a zero baseline. This simplification may improve the calibration performance by reducing the effect of the noise [18,19].

Although the use of such linear calibration techniques is appealing for its simplicity and ease of implementation, it introduces several critical aspects that, together with uncertainty on the measured sensor output and BG references, are often a cause of CGM sensor inaccuracy. Two examples of sensor inaccuracy are illustrated in Fig. 9.3, where two representative CGM profiles (continuous lines) acquired by the Dexcom G4 Platinum (Dexcom Inc. San Diego, CA) CGM sensor, are compared with reference BG concentrations measured in parallel by gold-standard laboratory instruments (points). Two major causes of these deviations are discussed in the following section.

Critical aspects affecting calibration

A first aspect explaining the discrepancies evidenced in Fig. 9.3 is the distortion introduced by the blood glucose-to-interstitial glucose (BG-to-IG) kinetics. Indeed,

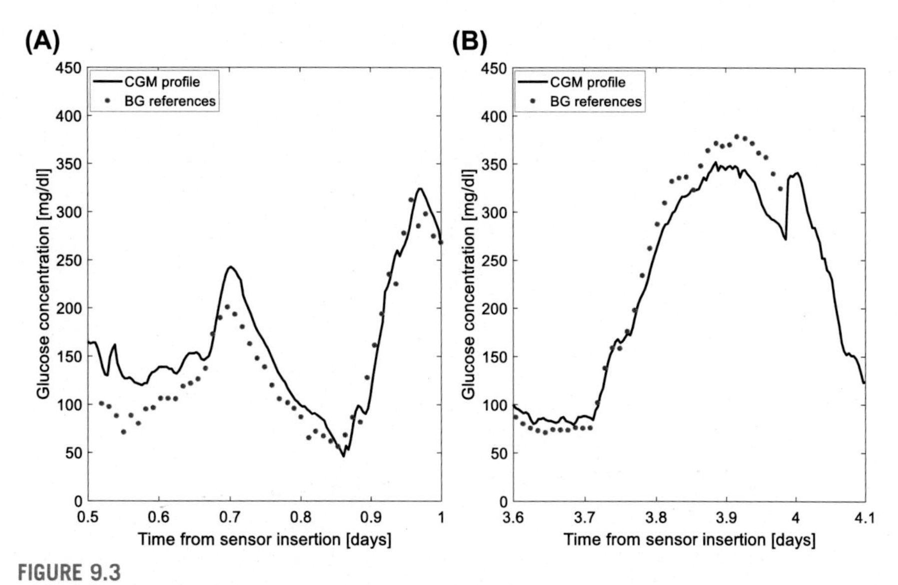

FIGURE 9.3

Examples in which the CGM sensor output (*continuous line*) (A) overestimates and (B) underestimates the reference blood glucose (*points*).

Taken from Acciaroli G, Vettoretti M, Facchinetti A, Sparacino G. Calibration of minimally invasive continuous glucose monitoring sensors: state-of-the-art and current perspectives. Biosensors 2018;8(1):24.

the needle sensor is inserted in the subcutaneous tissue and measures a current signal that is proportional to the glucose concentration in the interstitial fluid. This is due to the fact that, to reduce the invasiveness of CGM devices, sensors are placed in the subcutis and measure the glucose-related current signal from the interstitial fluid rather than directly from the blood. Thus, the two measurements available during the calibration process, that is, the electrical current signal measured by the sensor and the BG references acquired through fingerprick devices, belong to different physiological sites. A widely established description of the relationship between BG and IG is based on a two-compartment model [20,21], illustrated in Fig. 9.4, panel (A), where C_B is the blood glucose concentration, C_I the interstitial glucose concentration, R_a the rate of appearance of glucose in the blood, and k_{ij} $(i = 0, 1, 2$ and $j = 1, 2)$ the diffusion constants. The model of Fig. 9.4 can be described by the following differential equations:

$$\begin{aligned} \acute{C}_B(t) &= R_a + k_{12} \cdot C_I(t) - (k_{01} + k_{21}) \cdot C_B(t) \\ \acute{C}_I(t) &= k_{21} \cdot C_B(t) - (k_{02} + k_{12}) \cdot C_I(t) \end{aligned} \tag{9.8}$$

The parametrization $\tau = \frac{1}{k_{02}+k_{12}}$, $\alpha = \frac{k_{21}}{k_{02}+k_{12}}$ brings to the equivalent expression:

$$\acute{C}_I(t) = -\frac{1}{\tau} C_I(t) + \frac{\alpha}{\tau} C_B(t) \tag{9.9}$$

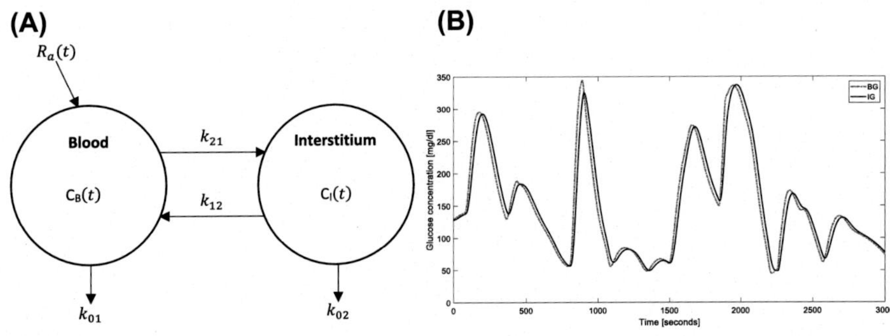

FIGURE 9.4

(A) Two-compartment model describing the BG-to-IG kinetics. R_a is the rate of appearance; k_{01}, k_{02}, k_{12}, k_{21} are rate constants. The time constant of the BG-to-IG system is $\tau = \frac{1}{k_{02} + k_{12}}$. (B) Representative blood glucose (*dashed line*) and interstitial glucose (*continuous line*) concentration profiles simulated assuming $\tau = 11$ min.

Adapted from Acciaroli G, Vettoretti M, Facchinetti A, Sparacino G. Calibration of minimally invasive continuous glucose monitoring sensors: state-of-the-art and current perspectives. Biosensors 2018;8(1):24.

where τ is the diffusion time constant and α is the system gain. In steady-state conditions, that is, when concentrations in the two compartments can be considered constant, the following relations hold:

$$\begin{gathered}\acute{C}_I = -\frac{1}{\tau}C_I + \frac{\alpha}{\tau}C_B = 0 \\ \frac{1}{\tau}C_I = \frac{\alpha}{\tau}C_B \\ C_I = \alpha \cdot C_B \\ \alpha = 1\end{gathered} \tag{9.10}$$

Substituting $\alpha = 1$ in Eq. (9.9),

$$\acute{C}_I(t) = -\frac{1}{\tau}C_I(t) + \frac{1}{\tau}C_B(t) \tag{9.11}$$

Moreover, the explicit solution of Eq. (9.11) is

$$C_I(t) = C_B(t) \otimes \frac{1}{\tau}e^{-\frac{t}{\tau}} \tag{9.12}$$

Thus, the interstitial concentration $C_I(t)$ can be seen as the output of a linear time-invariant system whose impulse response is

$$h(t) = \frac{1}{\tau}e^{-\frac{t}{\tau}} \tag{9.13}$$

Given the low-pass filtering nature of the system, the IG signal is a smoothed and delayed version of the BG concentration [22]. An example is reported in Fig. 9.4

panel (B), where the IG profile (obtained by convolving a given, simulated, BG profile with a single exponential with $\tau = 11$ min) shows both amplitude attenuation and phase delay compared to the BG profile. Notably, τ shows inter- and intrasubject variability and its numerical identification requires a suitable collection of both BG and IG samples. Published values of the time constant τ range from 6 to 15 min [21]. In practice, the BG-to-IG time constant τ is treated as a user parameter, but its role in the calibration process needs to be carefully considered [23,24].

A second critical aspect behind the differences pointed out in Fig. 9.3 is related to the time variability of sensor sensitivity. The raw electrical current signals acquired by CGM sensors often exhibit a nonphysiological drift, especially on the first day after sensor insertion. An example of nonphysiological drift observed in a raw CGM signal acquired by the Dexcom G4 Platinum CGM sensor is depicted in Fig. 9.5, where the continuous line represents the electrical current signal (in units not specified by the manufacturer) and the dashed line shows the drift. This phenomenon is related to a variation of sensor sensitivity after its insertion in the body when the sensor membrane enters in contact with the biological environment and undergoes the immune system reaction [25,26]. The calibration model has to properly compensate for such time variability, which is often nonlinear.

Finally, a third critical aspect is related to the low number of BG samples that are usually available as references for calibration and to the fact that those measures are affected by noise. Indeed, patients usually acquire only a few (e.g., 2–3) SMBG

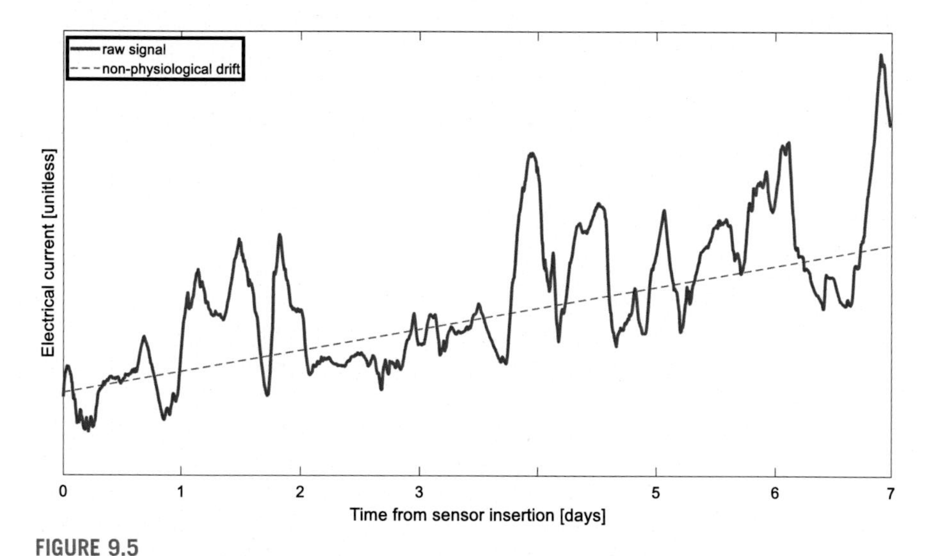

FIGURE 9.5

Representative raw CGM sensor signal (*continuous line*, units not specified by the manufacturer) that exhibits a nonphysiological drift (*dashed line*) due to the time variability of sensor sensitivity.

Taken from Acciaroli G, Vettoretti M, Facchinetti A, Sparacino G. Calibration of minimally invasive continuous glucose monitoring sensors: state-of-the-art and current perspectives. Biosensors 2018;8(1):24.

samples per day. The calibration process should be able to guarantee accurate glucose estimations in this data-poor scenario.

Several techniques have been proposed to deal with these issues affecting CGM sensor calibration. The next section presents and discusses the most recent algorithms proposed in the literature.

State-of-art calibration algorithms and today's challenges

Simple heuristic to deal with the BG-IG system

One of the major limitations of the calibration linear regression techniques presented in Section Problem statement is that they all neglect the time lag between the BG and the raw sensor signal, which can lead to a suboptimal estimation of the parameters of the calibration function. Therefore, most of the calibration algorithms developed by the scientific community included more or less sophisticated approaches to overcome this limitation and take BG-to-IG dynamics into account. The first simple approach is to require calibration of the sensor when glucose is relatively stable. This approach can be applied to any calibration algorithm. The rationale of this heuristic is that, in such a condition, BG and IG concentrations should be at equilibrium and, thus, the estimation of the linear regression parameters should not be influenced by not considering the BG-to-IG dynamics [27]. Following this rationale, Aussedat et al. [28] developed an automated algorithm that requests sensor calibration only when a window of stable signal is detected, that is, when the sensor signal has not changed by more than 1% over a 4-min window, and when the raw current value for the second calibration point differs from the first by $\geq$2 nA. The study proved that performing calibrations during periods of relative glucose stability minimizes the difference between BG references and raw sensor measurements due to the BG-to-IG kinetics.

Kalman filter-based approaches

More sophisticated model-based approaches to account for the BG-to-IG dynamics have been developed relying on Kalman filter theory. In particular, Knobbe et al. [29] proposed a five-state extended Kalman filter, which estimates subcutaneous glucose levels, BG levels, time lag between the sensor measured subcutaneous glucose and BG, time-rate-of-change of the BG level, and the subcutaneous glucose sensor scale factor [30]. In this study, BG levels are reconstructed in continuous time from CGM measurements, employing a state-space Bayesian framework with a priori knowledge of unknown variables. A direct application of a Kalman filter to improve CGM sensor accuracy was proposed by Kuure-Kinsey and colleagues [31], employing a dual-rate Kalman filter and exploiting sparse SMBG measurements to estimate the sensor sensitivity in real time. Although designed for real-time glucose and its rate of change estimation, the algorithm does not account for BG-to-IG kinetics. A further development, with a direct application to the calibration problem and incorporation of the BG-to-IG dynamic model, was given by Facchinetti et al. [32]. The authors proposed an extended Kalman filter method that works in cascade to the standard device calibration to enhance sensor accuracy. By taking into account BG-to-IG kinetics, using a model to describe the variability of sensor sensitivity, and exploiting four BG reference samples per day, the method significantly improves CGM accuracy when applied to synthetic data. However, its real-time implementation is not straightforward, requiring the knowledge of the variances of both state and measurement error processes, as well as an initial burn-in interval.

Methods relying on autoregressive models

Another approach for real-time glucose estimation based on autoregressive (AR) models was proposed by Leal et al. [33]. The study used AR models to estimate BG from raw CGM measurements. Data acquired from 18 T1D patients were used to train a population AR model, which was then incorporated into a calibration algorithm for real-time BG estimation. The raw sensor signal, used as the independent variable, and the BG concentration, considered as a dependent variable, were both normalized based on the maximum range of the available signals. The best overall estimated model, with a third-order Box–Jenkins structure and fixed parameters, enhanced CGM performance, especially in hypoglycemia detection. Significant improvement in hypoglycemia detection was also obtained by the same authors in another study performed on 21 patients with T1D where a new linear regression algorithm with enhanced offset estimation was proposed [34].

A calibration method integrating several local dynamics models

Barcelo-Rico and colleagues proposed an alternative calibration algorithm based on a dynamic global model of the relationship between BG and interstitial CGM signal [35]. The algorithm integrates several local dynamic models, each one representing a different metabolic condition and/or sensor-subject interaction. The local models are then weighted and added to compose the global calibration model. Inputs of the model are the signal measured by the sensor and other signals containing information relevant to glucose dynamics, which are normalized in magnitude using population parameters. The algorithm showed improvements in CGM sensor accuracy, although it was tested on only eight healthy subjects and a more extensive assessment on the diabetic population would be needed to confirm the findings. Further development of the algorithm was proposed in Ref. [36], where an adaptive scheme is used to estimate a patient's normalization parameters in real-time instead of using simple population parameters. The results on 30 virtual patients showed that the adaptation of normalization parameters further improved the performance of the algorithm, as they were able to compensate for sensor sensitivity variations.

Two approaches to optimize the computational complexity

Most of the algorithms proposed for improving CGM performance employ sophisticated models and signal processing features that, although still allowing the implementation on wearable devices/smartphones, increase the computational complexity and processing delay compared to the simple linear regression techniques. With the aim of reducing the delay due to signal processing, Mahmoudi et al. proposed a multistep calibration algorithm based on rate-limiting filtering, selective smoothing, and robust regression [37]. The rate-limiting filter limits the rate of change if a physiological threshold is exceeded; the selective smoothing is applied if the signal is noisy, that is, if the number of zero crossings of the signal first-order differences exceeds a predefined threshold; the robust regression then converts the raw measured current to BG levels using reference SMBG measurements (for a maximum of four references per day). The application of the filtering step to only the noisy parts of the signal lowered the delay introduced by the signal processing of the CGM profile.

Another approach that has low computational complexity as a major strength was proposed by Kirchsteiger and colleagues employing linear matrix inequalities techniques, resulting in convex optimization problems of low complexity [38,39]. The authors proposed two different parametric descriptions of the relationship between IG and BG and a constructive algorithm to adaptively estimate the unknown parameters. The algorithm explicitly considers the measurement uncertainty of the device used to collect the calibration measurements, which was first pointed out by Choleau and colleagues [18]. Moreover, the algorithm embeds an automatic feature to detect fingerprick measurements, which are not suitable to be used for calibration.

Deconvolution-based Bayesian approach

The uncertainty in the reference SMBG samples used for calibration is a key issue in the development of robust calibration algorithms. The real-time deconvolution-based approach proposed by Guerra et al. [40] demonstrated its robustness against both temporal misplacement of the SMBG references and uncertainty in the BG-to-IG kinetics model. The authors proposed a real-time signal-enhancement module to be applied to the CGM sensor output to improve the accuracy of the device. The algorithm compensates the distortion due to the BG-to-IG dynamic by means of regularized deconvolution [41] and relies on a linear regression model that is updated each time a pair of SMBG references is collected. Significant accuracy improvements were observed both on simulated and real datasets. The deconvolution-based approach of [40] was further developed in Refs. [42,43], where it was directly applied to the raw measured signal rather than in cascade to the CGM sensor output. The algorithm fits the raw current signal against BG references (collected twice a day) using a time-varying linear calibration function whose parameters are identified in the Bayesian framework using a priori knowledge on their statistical distribution. The BG-to-IG kinetics is compensated, as in Ref. [40], via nonparametric deconvolution. Results showed significant accuracy improvements compared to the manufacturer calibration.

Recursive approaches exploiting past CGM data

Current CGM products are available for continuous use and are replaced after several days. However, none of the methods discussed so far have embedded any features able to capture this essential cyclic nature by exploiting, for example, the data from prior weeks to better calibrate new CGM data. The first attempt in this direction was made by Lee and colleagues in Ref. [44], where a run-to-run strategy that personalizes sensor calibration parameters using data from previous weeks' use was proposed. Before each weekly new sensor insertion, the algorithm minimizes a cost function that penalizes differences between fingerprick reference values and CGM output of previous weeks. Repeated iterations of the run-to-run procedure demonstrated improved performance on synthetic data (summed square error reduced by 20% after 2 weeks, and up to 50% after 6 weeks). On the same line, another calibration algorithm, employing a time-varying linear calibration function as in Ref. [42], was augmented with a weekly updating feature for parameter optimization [45]. The algorithm estimates the calibration parameters through the recursive least squares to fit SMBG measurements taken approximately every 12 h. Then, personalized calibration parameters are optimized after the first week of use using past data, employing a forgetting factor to give more weight to the most recent data.

Today's challenges for CGM calibration algorithms

The literature calibration algorithms discussed earlier in the present section showed, in general, several performance improvements compared to the simple linear regression methods described in Section Problem statement and implemented in the first commercialized CGM sensors. However, none of them explicitly aimed at enhancing sensor accuracy while reducing, at the same time, the frequency of calibrations, that is, the number of SMBG fingerprick measurements needed as input to the calibration algorithm, which is an obvious reason of discomfort for the patients.

To pursue this objective, the use of the Bayesian estimation in the calibration process, as proposed in Refs. [42,43], appears the most promising technique. Indeed, by setting the calibration problem in the Bayesian framework, the information brought by additional BG references could be substituted by a priori knowledge on calibration parameters derived from ad hoc training sets, allowing, in principle, the reduction of calibration frequency without scarifying sensor accuracy. The following section gives a more detailed description of the application of the Bayesian estimation to the calibration problem.

The Bayesian approach applied to the calibration problem

The key aspect of Bayesian estimation is that, in addition to experimental data, it also considers statistical expectation on the unknown model parameters, usually called "prior" (see Ref. [46] for a comprehensive review on the topic). With reference to the general calibration function formulation of Eq. (9.1), this means that there is some expectation on the parameters $\boldsymbol{P}$, which is usually expressed as Gaussian distribution described by its mean and standard deviation. The Bayesian estimation can be used to estimate the parameters $\boldsymbol{P}$ of a generic calibration function f by using the available data, $u(t)$, and the prior information on $\boldsymbol{P}$.

In the following subsections, we will give an example of the implementation of a calibration algorithm that relies on the Bayesian estimation (see Refs. [47,48] for an extensive description of the method and relative results).

Description of a Bayesian calibration algorithm

Let us define the calibration algorithm as the series of operations that process the electrical current signal $y_I(t)$, measured by the CGM sensor, and the SMBG measurements, acquired by the fingerprick device, to obtain the IG profile $u_I(t)$. The two measurements, $y_I(t)$ and SMBG, belong to different physical domains, that is, the current and the glucose domain respectively, as well as to different physiological sites. SMBG measures glucose concentration into the blood, while the sensor current reflects glucose concentration in the interstitium. Thus, to calibrate a CGM sensor, a model describing the relationship between the two measurements is needed.

Letting $u_B(t)$ be the glucose concentration in blood (whose samples are observable by SMBG), a widely established description of the relation between BG and IG profiles is based on a two-compartment model [21] (see Section Critical aspects affecting calibration), in which the glucose concentration in the interstitial compartment can be described by:

$$\tau \cdot \frac{d}{dt} u_I(t) = -u_I(t) + u_B(t) \tag{9.14}$$

where τ is the equilibrium time-constant between plasma and interstitium. In the literature, the value of τ exhibits significant intersubject variability but, in a given subject, it is commonly assumed to be time invariant [21,23,49,50]. According to Eq. (9.14), the IG profile can be interpreted as the output of a first-order linear dynamic system, having the BG profile as input (see first block in the schematic diagram of Fig. 9.6), and $h(t)$, as defined in Eq. (9.13), as impulse response.

The low-pass filtering nature of $h(t)$ causes $u_I(t)$ to be a distorted version of $u_B(t)$, presenting both amplitude attenuation and phase delay. Following the cascade of blocks of Fig. 9.6, the IG profile $u_I(t)$ is the input of the calibration function which produces as output a current signal referred to the interstitium, $y^{nf}(t)$, which once corrupted by additive noise $w(t)$ finally produces the sensor signal $y_I(t)$. To describe the transformation between glucose and current profiles, let us consider the following model:

$$y_I(t) = [u_I(t) + b] \cdot s(t) + w(t) \tag{9.15}$$

where b is the baseline of the glucose profile and $s(t)$ represents the sensitivity of the sensor, described by the model:

$$s(t) = s_1 \cdot \alpha(t) + s_2 \cdot \beta(t) + s_3 \tag{9.16}$$

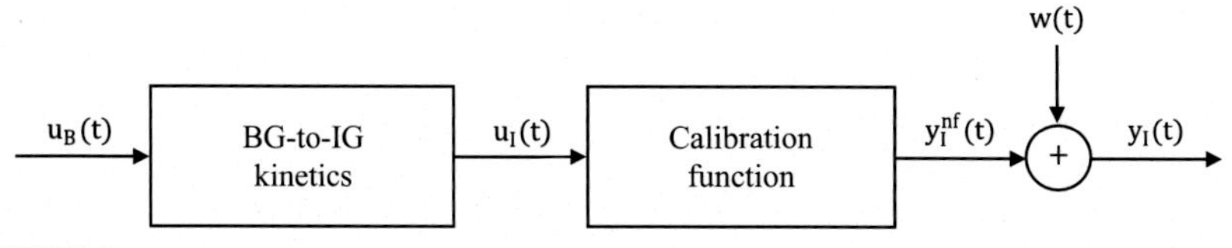

FIGURE 9.6

Schematic representation of the dynamic system relating the blood glucose $u_B(t)$, whose samples are given by SMBG measurements, to the interstitial current $y_I(t)$, measured by the sensor. The intermediate variables $u_I(t)$, and $y_I^{nf}(t)$, represent, respectively, interstitial glucose and noise-free interstitial current. The variable $w(t)$ represents the measurement noise.

Taken from Acciaroli G, Vettoretti M, Facchinetti A, Sparacino G, Cobelli C. Reduction of blood glucose measurements to calibrate subcutaneous glucose sensors: a Bayesian multiday framework. IEEE Transactions on Biomedical Engineering 2018;65(3):587–595.

In Eqs. (9.15) and (9.16), $\alpha(t)$ and $\beta(t)$ are fixed time-domain functions that describe the drift of the specific sensor (see Section Critical aspects affecting calibration), whereas b, s_1, s_2, and s_3 are the model parameters, undergoing the following constraints:

$$s_1, s_2, s_3 > 0; \quad \frac{s_1}{s_2} = \varphi \tag{9.17}$$

where φ is a fixed value (see Section Implementation for details). As a result, considering the dependence between sensitivity parameters ($s_1 = \varphi \cdot s_2$), the final parameters vector is as follows:

$$\boldsymbol{p} = [b, s_2, s_3]^T \tag{9.18}$$

In comparison with the other approaches proposed in the literature, the peculiarity of the model described by Eq. (9.15) is its temporal domain of validity, which is the entire monitoring period, at variance with the models described in Section Deconvolution-based Bayesian approach, whose domains of validity were restricted to the time window between two consecutive calibrations.

To note that, for the sake of method generality, in Eq. (9.16) the functional form and corresponding parameters of $\alpha(t)$ and $\beta(t)$ are intentionally treated as being device dependent. Indeed, optimizing them is done as part of industrial device development. In general, any time-domain function (such as linear, logarithmic, polynomial, and exponential) able to capture the sensor drift (as described in Section Critical aspects affecting calibration) could be embedded in Eq. (9.16).

Estimation of model parameters

Each time a new SMBG is acquired for calibration at time t_i, $i = 1, 2, \ldots, M$ (where M represents the total number of SMBG samples used for calibration), the set of parameters $\boldsymbol{p}$ is updated by exploiting the new measure $u_B(t_i)$ and all previously acquired SMBG samples. In particular, let us consider the following relation, expressed in the vector form:

$$\boldsymbol{u_B} = \widehat{\boldsymbol{u}}_{\boldsymbol{B}}(\boldsymbol{p}) + \boldsymbol{w} \tag{9.19}$$

where $\boldsymbol{u_B}$ is the $i \times 1$ vector containing the SMBG samples acquired at calibration times t_j, $j = 1, \ldots, i$ $(i = 1, \ldots, M)$:

$$\boldsymbol{u_B} = [u_B(t_1), \ldots, u_B(t_i - 1), u_B(t_i)]^T \tag{9.20}$$

$\widehat{\boldsymbol{u}}_{\boldsymbol{B}}$ is the $i \times 1$ vector (function of the model parameters), obtained transforming the $i \times 1$ vector $\boldsymbol{y_I}$, containing $y_I(t_j), j = 1, \ldots, i$, into BG values. Both the BG-to-IG kinetics and the calibration model are considered in the transformation (as discussed later). The $i \times 1$ vector $\boldsymbol{w}$ represents the error. Note that the length i of the vectors increases of one unit each time a new SMBG is acquired for calibration, as all previous measurements are anyway considered.

The unknown parameters vector $\boldsymbol{p}$ is estimated by exploiting the data contained in $\boldsymbol{u_B}$ and $\boldsymbol{y_I}$ in addition to some a priori knowledge on the distribution of $\boldsymbol{p}$, derived from a data training set (details in Section Implementation). In particular, the a priori distribution of the parameters vector $\boldsymbol{p}$ has mean $\boldsymbol{\mu_p}$ and covariance matrix $\boldsymbol{\Sigma_p}$. The error vector $\boldsymbol{w}$ is assumed to contain white noise samples, uncorrelated from $\boldsymbol{p}$, with zero mean and diagonal covariance matrix $\boldsymbol{\Sigma_w}$. The error variance is assumed constant over time, that is, $\boldsymbol{\Sigma_w} = \sigma_w^2 \cdot \boldsymbol{I}$ and σ_w^2 is estimated from the training set (details in Section Implementation).

The Bayesian maximum a posteriori estimate of $\boldsymbol{p}$ is obtained by solving the following optimization problem:

$$\widehat{\boldsymbol{p}} = \underset{\boldsymbol{p}}{\operatorname{argmin}}[\boldsymbol{u_B} - \widehat{\boldsymbol{u}}_{\boldsymbol{B}}(\boldsymbol{p})]^T \boldsymbol{\Sigma}_{\boldsymbol{w}}^{-1} [\boldsymbol{u_B} - \widehat{\boldsymbol{u}}_{\boldsymbol{B}}(\boldsymbol{p})] + (\boldsymbol{\mu_p} - \boldsymbol{p})^T \boldsymbol{\Sigma}_{\boldsymbol{p}}^{-1} (\boldsymbol{\mu_p} - \boldsymbol{p}) \tag{9.21}$$

which, given the presence of nonlinearities, does not have a closed-form solution. Thus, the estimate $\widehat{\boldsymbol{p}}$ is found by looking iteratively into the parameter space. The iterative procedure, schematically described by the diagram of Fig. 9.7, is summarized in five steps:

- (i) initialization of parameter vector **p**;
- (ii) estimation of IG profile $\widehat{u}_I(t)$ according to the calibration model of Eq. (9.15);
- (iii) estimation of BG profile, $\widehat{u}_B(t)$ accounting for the distortion introduced by the BG-to-IG kinetics, through nonparametric deconvolution;
- (iv) matching between SMBG measurements and $\widehat{u}_B(t)$;
- (v) update of the parameter vector for the next iteration.

The following subsections describe individually each of these five steps.

Step 0: parameter initialization

At the first iteration of the ith calibration ($i = 1, \ldots, M$) the parameter vector is initialized to the mean value of the prior distribution, $\boldsymbol{p}_0 = \boldsymbol{\mu_p}$

Step 1: use of calibration model

At each iteration k ($k = 0, 1, \ldots, Niter$), the interstitial glucose $u_I(t)$ is estimated from the current signal $y_I(t)$ by inverting Eq. (9.15), which depends on the parameter vector $\boldsymbol{p_k}$:

$$\widehat{u}_I(t, \boldsymbol{p_k}) = \frac{y_I(t)}{s(t, s_2, s_3)} - b \tag{9.22}$$

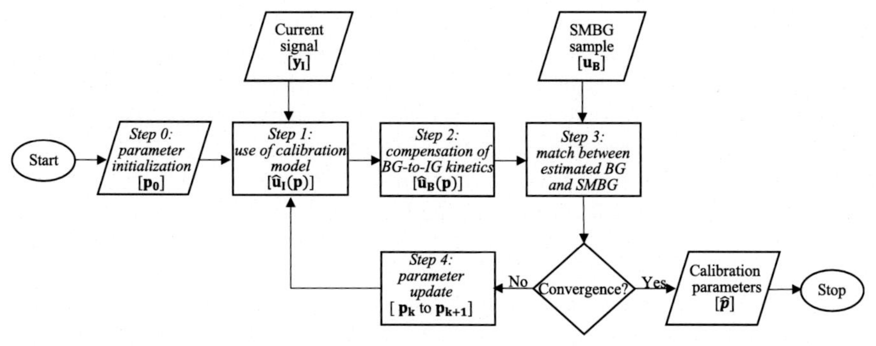

FIGURE 9.7

Flowchart of the iterative procedure for the estimation of the calibration model parameters (*parallelograms* denote input/output blocks, whereas the *diamond* denotes the decision block). Starting from initial vector p_0, at each iteration k the parameter vector is updated, according to the input values y_I (current signal) and u_B (blood glucose samples), using the calibration model of Eq. (9.15) and compensating for the blood-to-interstitial glucose kinetics.

Taken from Acciaroli G, Vettoretti M, Facchinetti A, Sparacino G, Cobelli C. Reduction of blood glucose measurements to calibrate subcutaneous glucose sensors: a Bayesian multiday framework. IEEE Transactions on Biomedical Engineering 2018;65(3):587–595.

Step 2: compensation of BG-to-IG kinetics

The IG profile obtained at Step 1, $\hat{u}_I(t, \boldsymbol{p_k})$, cannot be directly matched to SMBG measures, $u_B(t)$. Indeed, the distortion induced by BG-to-IG kinetics needs to be compensated first. As already discussed in Section Critical aspects affecting calibration, the IG profile can be seen as the output of a first-order linear dynamic system whose impulse response is given by Eq. (9.13) and whose input is the BG profile. Thus, the estimation of $\hat{u}_B(t, \boldsymbol{p_k})$ from $\hat{u}_I(t, \boldsymbol{p_k})$ is an inverse problem that can be solved by deconvolution. In particular, to reconstruct the BG profile from the IG profile a nonparametric stochastic approach [41] is used.

For computational reasons, the deconvolution is applied to temporal windows containing the time instant $t_j, j = 1, ..., i$ at which each SMBG is acquired. Practically, for each of the i BG measures collected in vector $\boldsymbol{u_B}$, a time window Λ from $t_i - 100$ to $t_i + 5$ min is considered. Letting $\boldsymbol{u_I}(\Lambda)$ be the $n \times 1$ vector containing the IG measures estimated at the previous step at the sampling instants lying in Λ, a uniform sampling grid, with a 5-min step, can be defined: $\Omega_s = t_1, t_2, \dots, t_n$. In addition, $\boldsymbol{w}$ is defined as the $n \times 1$ vector of measurement error $w(t)$ at time instants in Ω_s, assumed to have zero mean and covariance matrix $\boldsymbol{\Sigma_w} = \sigma^2 \boldsymbol{R}$, with σ^2 unknown constant and $\boldsymbol{R} n \times n$ known matrix whose structure reflects expectations on measurement error variance (here, $\boldsymbol{R} = \boldsymbol{In}$, as the error samples are assumed to be uncorrelated from the current signal and with variance constant over time). The vector $\boldsymbol{u_B}(\Lambda)$ is defined as the $N \times 1$ unknown vector containing samples of $u_B(t)$ at time instants on a virtual grid $\boldsymbol{v} = tv_1, tv_2, \dots, tv_N$, which is independent of and usually denser than Ω_s (here a uniform 1-min step is used). The virtual grid allows us to obtain a denser profile, which can be more easily matched with SMBG samples (see Step 3). Moreover, Σ_v starts from $t_1 - 100$ min, to allow initial condition transient to vanish, so that the reconstruction of $u_B(t)$ is not altered in the window of interest Λ.

Once all variables have been defined, having Ω_s and Ω_v both uniform, with $\Omega_s \subseteq \Omega_v$, the following matrix equation can be written as follows:

$$\boldsymbol{u}_{I(\Lambda)} = \boldsymbol{H} \cdot \boldsymbol{u}_{B(\Lambda)} + \boldsymbol{w} \tag{9.23}$$

where $\boldsymbol{H}$ is the $n \times N$ matrix obtained downsampling the $N \times N$ transfer matrix $\boldsymbol{Hv}$ of the BG-to-IG system, maintaining only the rows correspondent to sampling instants in Σ_s. As vector $\boldsymbol{u}_{B(\Lambda)}$ contains samples of a BG profile, which is a biological signal expected to have a certain smoothness, a double integrated white noise model [41] of unknown variance λ^2 is chosen to describe entries of $\boldsymbol{u}_{B(\Lambda)}$. Thus, its covariance matrix is

$$\Sigma_{\boldsymbol{u}_B(\Lambda)} = \lambda^2 \left(\boldsymbol{F}^T \boldsymbol{F}\right)^{-1} \tag{9.24}$$

with $\boldsymbol{F}$ $N \times N$ Toeplitz lower-triangular matrix having $[1, -2, 1, 0, \ldots, 0]T$ as the first column. Assuming that $\boldsymbol{u}_{B(\Lambda)}$ and $\boldsymbol{w}$ are uncorrelated, the following quadratic optimization problem corresponds to the linear minimum error variance Bayesian estimate of $\boldsymbol{u}_{B(\Lambda)}$:

$$\widehat{\boldsymbol{u}}_{B(\Lambda)} = \underset{\boldsymbol{u}_{B(\Lambda)}}{\operatorname{argmin}} \left(\boldsymbol{u}_{I(\Lambda)} - \boldsymbol{H}\boldsymbol{u}_{B(\Lambda)}\right)^T \boldsymbol{R}^{-1} \left(\boldsymbol{u}_{I(\Lambda)} - \boldsymbol{u}_{B(\Lambda)}\right) + \gamma \boldsymbol{u}_{B(\Lambda)}^T \boldsymbol{F}^T \boldsymbol{F} \boldsymbol{u}_{B(\Lambda)} \tag{9.25}$$

where parameter $\gamma = \frac{\sigma^2}{\lambda^2}$, estimated by maximum likelihood [41] represents the regularization term that balances the data fit with the smoothness of the estimated profile. The optimization problem of Eq. (9.25) admits a closed-form solution, expressed as follows:

$$\widehat{\boldsymbol{u}}_{B(\Lambda)} = \left(\boldsymbol{H}^T \boldsymbol{R}^{-1} \boldsymbol{H} + \gamma \boldsymbol{F}^T \boldsymbol{F}\right)^{-1} \boldsymbol{H}^T \boldsymbol{R}^{-1} \boldsymbol{u}_{I(\Lambda)} \tag{9.26}$$

For every SMBG measurement in vector $\boldsymbol{u}_B$ (see Eq. 9.20), the BG profile $\widehat{\boldsymbol{u}}_{B(\Lambda)}$, which depends on parameter vector $\boldsymbol{p}_k$, is estimated inside the window Λ that contains the time instant at which the SMBG sample t_j, $j = 1, \ldots, i$ is acquired.

Step 3: match between estimated BG and available SMBG

For each SMBG sample in vector $\boldsymbol{u}_B$, acquired at time t_j, $j = 1, \ldots, i$, the corresponding estimated value of $\widehat{\boldsymbol{u}}_{B(\Lambda)}$ at time t_j is considered, by exploiting the vector $\widehat{\boldsymbol{u}}_B(\boldsymbol{p}_k)$ used in Eq. (9.21):

$$\widehat{\boldsymbol{u}}_B(\boldsymbol{p}_k) = \left[\widehat{\boldsymbol{u}}_B(t_1, \boldsymbol{p}_k), \ldots, \widehat{\boldsymbol{u}}_B(t_{i-1}, \boldsymbol{p}_k), \widehat{\boldsymbol{u}}_B(t_i, \boldsymbol{p}_k)\right]^T \tag{9.27}$$

Step 4: parameter update

At each iteration k, the parameter vector $\boldsymbol{p}_k$ is updated to a new set of values, $\boldsymbol{p}_{k+1}$, using the Nelder–Mead simplex algorithm, as described in Ref. [51].

Steps 1–4 are reiterated until one of the following stopping criteria occurs: (i) the step size in parameters update is smaller than a fixed tolerance (e.g., 10^{-6}); (ii) the relative change in the value of the objective function is lower than a fixed tolerance (e.g., 10^{-6}); (iii) the algorithm reaches the maximum number of iterations (e.g., 10^4).

Calibration of the current signal

For each of the M SMBG samples used for calibration, the parameter vector $\widehat{\boldsymbol{p}}$, estimated from Eq. (9.21) by following the five-step procedure described earlier, is used to calibrate in real time the electrical current signal $y_I(t)$, by inverting the model of Eq. (9.15):

$$z(t,\widehat{\boldsymbol{p}}) = \frac{y_I(t)}{s(t,\widehat{s_2},\widehat{s_3})} - \widehat{b} \tag{9.28}$$

In particular, as the SMBG samples are acquired at times ti, $i = 1, 2, \ldots, M$, the parameter estimated at the ith calibration is used to calibrate the current signal from $ti + 5$ to $ti + 1 + 5$ min. Indeed, the deconvolution window Λ is defined to end 5 min after the reference time of the BG measurement (to avoid edge effects), thus introducing the need to wait 5 min from any ti before starting a new calibration.

Example of implementation

The algorithm presented in Section Description of a Bayesian calibration algorithm is here applied to a set of data acquired by the Dexcom G4 Platinum (DG4P) CGM sensor (Dexcom Inc., San Diego, CA).

Dataset description

Data were collected during a multicenter pivotal study involving 72 diabetic patients (60 subjects with T1D, 12 subjects with T2D) wearing the DG4P sensor for a 7-day period [52]. In total, a pool of 108 datasets was available (36 subjects wore two sensors), each one including the raw electrical current signal and the CGM profile (mg/dL) originally calibrated by the manufacturer using SMBG measurements (mg/dL)—also available in the datasets. In addition, on days 1, 4 and 7, subjects underwent 12-h clinical sessions during which their BG concentration was monitored, every 15 min, by a reliable laboratory method, the Yellow Springs Instruments Glucose Analyzer (YSI, Yellow Spring, OH).

An example of the dataset is reported in Fig. 9.8, where the top panel shows the current signal and the SMBG samples used by the manufacturer for calibration, whereas the bottom panel represents the correspondent CGM profile and YSI references used for accuracy assessment.

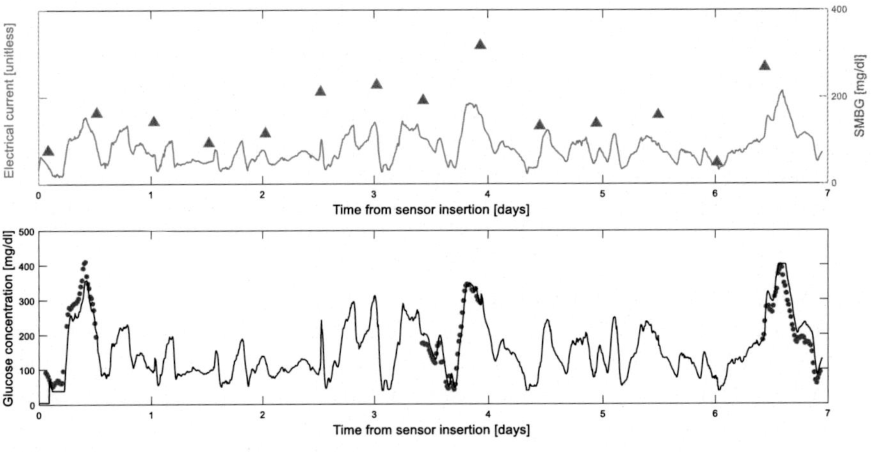

FIGURE 9.8

Representative Dexcom G4 Platinum (DG4P) dataset. Top panel: raw, unprocessed, current signal (*blue line*) and calibration SMBG references (*orange triangles*). Bottom panel: CGM profile as originally calibrated by the manufacturer (*black line*) and laboratory references for accuracy assessment (*red points*).

Implementation

The algorithm is assessed using a ninefold cross-validation technique. The database under analysis is divided into nine groups of 12 datasets each. Iteratively, one group was the test set used to evaluate the method, whereas all other groups formed the training set on which the priors for the calibration parameters are derived.

Prior derivation

The a priori information on calibration model parameters is derived by identifying the following nonlinear model on the training dataset:

$$y_I(t) = [(u_B(t) \otimes h(t)) + b] \cdot s(t) \tag{9.29}$$

where $\otimes$ stands for convolution, $h(t)$ is as in Eq. (9.13), $s(t)$ is defined by Eq. (9.16) and depends on parameters $s1$, $s2$, and s_3. The unknown parameters s_1, s_2, s_3, b, and τ are estimated by nonlinear least squares. The input of the identification procedure is a smoothed BG profile $u_B(t)$, obtained from YSI references using a stochastic Bayesian smoother [41] and the output is the electrical current signal $y_I(t)$. The identification process is performed on each of the N_t time series of the training set, thus obtaining N_t values for each parameter. In particular, defining $\Gamma = \tau_1, ..., \tau_{N}t$ the values of parameter τ, a Bayesian prior is built assuming a priori distribution with mean $\mu\tau$ and variance $\sigma\tau 2$, determined from samples in Γ:

$$\mu_\tau = \frac{1}{N_t}\sum_{k=1}^{N_t} t_k$$

$$\sigma_\tau^2 = \frac{1}{N_t - 1}\sum_{k=1}^{N_t} (\tau_k - \mu_\tau)^2 \tag{9.30}$$

Similarly, a Bayesian prior is built for the parameter vector $\boldsymbol{p}$.

In the online working modality, as the number of BG measurements available was not sufficient to estimate an individual value for τ, we fixed its value to the prior mean μ_τ obtained from the training set at each iteration of the cross-validation. Regarding the other parameters, given the stability of the ratio between parameters s_1 and s_2, to facilitate model identifiability in online modality (when only a few BG references are available) the ratio $\frac{s_1}{s_2}$ was fixed to the constant φ. The value of φ is estimated, at each iteration of the cross-validation, from the mean value of its distribution on the corresponding training set of size N_t. Thus, Eq. (9.18) represents the final parameter vector. Defining $\boldsymbol{P} = p_1, ..., p_{N_t}$ the samples of the parameters vector $\boldsymbol{p}$, a Bayesian prior is built by assuming a distribution with prior mean $\boldsymbol{\mu_p}$ and prior covariance matrix $\boldsymbol{\Sigma_p}$, where the ith element of $\boldsymbol{\mu_p}$, μ_i, and the ijth element of $\boldsymbol{\Sigma_p}$, σ_{ij}, are defined as follows:

$$\mu_i = \frac{1}{N_t}\sum_{k=1}^{N_t} p_{k,i}$$
$$\sigma_{i,j} = \frac{1}{N_t - 1}\sum_{k=1}^{N_t}\left(p_{k,i} - \mu_i\right)\left(p_{k,j} - \mu_j\right) \tag{9.31}$$

Training set data are used also to estimate the error covariance matrix $\boldsymbol{\Sigma_w}$, used in Eq. (9.21). The error variance is assumed constant over time and its value is estimated from the distribution of the differences between SMBG measurements and the correspondent calibrated values (in mg/dL) given by the manufacturer. In particular, for all SMBG samples available in the training set the correspondent CGM calibrated value is matched following the procedure described in Section Description of a Bayesian calibration algorithm and the difference between the two measurements is computed. The error variance is thus obtained, at each cross-validation iteration, from the distribution of the SMBG−CGM error on the training set.

Calibration scenarios

The calibration algorithm is applied to the test set by simulating an online working modality. In particular, for each sensor in the test group, the raw current signal $y_I(t)$ is calibrated by exploiting a set of BG references provided by SMBG measurements. Different calibration schedules are tested, to assess the accuracy of the calibrated profiles using a different number of SMBG samples per day, that is, varying the frequency at which parameters of the calibration model are updated. In particular, apart from the first calibration, which is always performed about 2 h after sensor insertion (exploiting a pair of SMBG samples acquired a few minutes of distance from each other), the following schedules are tested:

- one calibration about every day;
- one calibration about every 2 days;
- one calibration about every 4 days.

Performance assessment

The performance of the calibration algorithm is assessed by comparing the CGM calibrated profiles under test and the YSI laboratory references. First, the two measurements are matched in time. Then, accuracy is quantified by computing the following three metrics:

- Mean absolute relative difference (MARD) between the calibrated CGM profile and the YSI measurements
- Percentage of accurate glucose estimates (PAGE), that is, the percentage of estimated glucose values falling within either 20 mg/dL from the relative YSI reference if YSI is lower than 80 mg/dL or within 20% of the relative YSI reference if YSI is above 80 mg/dL.
- Percentage of CGM-YSI pairs lying in the "A" zone of the Clarke error grid (CEG-A). The CEG is an error grid divided into five zones indicating the accuracy of BG estimates generated by meters as compared to a reference value. In particular, zone "A" indicates accurate glucose results that do not lead to subsequent wrong or dangerous treatments.

These three metrics are used to assess the accuracy of both the CGM profiles as originally calibrated by the manufacturer and the CGM profiles as calibrated by the Bayesian algorithm discussed here.

For each CGM profile under test, a subject-level analysis is performed by computing the performance metrics for each dataset and then taking the population statistics. The population performance indexes are obtained by computing the mean and standard deviation of the metrics obtained in each dataset for normally distributed metrics and the median and interquartile range of the metrics obtained in each dataset for nonnormally distributed metrics. Normality is assessed for each metric by the Lilliefors test.

The statistical significance of the differences in performance metrics obtained with the new and the original calibration is determined by the Wilcoxon signed-rank test, a nonparametric-paired statistical test on the median of the performance metrics distributions. In particular, we tested the null hypothesis "the median difference between the paired values of the two groups is zero" with a significance level of 0.05 on the performance metrics distributions obtained with the original calibration against the new algorithm (for different calibration frequencies).

Results

The calibrated profiles obtained with the three scenarios listed earlier and the original manufacturer CGM output are shown, for a representative subject of the database, in Fig. 9.9, where also YSI references (not used by the calibration procedure) are reported. Top panel refers to the original manufacturer calibration (performed twice per day), whereas the other panels to the Bayesian calibration algorithm and, in particular, from top to bottom, to the one-every-day, one-every-two-days, and one-every-four-days calibration scenarios. The calibrated profiles obtained with the Bayesian calibration algorithm are, independently from the

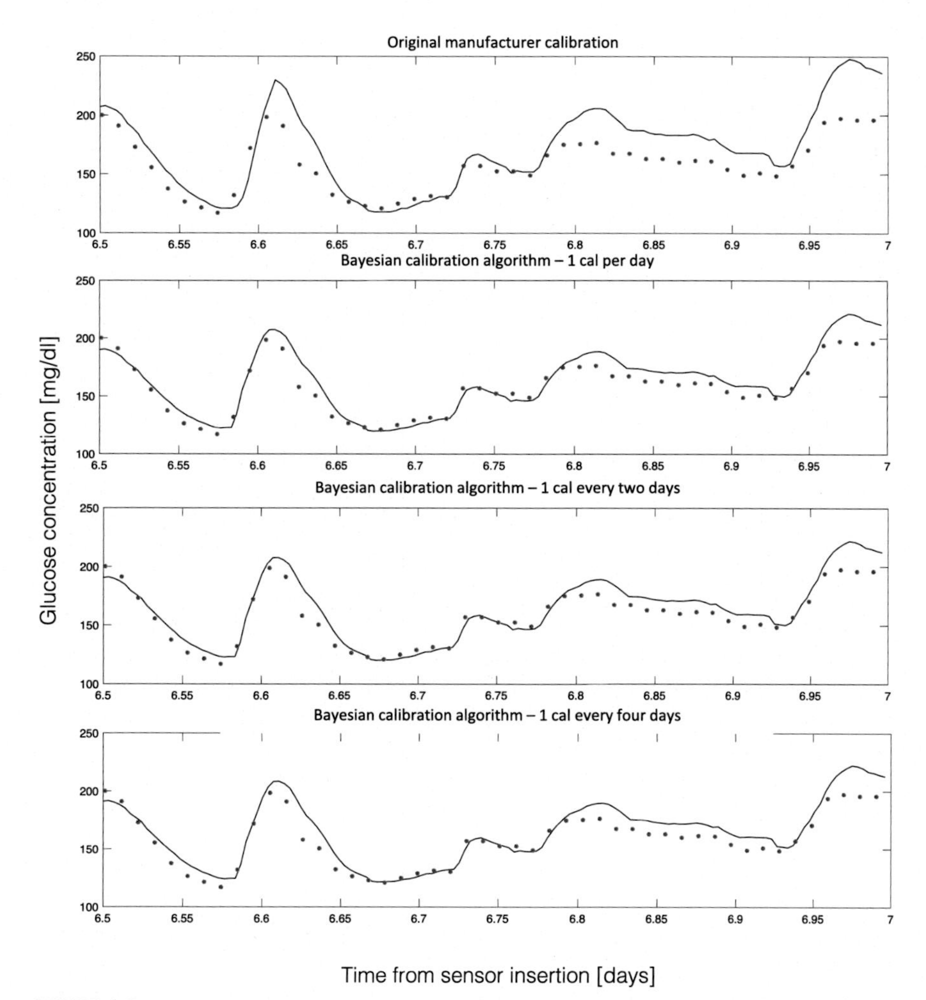

FIGURE 9.9

Calibrated CGM profiles (*continuous lines*) and laboratory references for accuracy assessment (*points*) for a representative subject during an in-clinic session on day 7. From top to bottom: original manufacturer calibration (performed two times per day), Bayesian calibration algorithm with one calibration per day, one calibration every 2 days and one calibration every 4 days.

Adapted from Acciaroli G, Vettoretti M, Facchinetti A, Sparacino G, Cobelli C. Reduction of blood glucose measurements to calibrate subcutaneous glucose sensors: a Bayesian multiday framework. IEEE Transactions on Biomedical Engineering 2018;65(3):587–595.

number of calibrations, more accurate than the original CGM output given by the manufacturer. In particular, MARD, PAGE, and CEGA-A resulted, respectively, 15.04%, 74.26%, 66.18% (original manufacturer calibration, on average two calibrations per day), 8.85%, 97.06%, 94.12% (Bayesian algorithm, one calibration

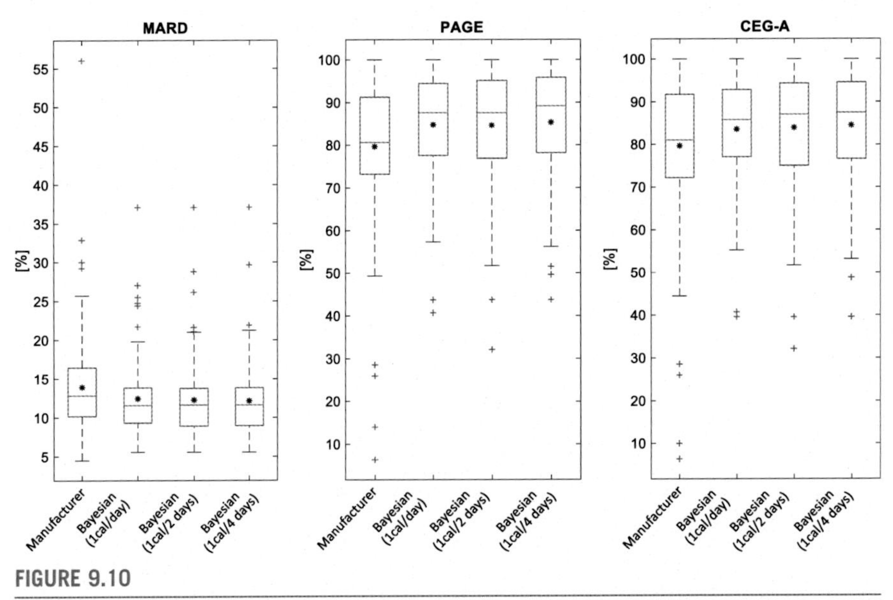

FIGURE 9.10

Boxplot of performance metrics. From left to right, MARD, PAGE, and percentage of points in A zone of the CEG obtained on the test sets for the original calibration (performed twice per day) and for the Bayesian calibration algorithm with three different calibration frequencies (one calibration per day, one calibration every 2 days and one calibration every 4 days). The stars represent the mean values of each distribution.

Adapted from Acciaroli G, Vettoretti M, Facchinetti A, Sparacino G, Cobelli C. Reduction of blood glucose measurements to calibrate subcutaneous glucose sensors: a Bayesian multiday framework. IEEE Transactions on Biomedical Engineering 2018;65(3):587–595.

per day), 8.94%, 97.06%, 92.65% (Bayesian algorithm, one calibration every 2 days), and 10.15%, 93.38%, 87.50% (Bayesian algorithm, one calibration every 4 days).

The results on the entire dataset are reported via boxplot in Fig. 9.10, where the distributions of the three performance metrics, both for the original calibration and for the Bayesian algorithm (with different calibration frequencies), are shown. In general, the Bayesian algorithm appears to be more accurate than the original manufacturer calibration, for all the considered metrics and independently from the frequency of calibrations. Indeed, the metrics obtained for the Bayesian calibration algorithm show distributions concentrated at lower MARD values and higher PAGE and CEGA-A values with respect to the original manufacturer calibration. In particular, the Bayesian algorithm with one calibration every 4 days compared to the original manufacturer calibration (on average two calibrations per day) shows 11.62% MARD (vs. 12.83%), 89.20% PAGE (vs. 80.62%), and 87.5% CEGA-A (vs. 81%).

Numeric values of performance metrics and relative statistical analysis results are reported in Table 9.1. Columns 2–5 report, respectively, the median values of

Table 9.1 Performance metrics (median values).

Metric	Original calibration	Bayesian calibration (1 cal/day)	Bayesian calibration (1 cal/2 days)	Bayesian calibration (1 cal/4 days)	p-value (original vs. Bayesian 1 cal/day)	p-value (original vs. Bayesian 1 cal/2 days)	p-value (original vs. Bayesian 1 cal/4 days)
MARD	12.83	11.59	11.63	11.62	0.0122	0.0025	0.0058
PAGE	80.62	87.63	87.59	89.20	0.0017	0.0010	0.0012
CEGA-A	81.00	85.81	87.07	87.50	0.0442	0.0092	0.0112

the performance metrics obtained with the original calibration, the Bayesian algorithm with one calibration per day, one calibration every 2 days and one calibration every 4 days. Columns 6–8 report the corresponding p-values by the Wilcoxon signed-rank test. The indexes reported in Table 9.1 show that the Bayesian calibration algorithm improves sensor accuracy compared to manufacturer calibration, independently from the frequency of calibrations. The improvement achieved with the Bayesian algorithm is statistically significant ($P < .05$) for all the considered metrics and for all the calibration frequencies tested. In addition, no statistically significant difference ($P > .05$) is found between the performance metrics distributions obtained with the Bayesian algorithm by using different calibration frequencies, that is, one per day versus one every 2 days versus one every 4 days (*P* not shown).

It is interesting to consider also an extreme scenario where no calibration, except the initial one, is performed. In this zero calibrations scenario, the calibration parameters are estimated about 2 h after sensor insertion and used for the entire monitoring session without any further update. As the estimation is performed in the Bayesian setting exploiting only a pair of SMBG samples, it strongly relies on prior information. In this case, the following median values are obtained: 12.03% MARD, 85.67% PAGE, and 85.62% CEG-A. Although these indexes are still slightly better than those of the original manufacturer calibration (see Table 9.1), no statistically significant differences are observed. Indeed, the larger the number of calibration references (as moving from day 1 to day 7), the more the posterior distributions differentiate from the priors. This phenomenon is depicted in Fig. 9.11, where the posterior distributions of the three calibration parameters are reported, centered with respect to the a priori expected values. We can observe that, not surprisingly, as moving from day 1 to day 7, that is, as the number of calibration references increases, the posterior distributions differentiate from the priors (this is more evident for parameters b and s_2). Notably, the prior distributions are quite flat, much more than the posterior distributions. This suggests that the information brought into the Bayesian estimation process by the calibration references plays a key role in determining the parameter values and that the use of only one calibration reference would let the estimate to excessively rely on prior information.

Conclusions

Most of the commercially available CGM sensors need to be calibrated to convert the raw measurements to glucose values. To preserve sensor accuracy, manufacturer instructions recommend a calibration at least every 12 h. Simple linear regression techniques have been extensively employed for calibration since the commercialization of the first CGM devices. Although their simplicity and ease of implementation in wearable devices represent the fundamental strength of these approaches, sensor inaccuracy problems and the need for frequent recalibrations called for the development of more sophisticated techniques.

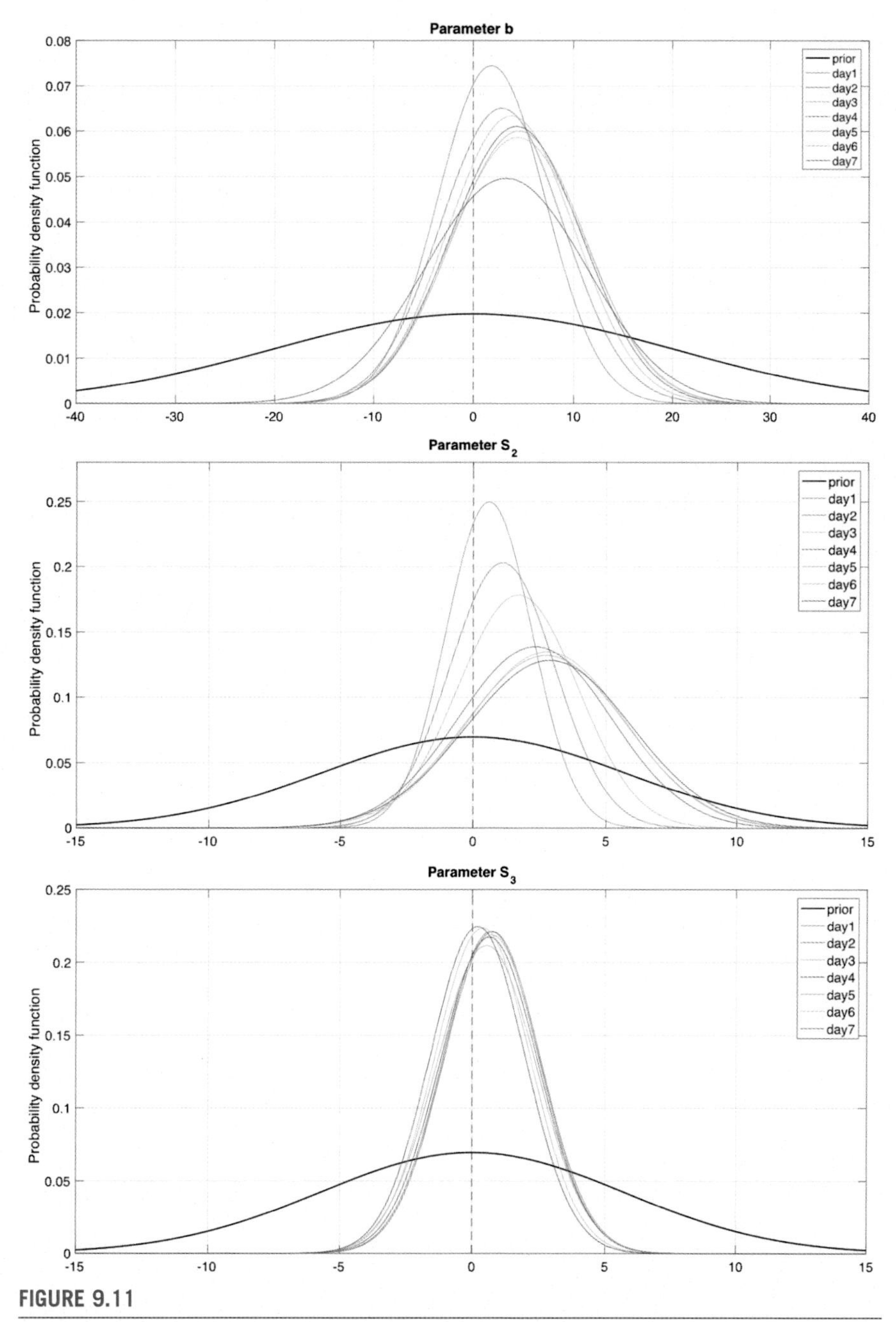

FIGURE 9.11

Posterior distributions of the three calibration parameters as moving from day 1 to day 7 (each day is depicted with a different color, as reported in the legend) and respective a priori distributions (*black curve*). From top to bottom: parameter b, s_2, and s_3. All distributions have been centered with respect to the a priori expected values.

The proved efficacy of some recently proposed techniques in improving sensor performance predicts that, in the following years, more sophisticated algorithms will be integrated into next-generation CGM systems to support the increasing sensors' lifetimes and performance requirements and to reduce calibration need.

References

[1] Wang J. Electrochemical glucose biosensors. Chemical Reviews 2008;108(2):814–25.
[2] McGarraugh G. The chemistry of commercial continuous glucose monitors. Diabetes Technology and Therapeutics 2009;11(Suppl. 1):S17–24.
[3] Rossetti P, Bondia J, Vehi J, Fanelli CG. Estimating plasma glucose from interstitial glucose: the issue of calibration algorithms in commercial continuous glucose monitoring devices. Sensors 2010;10:10936–52.
[4] Bequette BW. Continuous glucose monitoring: real-time algorithms for calibration, filtering, and alarms. Journal of Diabetes Science and Technology 2010;4(2):404–18.
[5] Sparacino G, Facchinetti A, Cobelli C. Smart continuous glucose monitoring sensors: on-line signal processing issues. Sensors 2010;10(7):6751–72.
[6] Lodwing V, Heinemann L. Continuous glucose monitoring with glucose sensors: calibration and assessment criteria. Diabetes Technology and Therapeutics 2003;5:572–86.
[7] Dexcom G4 platinum continuous glucose monitoring system user's guide. Available from: https://s3-us-west-2.amazonaws.com/dexcompdf/LBL012528 +Rev+004+User \T1\textquoterights+Guide%2C+G4+PLATINUM+with+Share+US+Web+with+cover.pdf. [Accessed 10 01 2018].
[8] Dexcom G5 mobile continuous glucose monitoring system user guide. Available from: https://s3-uswest-2.amazonaws.com/dexcompdf/LBL013990-REV003-G5-Mobile-User-Guide-NA-Android-US.pdf. [Accessed 10 01 2018].
[9] Medtronic diabetes. Sensors & transmitters. Calibrating your sensor. Available from: https://www.medtronicdiabetes.com/customer-support/sensors-and-transm itters-support/calibration-sensor. [Accessed 10 01 2018].
[10] Cappon G, Acciaroli G, Vettoretti M, Facchinetti A, Sparacino G. Wearable continuous glucose monitoring sensors: a revolution in diabetes treatment. Electronics 2017;6(3): 65.
[11] Vettoretti M, Cappon G, Acciaroli G, Facchinetti A, Sparacino G. Continuous glucose monitoring: current use in diabetes management and possible future applications. Journal of Diabetes Science and Technology 2018;12(5):1064–71.
[12] Acciaroli G, Vettoretti M, Facchinetti A, Sparacino G. Calibration of minimally invasive continuous glucose monitoring sensors: state-of-the-art and current perspectives. Biosensors 2018;8(1):24.
[13] Panteleon AE, Rebrin K, Steil GM. The role of the independent variable to glucose sensor calibration. Diabetes Technology and Therapeutics 2003;5(3):401–10.
[14] Mastrototaro J, Gross T, Shin J. Glucose monitor calibration methods. Available from: https://patents.google.com/patent/US6424847B1/en. [Accessed 11 12 2018].
[15] Shin J, Holtzclaw K, Dangui N, Kanderian J, Mastrototaro SJ, Hong P. Real time self-adjusting calibration algorithm. Available from: https://patentscope.wipo.int/search/en/detail.jsf?docId=WO2005065542. [Accessed 11 12 2018].

[16] Kamath A, Simpson P, Brauker J, Goode PV. Calibration techniques for a continuous analyte sensor. Available from: https://patents.google.com/patent/US8386004 [Accessed 11 12 2018].

[17] Budiman E. Method and device for providing offset model based calibration for analyte sensor. Available from: https://patents.google.com/patent/US8224415. [Accessed 11 12 2018].

[18] Choleau C, Klein JC, Reach G, Aussedat B, Demaria-Pesce D, Wilson GS, Gifford R, Ward WK. Calibration of a subcutaneous amperometric glucose sensor implanted for 7 days in diabetic patients: part 2. Superiority of the one-point calibration method. Biosensors and Bioelectronics 2002;17(8):647–54.

[19] Mahmoudi Z, Johansen MD, Christiansen JS, Hejlesen O. Comparison between one-point calibration and two-point calibration approaches in a continuous glucose monitoring algorithm. Journal of Diabetes Science and Technology 2014;8(4):709–19.

[20] Rebrin K, Steil GM, Van Antwerp WP, Mastrototaro JJ. Subcutaneous glucose predicts plasma glucose independent of insulin: implications for continuous monitoring. American Journal of Physiology 1999;277(3):561–71.

[21] Schiavon M, Dalla Man C, Dube S, Slama M, Kudva YC, Peyser T, Basu A, Basu R, Cobelli C. Modeling plasma-to-interstitium glucose kinetics from multitracer plasma and microdialysis data. Diabetes Technology and Therapeutics 2015;17(11):825–31.

[22] Keenan DB, Mastrototaro JJ, Voskanyan G, Steil GM. Delays in minimally invasive continuous glucose monitoring devices: a review of current technology. Journal of Diabetes Science and Technology 2009;33(55):1207–14.

[23] Rebrin K, Sheppard NF, Steil GM. Use of subcutaneous interstitial fluid glucose to estimate blood glucose: revisiting delay and sensor offset. Journal of Diabetes Science and Technology 2010;4(5):1087–98.

[24] Facchinetti A, Sparacino G, Cobelli C. Reconstruction of glucose in plasma from interstitial fluid continuous glucose monitoring data: role of sensor calibration. Journal of Diabetes Science and Technology 2007;1(5):617–23.

[25] Helton KL, Ratner BD, Wisniewski NA. Biomechanics of the sensor-tissue interface—effects of motion, pressure, and design on sensor performance and the foreign body response—Part I: theoretical framework. Journal of Diabetes Science and Technology 2011;5(3):632–46.

[26] Klueh U, Liu Z, Feldman B, Henning TP, Cho B, Ouyang T, Kreutzer D. Metabolic biofouling of glucose sensors in vivo: role of tissue microhemorrhages. Journal of Diabetes Science and Technology 2011;5(3):583–95.

[27] Diabetes Research In Children Network (Direcnet) Study Group, Buckingham BA, Kollman C, Beck R, Kalajian A, Fiallo-Scharer R, Tansey MJ, Fox LA, Wilson DM, Weinzimer SA, Ruedy KJ, Tamborlane WV. Evaluation of factors affecting CGMS calibration. Diabetes Technology and Therapeutics 2006;8(3):318–25.

[28] Aussedat B, Thomé-Duret V, Reach G, Lemmonier F, Klein JC, Hu Y, Wilson GS. A user-friendly method for calibrating a subcutaneous glucose sensor-based hypoglycaemic alarm. Biosensors and Bioelectronics 1997;12(11):1061–71.

[29] Knobbe EJ, Lim WL, Buckingham BA. Method and apparatus for real-time estimation of physiological parameters. Available from: https://patents.google.com/patent/US6572545B2/en. [Accessed 11 12 2018].

[30] Knobbe EJ, Buckingham B. The extended Kalman filter for continuous glucose monitoring. Diabetes Technology and Therapeutics 2005;7(1):15–27.

[31] Kuure-Kinsey M, Palerm CC, Bequette BW. A dual-rate Kalman filter for continuous glucose monitoring. Conference Proceedings IEEE Engineering in Medicine and Biology 2006;1:63–6.

[32] Facchinetti A, Sparacino G, Cobelli C. Enhanced accuracy of continuous glucose monitoring by online extended Kalman filtering. Diabetes Technology and Therapeutics 2010;12(5):353–63.

[33] Leal Y, Garcia-Gabin W, Bondia J, Esteve E, Ricart W, Fernandez Real JM, Vehi J. Real-time glucose estimation algorithm for continuous glucose monitoring using autoregressive models. Journal of Diabetes Science and Technology 2010;4(2):391–403.

[34] Leal Y, Garcia-Gabin W, Bondia J, Esteve E, Ricart W, Fernández Real JM, Vehí J. Enhanced algorithm for glucose estimation using the continuous glucose monitoring system. Medical Science Monitor 2010;16(6):51–8.

[35] Barceló-Rico F, Bondia J, Díez JL, Rossetti P. A multiple local models approach to accuracy improvement in continuous glucose monitoring. Diabetes Technology and Therapeutics 2012;14(1):74–82.

[36] Barcelo-Rico F, Diez JL, Rossetti P, Vehi J, Bondia J. Adaptive calibration algorithm for plasma glucose estimation in continuous glucose monitoring. IEEE Journal of Biomedical and Health Informatics 2013;17(3):530–8.

[37] Mahmoudi Z, Johansen MD, Christiansen JS, Hejlesen OK. A multistep algorithm for processing and calibration of microdialysis continuous glucose monitoring data. Diabetes Technology and Therapeutics 2013;15(10):825–35.

[38] Kirchsteiger H, Zaccarian L, Renard E, Del Re L. A novel online recalibration strategy for continuous glucose measurement sensors employing LMI techniques. Conference Proceedings IEEE Engineering in Medicine and Biology 2013;1:3921–4.

[39] Kirchsteiger H, Zaccarian L, Renard E, Del Re L. LMI-based approaches for the calibration of continuous glucose measurement sensors. IEEE Journal of Biomedical and Health Informatics 2015;19(5):1697–706.

[40] Guerra S, Facchinetti A, Sparacino G, De Nicolao G, Cobelli C. Enhancing the accuracy of subcutaneous glucose sensors: a real-time deconvolution-based approach. IEEE Transactions on Biomedical Engineering 2012;59(6):1658–69.

[41] De Nicolao G, Sparacino G, Cobelli C. Nonparametric input estimation in physiological systems: problems, methods, and case studies. Automatica 1997;33(5):851–70.

[42] Vettoretti M, Facchinetti A, Del Favero S, Sparacino G, Cobelli C. Online calibration of glucose sensors from the measured current by a time-varying calibration function and Bayesian priors. IEEE Transactions on Biomedical Engineering 2016;63(8):1631–41.

[43] Acciaroli G, Vettoretti M, Facchinetti A, Sparacino G, Cobelli C. From two to one per day calibration of Dexcom G4 Platinum by a time-varying day-specific Bayesian prior. Diabetes Technology and Therapeutics 2016;18(8):472–9.

[44] Lee JB, Dassau E, Doyle FJ. A run-to-run approach to enhance continuous glucose monitor accuracy based on continuous wear. IFACPapersOnLine 2015;48(20):237–42.

[45] Zavitsanou S, Lee JB, Pinsker JE, Church MM, Doyle FJ, Dassau E. A personalized week-to-week updating algorithm to improve continuous glucose monitoring performance. Journal of Diabetes Science and Technology 2017;11(6):1070–9.

[46] Magni P, Sparacino G. Parameter estimation. In: Carson E, Cobelli C, editors. Modeling methodology for physiology and medicine. 2nd ed. Oxford: Elsevier; 2013.

[47] Acciaroli G, Vettoretti M, Facchinetti A, Sparacino G, Cobelli C. Reduction of blood glucose measurements to calibrate subcutaneous glucose sensors: a Bayesian multiday framework. IEEE Transactions on Biomedical Engineering 2018;65(3):587–95.

[48] Acciaroli G, Vettoretti M, Facchinetti A, Sparacino G. Toward calibration-free continuous glucose monitoring sensors: Bayesian calibration approach applied to next-generation Dexcom technology. Diabetes Technology and Therapeutics 2018;20(1):59−67.

[49] Cobelli C, Schiavon M, Dalla Man C, Basu A, Basu R. Interstitial fluid glucose is not just a shifted-in-time but a distorted mirror of blood glucose: insight from an in silico study. Diabetes Technology and Therapeutics 2016;18(8):505−11.

[50] Basu A, Dube A, Veettil S, Slama M, Kudva JC, Peyser T, Carter RE, Cobelli C, Basu R. Time lag of glucose from intravascular to inter- stitial compartment in type 1 diabetes. Journal of Diabetes Science and Technology 2015;9(1):63−8.

[51] Lagarias JC, Reeds JA, Wright MH, Wright PE. Convergence properties of the Nelder−Mead simplex method in low dimensions. SIAM Journal on Optimization 1998;9(1):112−47.

[52] Christiansen M, Bailey T, Watkins E, Liljenquist D, Price D, Nakamura K, Boock R, Peyser T. A new-generation continuous glucose monitoring system: improved accuracy and reliability compared with a previous-generation system. Diabetes Technology and Therapeutics 2013;15(10):881−8.

CHAPTER

CGM filtering and denoising techniques

10

Andrea Facchinetti, PhD, Giovanni Sparacino, PhD, Claudio Cobelli, PhD
Department of Information Engineering, University of Padova, Padova, Italy

Introduction

Continuous glucose monitoring (CGM) sensor data can be affected by several sources of error, for example, due to imperfect calibration, sensor physics, chemistry, and electronics, which can affect both accuracy and precision of CGM readings [1−6]. In particular, the CGM signal is also corrupted by a random noise component, which dominates the true signal at high frequency [7−9].

The presence of the random noise component on CGM data is evident from Fig. 10.1, which illustrates two representative time series (black line) measured in two diabetic subjects through a commercial CGM device, the Glucoday (Menarini, Firenze, Italy), a minimally invasive microdialysis sensor that provides glucose readings every 3 min (data taken from Ref. [10], where details on the sensor can be found). On the top panel, 1-day CGM data of the first representative subject shows are clearly corrupted by a large noise component. On the other hand, the bottom panel shows 1-day CGM data of a second representative subject, in which the noise variance seems, in general, smaller than for subject #10, even if large spurious spikes (possibly due to patient movements that episodically may perturb sensor behavior) are present.

Spurious spikes and oscillations can affect the performance of any algorithm based on CGM data for therapeutic decisions and suggestions. For instance, the generation of hypoglycemic and hyperglycemic alerts is strongly influenced by the CGM sensor's accuracy and precision, with a percentage of false alerts of the order of 50% in the worst cases [11,12]. Another example in which the random fluctuations around the actual CGM value could be critical is in the calculation of insulin boluses, where it can lead to under/overestimations of the insulin amount of the injected, with the possibility of dangerously increasing risks of hypo/hyperglycemia [13]. Finally, other CGM-based applications negatively affected by the presence of random noise include glucose predictors [14−16] and closed-loop control strategies [17,18]. In this chapter, we will deal with the problem of reducing the impact of random noise on CGM data.

Glucose Monitoring Devices. https://doi.org/10.1016/B978-0-12-816714-4.00010-7

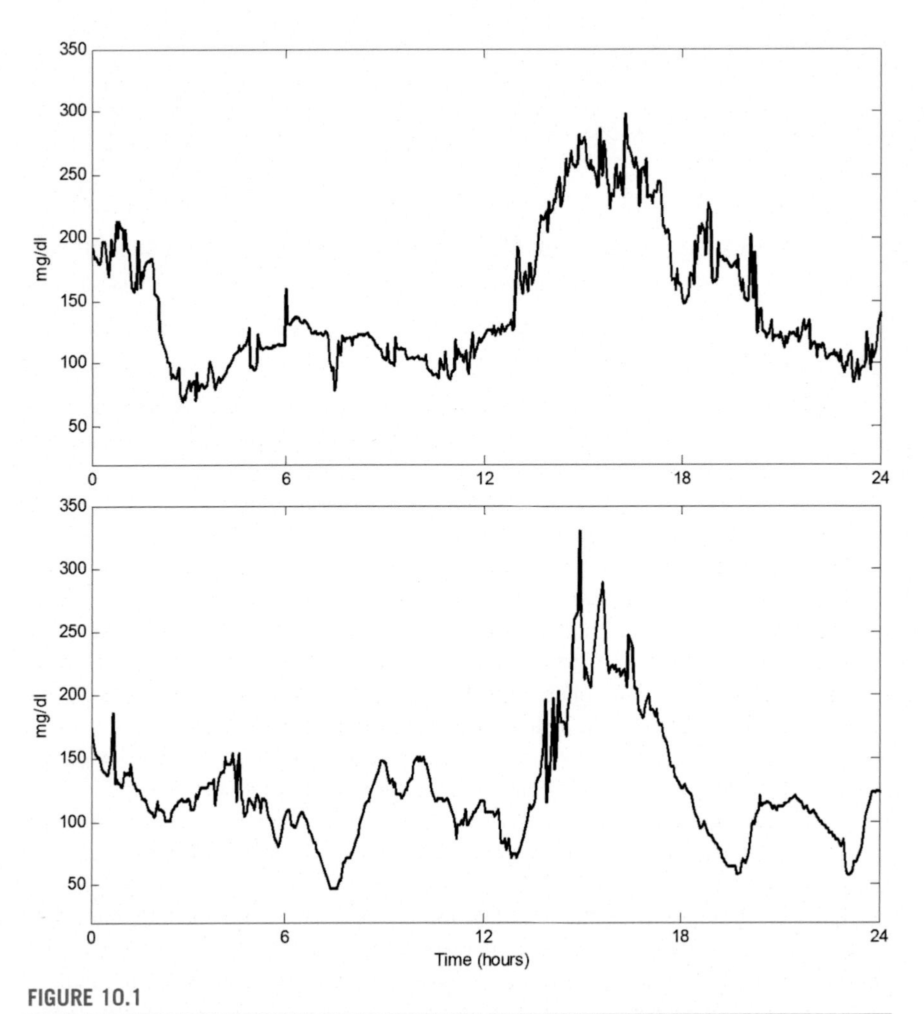

FIGURE 10.1

Two representative CGM time series measured for 24 h, at 3 min sampling rate, in two diabetic subjects by Glucoday. *Top*: low SNR condition. *Bottom*: high SNR condition.

Taken from Facchinetti A, Sparacino G, Cobelli C. An online self-tunable method to denoise CGM sensor data. IEEE Transactions on Biomedical Engineering 2010;57(3):634–641.

The denoising problem

In formal terms, if we consider the following equation

$$y(t) = u(t) + v(t) \tag{10.1}$$

where $y(t)$ is the glucose level measured at time t by the CGM sensor, $u(t)$ is the true, unknown, glucose level, and $v(t)$ is the random noise affecting it, supposed to be additive and uncorrelated from the useful signal, the purpose of denoising is recovering $u(t)$ from $y(t)$. Digital filtering is the most appropriate technology that can be used to enhance the quality of the CGM signal and reduce the random noise component [19–21]. Given the expected spectral characteristics of noise, for example, noise is white, (causal) low-pass filtering represents the most natural candidate to separate signal from noise in online applications [22]. One major problem in low-pass filtering is that, as signal and noise spectra normally overlap, it is not possible to remove the random noise $v(t)$ from the measured signal $y(t)$ without distorting the true signal $u(t)$. In particular, distortion results in a delay affecting the estimate $\hat{u}(t)$ with respect to the true $u(t)$: the more the filtering, the larger the delay. It is easily understood that having a consistently delayed, even if the less noisy version of CGM data could be useless in practice, for example, for the generation of timely hypoalerts. A clinically significant filtering issue is thus the establishment of a compromise between the regularity of $\hat{u}(t)$ and its delay with respect to the true $u(t)$.

Possible approaches to CGM denoising

Moving-average (MA) filtering is a first candidate approach to deal with CGM denoising. MA filters are commonly used in denoising in many applications, including processing in commercial CGM devices [21]. Briefly, having fixed the order, k, the output of the filter relative to the nth sample is given by a weighted sum of the last k measured samples

$$\hat{u}(n) = \frac{w_1 y(n) + w_2 y(n-1) + \cdots + w_k y(n-k+1)}{\sum_{i=1}^{k} w_i} \tag{10.2}$$

where $y(n)$ represents glucose of the nth sample. The parameters of the filter are the order k and the weights $w_1, \ldots, w_k$. The higher is k, the longer is the "memory" of the past data. Increasing k usually produces a more significant noise reduction and, at the same time, a larger signal distortion, for example, $\hat{u}(n)$ is significantly delayed, thus being unable to track fast changes of the true $u(n)$. Having fixed the order k, the weights $w_1, \ldots, w_k$ can be chosen in several ways. The most common strategy is an MA with exponential weights, where $w_i = \mu^i$, with μ (a real between 0 and 1) acting as a "forgetting factor" (the higher μ, the higher the memory of past data). The major weakness of MA is that, once weights have been chosen, it treats all the time series in the same way, irrespectively of possible differences of their signal-to-noise ratio (SNR) due to sensor and individual variability (see Fig. 10.1). As a consequence, a filter with fixed parameters is at risk of being suboptimal in denoising CGM data.

A different CGM denoising procedure, proposed by Chase et al. [7], was based on an integral-based fitting and filtering method. Even if the procedure can be used in real time during clinical trials, its major limitation is, in fact, that some of its components (e.g., the concentration of plasma insulin) cannot be identified if only CGM data are available. This hinders the possibility of using the method in daily-life conditions.

Finally, another proposed denoising procedure, in our opinion, more suited to CGM applications, resorts to Kalman filter (KF). Pioneering applications of KF to process CGM data were presented by Knobbe et al. [23], with the aim of reconstructing blood glucose concentration by employing a model of blood-to-interstitium glucose kinetics and blood glucose concentration references, by Palerm et al. [24,25], with the aim of predicting the glucose profile and detecting hypoglycemia, and by Kuure-Kinsey et al. [3], with the purpose of improving CGM calibration.

In the next section, we will provide an application of KF to the GCM denoising problem originally proposed by Facchinetti et al. [8,9].

CGM denoising by Kalman filter

Overview of the Kalman filter

Briefly, at discrete time, the KF is implemented by first-order difference equations that recursively estimate the unknown state vector $x(t)$ of a dynamic system exploiting vectors of noisy measurements $y(t)$ causally related to it [22,26]. The process update equation is given by:

$$x(t+1) = Fx(t) + w(t) \tag{10.3}$$

where $x(t)$ has in general size n, $w(t)$ is usually a zero-mean Gaussian noise vector (size n) with (unknown) covariance matrix Q (size $n \times n$), and F is a suitable matrix (size $n \times n$). The state vector $x(t)$ is linked to the measurement vector $y(t)$ (size m) by the equation:

$$y(t) = Hx(t) + v(t) \tag{10.4}$$

where $v(t)$ is the zero-mean Gaussian noise measurement error vector (size m) with (unknown) covariance matrix R, and which is uncorrelated with $w(t)$, and H is a suitable matrix (size $m \times n$). The linear minimum variance estimate of the state vector obtainable from the measurements $y(t)$ collected until time t is indicated by $\hat{x}(t|t)$ and can be computed by using the following linear equations:

$$\begin{cases} K_t = \left(FP_{t-1|t-1}F^T + Q\right)H^T\left(H\left(FP_{t-1|t-1}F^T + Q\right)H^T + R\right)^{-1} \\ \hat{x}(t|t) = Fx(t-1|t-1) + K_t(y(t) - H\hat{x}(t-1|t-1)) \\ P_{t|t} = (I - K_tH)\left(FP_{t-1|t-1}F^T + Q\right) \end{cases} \tag{10.5}$$

where $P_{t|t}$ (size $n \times n$) is the covariance matrix of the estimation error affecting $\hat{x}(t|t)$, K_t (size $n \times m$) is the Kalman gain matrix, and where $P_{0|0}$ and $\hat{x}(0|0)$ are the initial conditions. The Q and R matrices, that is, the process and the measurement noise covariance matrices (respectively), are key parameters in determining the performance of KF. Unfortunately, Q and R are usually unknown, or sometimes they are known except for a scale factor. The major problem of KF is the determination of Q and R, and, more specifically, of the so-called Q/R ratio [22,26]. This problem bears a close resemblance to determining the smoothing parameter in regularization methods [27–29].

Formulation as online self-tunable approach

A priori model for u(t)

Calling back the notation used in Eq. (10.1), if some a priori information on $u(t)$ is available, an "optimal" filter (i.e., determining the best trade-off between noise reduction and signal distortion) can be determined by embedding the filtering problem within a Bayesian context, where the Kalman filter (KF) can be adopted in the implementation step [22].

An a priori description of the unknown signal is however necessary. A simple but flexible way to model a smooth signal on a uniformly spaced discrete grid is to describe it as the realization of the multiple integrations of a white noise process. The choice of the number of integrators, to the best of our knowledge, cannot be easily addressed on firm theoretical basis. As a matter of fact, in the smoothing/regularization literature, the choice of m is normally left to the user or handled on empirical bases [30]. In our case, $u(t)$ can be reliably described as the double integration (the so-called integrated random walk model) [29], one has

$$u(t) = 2u(t-1) - u(t-2) + w(t) \tag{10.6}$$

where $w(t)$ is a zero-mean Gaussian noise with (unknown) variance equal to λ^2. The choice of $m=2$ integrators emerges from a simulation study using a cross-validation strategy similar to that of [27,30,31]. Bringing Eq. (11.6) into the state space, two state variables, that is, $x_1(t)=u(t)$ and $x_2(t)=u(t-1)$, are needed. Then, in Eq. (11.3) the state space vector at time t becomes $\boldsymbol{x}(t)=[x_1(t)\ x_2(t)]^T$, and F is consequently given by

$$F = \begin{bmatrix} 2 & -1 \\ 1 & 0 \end{bmatrix} \tag{10.7}$$

Reminding that the CGM measurement is the only output of the system, the measurement vector $y(t)$ of Eq. (10.4) becomes a scalar, and $H=[1\ 0]$. Updating Eq. (10.5) for the estimation of $\widehat{x}(t|t)$, $P_{t|t}$ becomes a 2×2 matrix (with $P_{0|0}=I_2$ and $\widehat{x}(0|0)=[y(0)\ y(-1)]^T$), K_t a 2×1 vector, and Q and R are

$$Q = \begin{bmatrix} \lambda^2 & 0 \\ 0 & 0 \end{bmatrix}, \quad R = \sigma^2 \tag{10.8}$$

To arrive at an estimate of glucose $\hat{u}(t)$, both λ^2 and σ^2 are required. In our problem, neither λ^2 nor σ^2 are known, and we need to estimate their values in real time from the data. An additional difficulty is that, as it appears from the two panels of Fig. 10.1, the SNR of CGM data can significantly vary from sensor to sensor and from individual to individual. This variability suggests that "universal" λ^2 and σ^2 estimates cannot be used for all individuals, but their values need to be individualized, calling for a real time and self-tunable parameter estimation procedure to make KF really online applicable.

Determination of λ^2 and σ^2

As seen previously, the λ^2 and σ^2 estimation problem (i.e., the Q/R ratio estimation) has been already faced in the literature [24,25]. However, in these studies, parameters are retrospectively tuned and are not individualized. To solve both these problems, the following two-step procedure for the estimation of λ^2 and σ^2 can be used.

Step 1

The first portion of each CGM time series is considered as a tuning interval, where the unknown parameters λ^2 and σ^2 are automatically estimated using a stochastically based smoothing criterion based on maximum before (ML). The tuning interval should contain a suitable number of CGM values, for example, some tens (hereafter we will call this number N). Briefly, approaching the problem of smoothing the data of the tuning interval in vector $\boldsymbol{y} = [y(1)\ y(2)\ \ldots\ y(N)]$ as a linear minimum variance estimation problem, and also defining $\boldsymbol{u} = [u(1)\ u(2)\ \ldots\ u(N)]$ and $\boldsymbol{v} = [v(1)\ v(2)\ \ldots\ v(N)]$, one has to solve:

$$\hat{\boldsymbol{u}} = \underset{\boldsymbol{u}}{\operatorname{argmin}}\left\{ (\boldsymbol{y}-\boldsymbol{u})^T \boldsymbol{B}^{-1} (\boldsymbol{y}-\boldsymbol{u}) + \frac{\sigma^2}{\lambda^2} \boldsymbol{u}^T \boldsymbol{L}^T \boldsymbol{L} \boldsymbol{u} \right\} \tag{10.9}$$

where the first term of the cost function on the right-hand side measures the fidelity to the data while the second term weights the roughness of the estimate, being $\boldsymbol{L}$ a square lower triangular Toeplitz matrix whose first column is $[1, -2, 1, 0, \ldots, 0]^T$. The estimate $\hat{\boldsymbol{u}}$ of Eq. (10.9) is given by

$$\hat{u} = \left(B^{-1} + \gamma L^T L\right)^{-1} B^{-1} y \tag{10.10}$$

with $\boldsymbol{B}$ squared N-size positive definite matrix expressing our prior knowledge on the structure of the autocorrelation of $\boldsymbol{v}$, assuming the covariance matrix of $\boldsymbol{v}$ depending on the scale factor σ^2, that is, $\boldsymbol{\Sigma}_v = \sigma^2 \boldsymbol{B}$, and with regularization parameter $\gamma = \sigma^2/\lambda^2$. The estimate of Eq. (11.10) can be interpreted as the linear minimum variance estimator of $\boldsymbol{u}$ given $\boldsymbol{y}$. Under Gaussianity assumptions, this linear estimator is optimal in a broad sense. When both σ^2 and λ^2 are unknown, the minimization problem of Eq. (10.9) should be solved for several trial values of the regularization parameter γ until:

$$\frac{\mathrm{WRSS}(\gamma)}{n - q(\gamma)} = \gamma \frac{\mathrm{WESS}(\gamma)}{q(\gamma)} \tag{10.11}$$

where $\mathrm{WRSS} = (\boldsymbol{y} - \hat{\boldsymbol{u}})^T \boldsymbol{B}^{-1} (\boldsymbol{y} - \hat{\boldsymbol{u}})$ (quadratic sum of the weighed residues), $\mathrm{WESS} = \hat{\boldsymbol{u}}^T \boldsymbol{F}^T \boldsymbol{F} \hat{\boldsymbol{u}}$ (quadratic sum of weighed estimates) and $q(\gamma) = \mathrm{trace}(\boldsymbol{B}^{1/2} (\boldsymbol{B}^{-1} + \gamma \boldsymbol{F}^T \boldsymbol{F})^{-1} \boldsymbol{B}^{-1/2})$ (equivalent degrees of freedom), k being the number of measured CGM samples in the selected tuning interval time window. As γ is determined, the estimate of σ^2 is given by

$$\hat{\sigma}^2 = \frac{\mathrm{WRSS}(\gamma)}{n - q(\gamma)} \tag{10.12}$$

The regularization criterion of Eq. (10.11) has interesting connections both with some average properties of linear minimum variance estimators [32] and with data before maximization [29] (for more details, we address the reader to the quoted papers).

Step 2

For the rest of the data, the values of λ^2 and σ^2 found in Step 1 are used inside Eqs. (10.5) and (10.8), allowing both real-time application of KF and individualization of KF parameters.

In silico assessment

To demonstrate the necessity of the individualization of filter parameters and the reliability of the new methodology, a Monte Carlo simulation study has been performed.

Accuracy in SNR determination

A reference noise-free 3-min sampled glucose profile (Fig. 10.2, top panel, thick gray line) was first created. Then, $N_{sim} = 300$ noisy time series have been generated by adding to the reference profile a zero-mean white Gaussian noise sequence [7] with variance σ^2 randomly sampled in the interval (1100) mg^2/dL^2. Two representative profiles with high and low SNR (obtained for realizations of σ^2 equal to 2 and 51, respectively) are shown in the middle and bottom panels of Fig. 10.2 (thick gray lines), respectively. In both panels, σ^2 and λ^2 estimates, obtained applying the procedure of described previously, are reported inside the gray box, which represents the 6 h tuning interval that has been selected for this example (here, the number of CGM samples is $N = 120$). Estimated σ^2 values are 2.4 and 46.2, very similar to true values. The KF output (tuned with parameters estimated in the 6 h window) is displayed by a thin black line. In these two realizations, the root means square error (RMSE) is equal to 1.2 and 4.5 mg/dL, respectively.

Results of the application of the criterion of Eq. (10.11) for all $N_{sim} = 300$ simulations show that the measurement noise variance σ^2 is estimated very well. Fig. 10.3 displays the comparison between true and estimated σ^2 values in $N_{sim} = 300$ runs, with a correlation coefficient $R^2 = 0.986$. Looking at the so-called *Q*/*R* ratio, estimated γ values are very different, with an average value of 21.6, and 10th and 90th percentile of 3.3 and 48.2, respectively. In addition, the estimation of the process noise variance λ^2 for all the $N_{sim} = 300$ realizations returned an average value of 1.5, with 10th and 90th percentile of 1.0 mg^2/dL^2 and 2.1 mg^2/dL^2, respectively (meaning that the variability of λ^2 is correctly estimated, irrespectively of the SNR).

Importance of filter parameters accuracy

Here we demonstrate the necessity of filter parameters individualization. The top panel of Fig. 10.4 displays a zoom of what happens if the signal of Fig. 10.2 (middle panel) is filtered using parameters obtained for the signal of Fig. 10.2 (bottom panel). As apparent, the use of suboptimal parameters, in this case, leads to oversmoothing, introducing possibly critical under/overshoots when signal derivative changes and large temporal delay (e.g., 6 min around hour 20.7). Conversely, if the signal of Fig. 10.2 (bottom panel) is filtered with the parameters obtained for

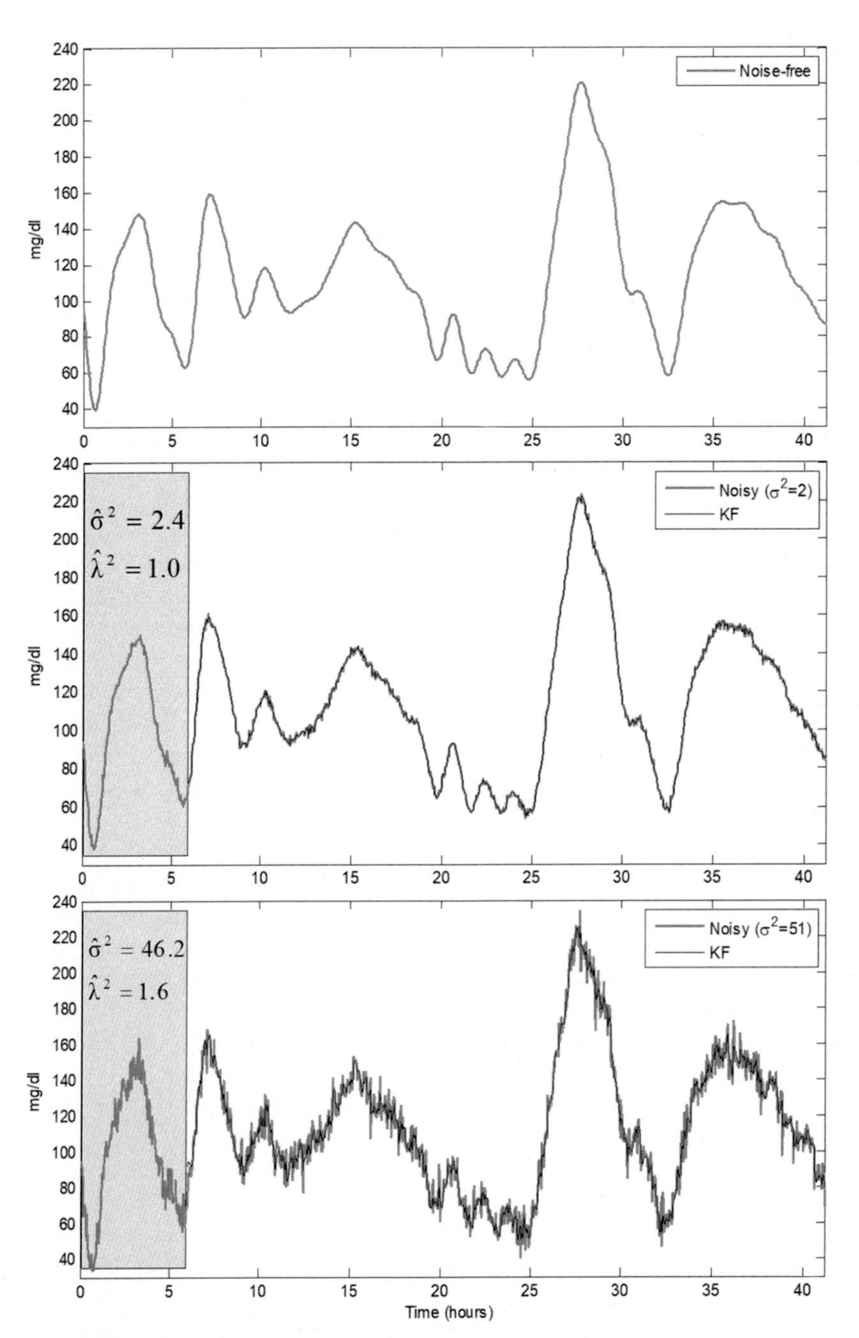

FIGURE 10.2

Simulated study. *Top:* noise-free CGM data (3 min sampling rate). *Middle*: representative high SNR ($\sigma^2 = 2$), noisy (*gray line*) versus KF (*thin black line*) time series. *Bottom*: representative low SNR ($\sigma^2 = 51$), noisy (*gray line*) versus KF (*thin black line*) time series. The gray box is the 6-h tuning interval (estimated parameters are reported inside).

Taken from Facchinetti A, Sparacino G, Cobelli C. An online self-tunable method to denoise CGM sensor data. IEEE Transactions on Biomedical Engineering 2010;57(3):634–641.

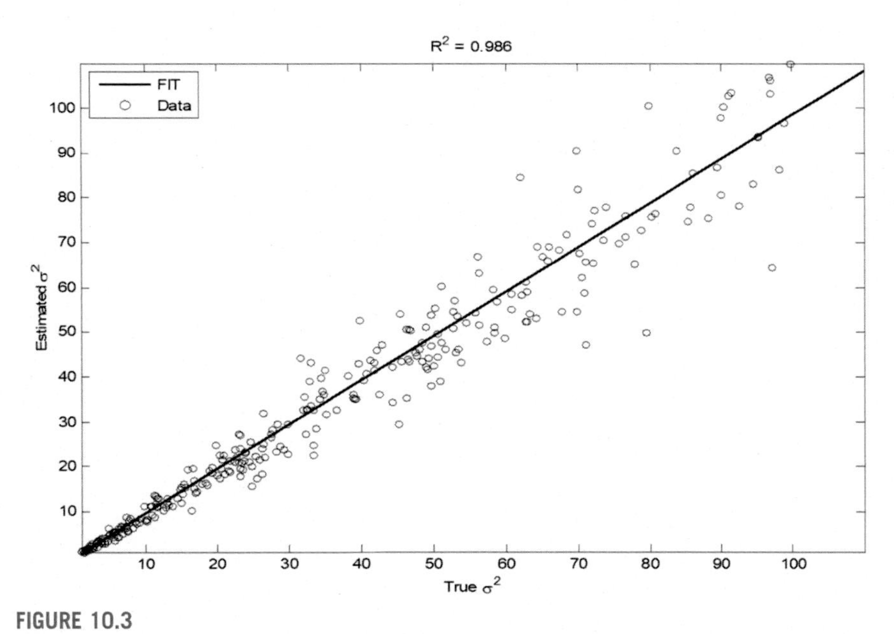

FIGURE 10.3

True versus estimated σ^2 values (*black circles*). The *black solid line* is the fit of the data ($R^2 = 0.986$).

the signal of Fig. 10.2 (middle panel), an undersmoothing situation is generated, with an RMSE increase of about 27% (5.7 vs. 4.5 mg/dL).

Comparison with MA

To further illustrate the novelty of the new KF methodology, its outcome in the $N_{sim} = 300$ simulations was compared with that of an MA filter with exponential fixed weights determined (after a preliminary study, here not documented for sake of space) by setting k and μ equal to 5 and to 0.65, respectively. The filter performance is quantitatively assessed by considering both the RMSE and the delay with respect to the original noise-free signal. Such a delay is measured by the index T, defined as the temporal shift (in minutes) that has to be applied to $\hat{u}$ to minimize the squared norm of the difference between $\hat{\boldsymbol{u}}$ and $\boldsymbol{y}$, that is,

$$T = \underset{T}{\operatorname{argmin}} \sum_{i} (y(i) - \hat{u}(i + T))^2 \tag{10.13}$$

Table 10.1 shows the average values of RMSE and T, together with their 10th and 90th percentile. Even if the RMSE is not significantly different (Wilcoxon rank-sum test, $P = .39$), the delay T is significantly lower (Wilcoxon rank-sum test, $P < .001$), with an average value that has been reduced by about 90% (0.4 vs. 3.5 min). In addition, the delay introduced by KF is, in the worst case, lower than the delay of MA in the best case (2.2 vs. 2.3 min, not documented here for sake of space). In summary, the performance of the new KF is quantitatively much better than a filter with fixed parameters, giving a similar estimation of the noise-free profile with a minimum delay.

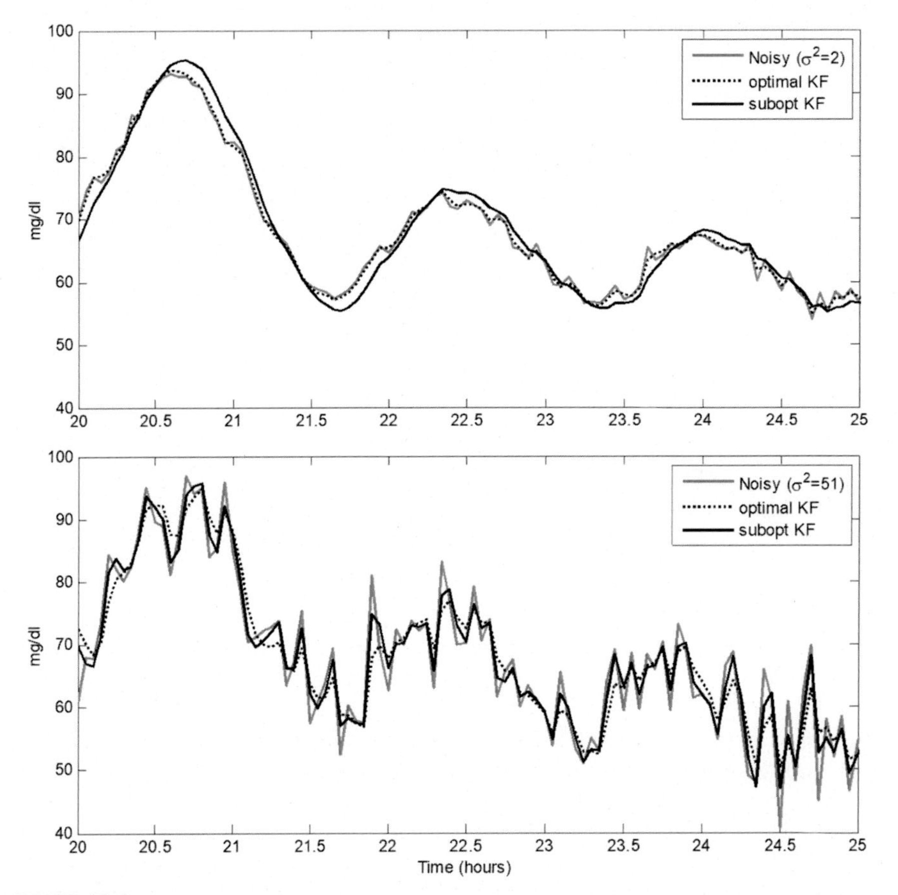

FIGURE 10.4

Suboptimal filtering. Noisy (*gray line*), optimal KF filtered (*black dashed line*), and suboptimal KF filtered (*black line*) time series. *Top:* oversmoothing case, real $\sigma^2 = 2$, used $\sigma^2 = 46.2$. *Bottom:* undersmoothing case, real $\sigma^2 = 51$, used $\sigma^2 = 2.4$.

Taken from Facchinetti A, Sparacino G, Cobelli C. An online self-tunable method to denoise CGM sensor data. IEEE Transactions on Biomedical Engineering 2010;57(3):634–641.

Table 10.1 Average values of T and RMSE, together with 10th and 90th percentiles, computed for MA and the new KF on all $N_{sim} = 300$ realizations.

	T (min)		RMSE	
	MA	**KF**	**MA**	**KF**
Average	3.5	0.4	3.6	3.5
10th perc	3.0	0.1	2.3	1.5
90th perc	4.0	1.0	5.0	5.4

Assessment on data

The database used for the test consists of 24 time series, taken from a larger study [10], collected in type-1 diabetic patients using the Glucoday system (Menarini Diagnostics, Firenze, Italy). Both MA and the new KF have been applied.

Before filtering, the time series were preprocessed through a simple causal nonlinear procedure, aimed at reducing the amplitudes of occasional nonphysiological spikes. In particular, each glucose sample is compared with the previous one, and, if the absolute difference (relative to the sampling period) is higher than the physiological limit of 4 mg/dL per minute [33], it is corrected accordingly. This hard-bounding procedure is similar to that employed within the Minimed CGMS device [34].

The performance of the two filtering approaches has been assessed by considering both the delay measured by index T of Eq. (10.13) and the regularity of the filtered signal (note that the RMSE as done previously in the simulation context), measured by the smoothness relative gain (SRG) index, defined as

$$\mathrm{SRG} = \frac{\mathrm{ESOD}(y) - \mathrm{ESOD}(\widehat{u})}{\mathrm{ESOD}(y)} \tag{10.14}$$

where ESOD(u) denotes the energy of the second-order differences of a time series u, a regularity index already proposed in a CGM prediction context in Ref. [14]. SRG is an index that varies between 0 and 1 and measures the relative amount of signal regularity introduced by (low-pass) filtering.

Fig. 10.5 shows the results of the application of both MA (black dotted line) and the new methodology (black solid line) on the same two representative real subjects illustrated in Fig. 10.1. To better highlight the most important features coming out from the comparison, two 6-h windows have been selected. For subject #10 (top panel) $\widehat{\sigma}^2$ results equal to 17.1 mg^2/dL^2, quantitatively confirming the presence of a rather low SNR, which could be also detected by eye inspection. KF produces a very good denoising, with $T = 4.6$ min lower than MA (where $T = 7.0$ min), and SRG = 0.91 higher than MA (where SRG = 0.90), meaning that it is able to perform a similar smoothing introducing less delay. For subject #8 (bottom panel), where the SNR appears lower than in subject #10 also by eye inspection, a lower value for the measurement noise variance is estimated ($\widehat{\sigma}^2 = 3.5$ mg^2/dL^2). From a quantitative point of view, KF gives a profile with SRG = 0.86 and $T = 1.4$ min, while with MA returns SRG = 0.91 and $T = 3.5$ min. Results highlight the fact that, in subject #8, MA clearly produces oversmoothing, while KF, thanks to the individualization of the parameters, correctly detects a high SNR. Table 10.2 reports mean (10th and 90th percentiles) values of T and SRG calculated on the 24 subjects of the dataset. On average, we can observe that the SRG has been reduced only by 0.03, while the delay T introduced by KF is significantly smaller (−35%) than MA ($P < .01$, Wilcoxon rank-sum test). Interestingly, the 10th and 90th percentiles of both T and SRG correspond to rather wide intervals, suggesting that KF, with parameters tuned according to the statistically based criterion and according to the individual

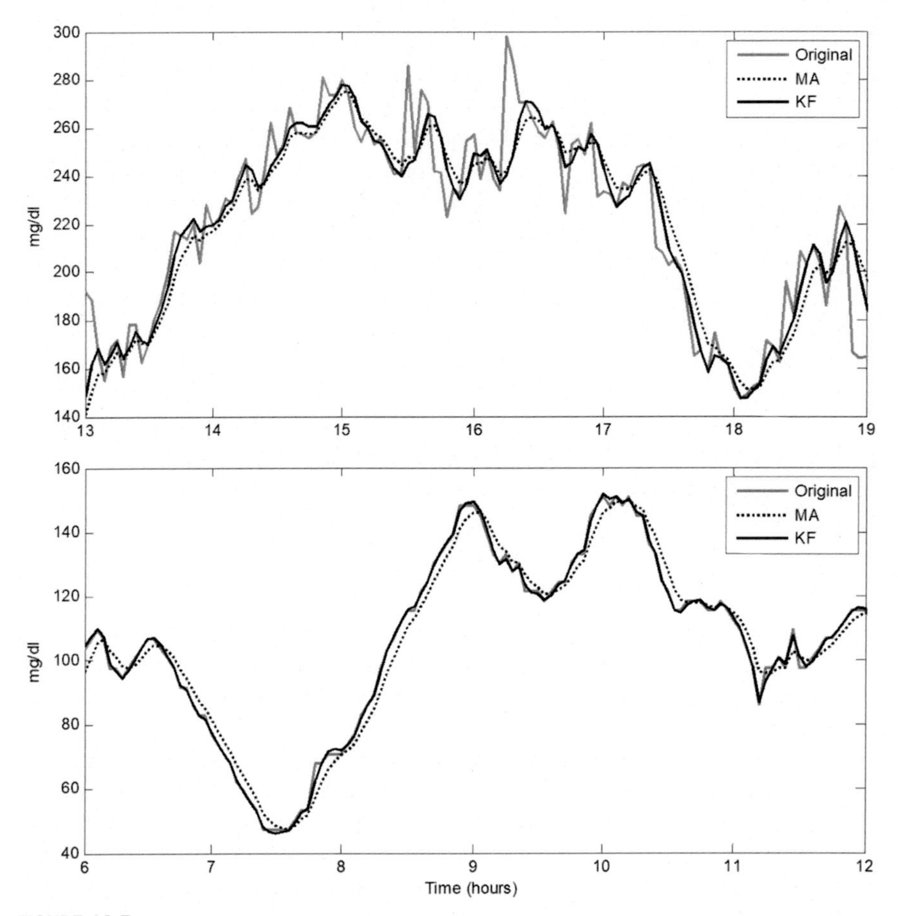

FIGURE 10.5

Two representative 6-h windows of real data. Original (*gray line*), MA filtered (*black dashed line*), and KF filtered (*black line*) time series. *Top*: subject #10, low SNR condition. *Bottom*: subject #8, high SNR condition.

Taken from Facchinetti A, Sparacino G, Cobelli C. An online self-tunable method to denoise CGM sensor data. IEEE Transactions on Biomedical Engineering 2010;57(3):634–641.

Table 10.2 Average, 10th and 90th percentiles values of T and SRG on the 24 real subjects. The last two columns show σ^2 and γ estimated in the burn-in interval.

	***T* (min)**		**SRG**			
Subject	**MA**	**KF**	**MA**	**KF**	σ^2	γ
Average	5.1	3.3	0.89	0.86	10.3	2.0
10th perc	3.5	0.7	0.88	0.75	3.5	0.2
90th perc	6.0	7.5	0.91	0.94	20.7	3.7

SNR, is able to tune the proper smoothing in different SNR conditions, and therefore it is an effective solution to problem of the SNR variability from individual to individual. As far as the estimation of σ^2 is concerned, it clearly appears from the 10th and 90th percentile values (3.5 and 20.7 mg^2/dL^2) that the measurement noise variance is very different from an individual to another, numerically resembling what has been observed by graphical inspection. Furthermore, as far as the regularization parameter γ is concerned, which we remind to be the so-called Q/R ratio, and which is estimated in the 6 h tuning interval, these values result very different between individuals. The fact is not surprising, resembling the observation on the need for filter parameters individualization made previously, in which more than one order of growth rate was detected. Quantitatively, on the real dataset, the difference between the maximum (11.45) to minimum (0.04) values is about three orders of growth rate. This confirms also on real data the necessity of parameter individualization to avoid suboptimal filtering.

Dealing with SNR intraindividual variability

In the presence of intraindividual variability of the SNR, the KF approach for denoising CGM data presented so far performed will result in suboptimal, with a portion of the signal that may result under- or oversmoothed. The existence of the intraindividual variability of the SNR is rather visible on the representative CGM time series, obtained with the Menarini Glucoday system and taken from Ref. [10], displayed in Fig. 10.6. As one can note by eye inspection, the noise component in the time

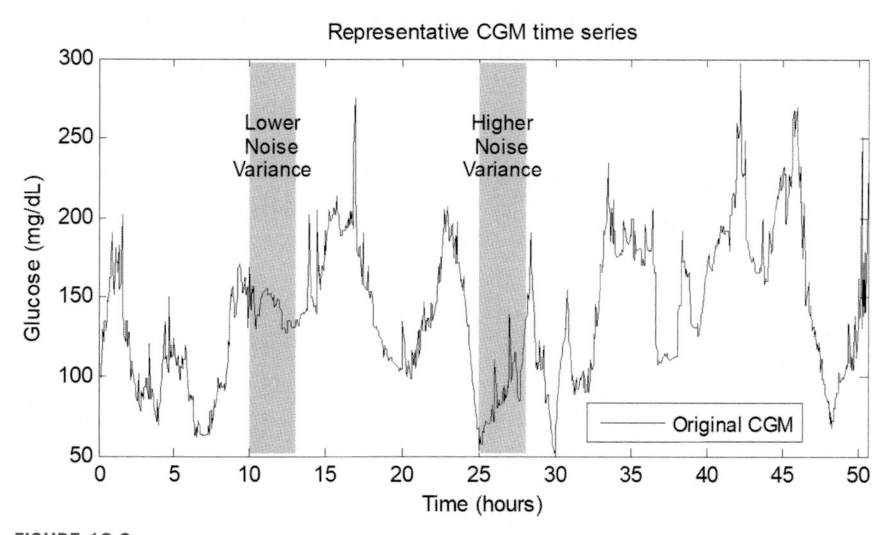

FIGURE 10.6

A representative CGM time series (*black line*) obtained with the Menarini Glucoday system and taken from Ref. [10]. Time intervals 10–13 h and 25–28 h (*gray areas*) show two situations of lower and higher noise variance, respectively.

Taken from Online denoising method to handle intraindividual variability of signal-to-noise ratio in continuous glucose monitoring. IEEE Transactions on Biomedical Engineering 2011;58(9):2664–71.

interval 25–28 h is greater than in 10–13 h. Therefore, different filtering would be needed for denoising these two portions of the same signal.

To deal with the intraindividual variability of the SNR, the KF proposed so far needs to be modified. In an online setting, the procedure of Eqs. (10.10)–(10.12), instead of been used only at the beginning of the monitoring, that is, at time t on the first time window containing N samples, should be repeated at time $t+1$, $t+2$, ..., with the N-size vector $\boldsymbol{y}$, $\boldsymbol{u}$, and $\boldsymbol{v}$ referred to a sliding temporal window of length N. Obviously, in practical applications, N determines also the length of a burn-in interval where no filtered data can be provided (in the previous application on Menarini Glucoday system the length of the initial tuning interval was set to 6 h, but this value may vary due to the sampling frequency of the CGM sensor).

Remark: so far, we assumed the measurement noise $v(t)$ of Eq. (10.1) to be white and Gaussian. In any case, the method we propose is general and flexible to these assumptions. As far as whiteness is concerned, this implies that in Eq. (10.10) the matrix $\boldsymbol{B} = \boldsymbol{I}_N$, where $\boldsymbol{I}_N$ is an N-size identity matrix, should be suitably modified to properly describe correlated noise, for example, autoregressive [4,35]. In addition, Gaussanity is not strictly required because it simply ensures the global optimality of the estimator in Eq. (10.9).

Conclusions

CGM data are affected by several sources of error, including bias errors due to imperfect/loss of calibration or to the physics/chemistry of the sensor, and random noise, which dominates the true signal at high frequency. Although calibration errors were discussed in Chapter 9 of this book, in this chapter we have discussed the denoising problem by online digital filtering. In particular, after having revised the major challenges of CGM filtering, that is, it is impossible to determine real-time filter parameters, and to adapt them to the individual SNR, we presented a KF-based methodology and assessed its performance on both Monte Carlo simulation and CGM data. The method has general applicability, also outside from the CGM context, and can be also used to quantify the variance of measurement noise on CGM data.

References

[1] Facchinetti A, Sparacino G, Cobelli C. Reconstruction of glucose in plasma from interstitial fluid continuous glucose monitoring data: role of sensor calibration. Journal of Diabetes Science and Technology 2007;1(5):617–23.

[2] Kovatchev B, Anderson S, Heinemann L, Clarke W. Comparison of the numerical and clinical accuracy of four continuous glucose monitors. Diabetes Care 2008;31(6): 1160–4.

[3] Kuure-Kinsey M, Palerm CC, Bequette BW. A dual-rate Kalman filter for continuous glucose monitoring. Conference of Proceedings IEEE Engineering in Medicine Biology Society 2006;1:63–6.

[4] Facchinetti A, Del Favero S, Sparacino G, Castle JR, Ward WK, Cobelli C. Modeling the glucose sensor error. IEEE Transactions on Biomedical Engineering 2014;61(3): 620–9.

[5] Vettoretti M, Cappon G, Acciaroli G, Facchinetti A, Sparacino G. Continuous glucose monitoring: current use in diabetes management and possible future applications. Journal of Diabetes Science and Technology 2018;12(5):1064–71.

[6] Cappon G, Acciaroli G, Vettoretti M, Facchinetti A, Sparacino G. Wearable continuous glucose monitoring sensors: a revolution in diabetes treatment. Electronics 2017;6(3).

[7] Chase JG, Hann CE, Jackson M, Lin J, Lotz T, Wong XW, Shaw GM. Integral-based filtering of continuous glucose sensor measurements for glycaemic control in critical care. Computer Methods and Programs in Biomedicine 2006;82(3):238–47.

[8] Facchinetti A, Sparacino G, Cobelli C. An online self-tunable method to denoise CGM sensor data. IEEE Transactions on Biomedical Engineering 2010;57(3):634–41.

[9] Online denoising method to handle intraindividual variability of signal-to-noise ratio in continuous glucose monitoring. IEEE Transactions on Biomedical Engineering 2011; 58(9):2664–71.

[10] Maran A, Crepaldi C, Tiengo A, Grassi G, Vitali E, Pagano G, Bistoni S, Calabrese G, Santeusanio F, Leonetti F, Ribaudo M, Di Mario U, Annuzzi G, Genovese S, Riccardi G, Previti M, Cucinotta D, Giorgino F, Bellomo A, Giorgino R, Poscia A, Varalli M. Continuous subcutaneous glucose monitoring in diabetic patients: a multicenter analysis. Diabetes Care 2002;25(2):347–52.

[11] Peyser TA, Nakamura K, Price D, Bohnett LC, Hirsch IB, Balo A. Hypoglycemic accuracy and improved low glucose alerts of the latest Dexcom G4 platinum continuous glucose monitoring system. Diabetes Technology and Therapeutics 2015;17(8): 548–54.

[12] Howsmon D, Bequette BW. Hypo- and hyperglycemic alarms: devices and algorithms. Journal of Diabetes Science and Technology 2015;9(5):1126–37.

[13] Cappon G, Marturano F, Vettoretti M, Facchinetti A, Sparacino G. In silico assessment of literature insulin bolus calculation methods accounting for glucose rate of change. Journal of Diabetes Science and Technology 2019;13(1):103–10.

[14] Sparacino G, Zanderigo F, Corazza S, Maran A, Facchinetti A, Cobelli C. Glucose concentration can be predicted ahead in time from continuous glucose monitoring sensor time-series. IEEE Transactions on Biomedical Engineering 2007;54(5):931–7.

[15] Zecchin C, Facchinetti A, Sparacino G, Cobelli C. How much is short-term glucose prediction in Type 1 diabetes improved by adding insulin delivery and meal content information to CGM data? A proof-of-concept study. Journal of Diabetes Science and Technology 2016;10(5):1149–60.

[16] Oviedo S, Vehi J, Calm R, Armengol J. A review of personalized blood glucose prediction strategies for T1DM patients. International Journal for Numerical Methods in Engineering 2017;33(6).

[17] Dadlani V, Pinsker JE, Dassau E, Kudva YC. Advances in closed-loop insulin delivery systems in patients with type 1 diabetes. Current Diabetes Reports 2018;18(10):88.

[18] Kovatchev B. Diabetes technology: monitoring, analytics, and optimal control. Cold Spring Harbor Perspectives in Medicine 2019;9(6). https://doi.org/10.1101/cshperspect.a034389. pii: a034389.

[19] Mastrototaro J. Glucose monitor calibration methods. 2002. US Patent No. 6424847.

[20] Feldman BJ, M. G.V.. Method of calibrating an analyte-measurement device, and associated methods, devices and systems. 2008. US Patent No. 0081969-A1.

[21] Simpson PC, Brister M, Wightlin M, Pryor J. Dual electrode system for a continuous analyte sensor. 2008. US Patent No. 0083617-A1.
[22] Anderson BDO, Moore JB. Optimal filtering. Dover Publications; 2005.
[23] Knobbe EJ, Buckingham B. The extended Kalman filter for continuous glucose monitoring. Diabetes Technology and Therapeutics 2005;7(1):15–27.
[24] Palerm CC, Willis JP, Desemone J, Bequette BW. Hypoglycemia prediction and detection using optimal estimation. Diabetes Technology and Therapeutics 2005;7(1):3–14.
[25] Palerm CC, Bequette BW. Hypoglycemia detection and prediction using continuous glucose monitoring-a study on hypoglycemic clamp data. Journal of Diabetes Science and Technology 2007;1(5):624–9.
[26] Grewal MS, Andrews AP. Kalman filtering: theory and practice using MATLAB. John Wiley and Sons; 2001.
[27] Hall P, Titterington DM. Bayesian "confidence intervals" for the cross-validate smoothing spline. Journal of the Royal Statistical Society Series B (Methodological) 1937;45: 133–50.
[28] Common structure of techniques for choosing smoothing parameters in regression problems. Journal of the Royal Statistical Society Series B (Methodological) 1987; 49(2):184–98.
[29] Nicolao GD, Sparacino G, Cobelli C. Nonparametric input estimation in physiological systems: problems, methods, and case studies. Automatica 1997;33(5):851–70.
[30] Wahba G, Wendelberger J. Some new mathematical methods for variational objective analysis using splines and cross validation. Monthly Weather Review 1980;108(8): 1122–43.
[31] Camber HA. Choice of an optimal shape parameter when smoothing noisy data. Communications in Statistics – Theory and Methods 1979;8(14):1425–35.
[32] Sparacino G, Cobelli C. A stochastic deconvolution method to reconstruct insulin secretion rate after a glucose stimulus. IEEE Transactions on Biomedical Engineering 1996; 43(5):512–29.
[33] Kovatchev BP, Clarke WL, Breton M, Brayman K, McCall A. Quantifying temporal glucose variability in diabetes via continuous glucose monitoring: mathematical methods and clinical application. Diabetes Technology and Therapeutics 2005;7(6): 849–62.
[34] Voskanyan G, Keenan DB, Mastrototaro JJ, Steil GM. Putative delays in interstitial fluid (ISF) glucose kinetics can be attributed to the glucose sensing systems used to measure them rather than the delay in ISF glucose itself. Journal of Diabetes Science and Technology 2007;1(5):639–44.
[35] Facchinetti A, Del Favero S, Sparacino G, Cobelli C. Model of glucose sensor error components: identification and assessment for new Dexcom G4 generation devices. Medical, and Biological Engineering and Computing 2015;53(12):1259–69.

CHAPTER

Retrofitting CGM traces

11

Simone Del Favero, PhD [1], Andrea Facchinetti, PhD [2], Giovanni Sparacino, PhD [2], Claudio Cobelli, PhD [2]

[1]*Assistant Professor, Department of Information Engineering, Padova, Italy;* [2]*Department of Information Engineering, University of Padova, Padova, Italy*

Introduction

Continuous glucose monitoring (CGM) technology has been constantly improving since its first appearance two decades ago, and CGM is now spreading in clinical practice [1]. CGM has been profitably used as an addition to self-monitoring blood glucose (SMBG) measurements to improve glucose control by using the glucose trends-in-time measured by the device to adjust insulin dosing in real time [2]. More recently, some CGM models have been approved by the US Food and Drug Administration (FDA) as a substitute to SMBG devices (nonadjunctive CGM use) insulin dosing [3]. Beside real-time use of CGM data for T1D treatment, the recorded CGM traces can be downloaded and used retrospectively for many purposes, for instance, to analyze glucose patterns and adjust standard therapy parameters (e.g., basal pattern, carbohydrate-to-insulin ratio, etc.) [4], to assess glucose control achieved in a clinical trial [5,6], to estimate physiological model parameters [7], and to identify glucose-insulin models [8,9].

In the present chapter, we consider the retrospective use of CGM data and review a "retrofitting" algorithm we originally proposed in Ref. [10], a technique designed to improve a posteriori both precision and accuracy of a CGM trace by using a few SMBG measurements collected in parallel to CGM. By merging information of CGM (high temporal resolution) and SMBG (sparse in time but more accurate than CGM), the retrofitting method produces a continuous-time BG profile which is more accurate than the original CGM data. Having a more accurate CGM trace is beneficial for the above-mentioned retrospective CGM applications.

Chapter organization

In section The retrofitting algorithm we review the retrofitting algorithm presented in Ref. [10].

In section Retrofitting outpatient study data we show that the retrofitting algorithm is very effective in enhancing precision and accuracy of Dexcom SEVEN PLUS CGM sensor, in the setup we encountered in some of our outpatient clinical

Glucose Monitoring Devices. https://doi.org/10.1016/B978-0-12-816714-4.00011-9

studies testing we conducted in 2012–14 [11–14], offering a relatively large number of highly accurate references (YSI) to retrospectively the CGM.

Then, in section Retrofitting real-life adjunctive data, we show that the retrofitting method is capable also to improve the accuracy of a newer and more accurate Dexcom sensor (Dexcom G5) that reached the 1-digit precision (currently one of the most accurate CGMs on the market) with data collected in real-life conditions.

Finally, in section Accuracy of retrofitted CGM versus number of references available, we investigate how the accuracy improvement granted by the retrofitting method is affected by the number of BG measurement available.

The retrofitting algorithm

Problem formulation

The retrofitting method reconstructs, with high temporal resolution, BG concentration profile, $bg(t)$ from CGM records, $cgm(t)$, i.e., measurements of the interstitial glucose concentration affected by noise and bias due to lack/loss of calibration. The method has also access to a few sparse but accurate BG reference measurements. Moreover, we assume that CGM calibration times are known. The signals $bg(t)$ and $cgm(t)$ are related by the model in Fig. 11.1 . The first block models the glucose transport between blood and interstitial fluid with a two-compartment model [15,16],

$$\tau \frac{\mathrm{d}}{\mathrm{d}t} ig(t) = -ig(t) + bg(t) \tag{11.1}$$

where $ig(t)$ is the interstitial fluid glucose concentration and τ is the diffusion time constant assumed to remain constant between two consecutive calibrations. CGM sensor measures glucose in the interstitial fluid producing a current signal, converted back to a glucose concentration by calibration. Due to uncertainties in the calibration process and to transduction sensitivity drifts, for sake of simplicity referred to as

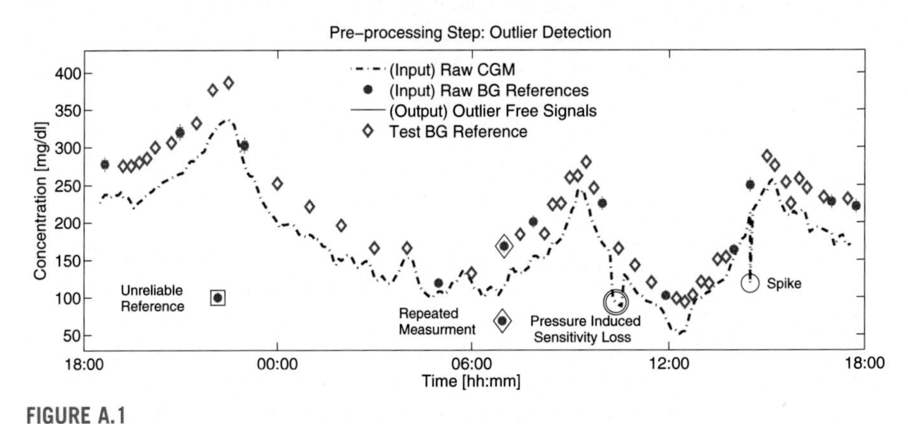

FIGURE A.1

Data preprocessing Step for the representative subject of Fig. 11.2.

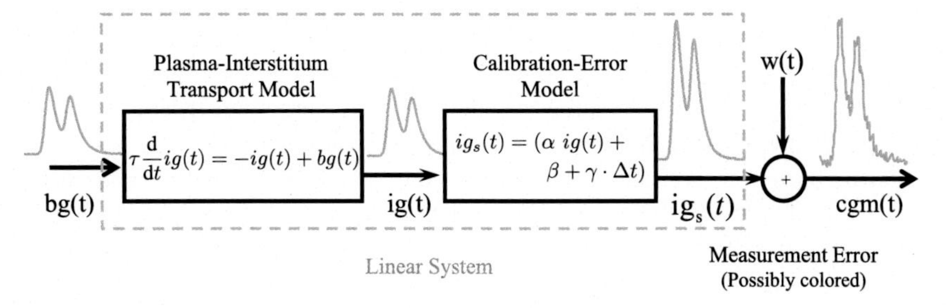

FIGURE 11.1

The semiblind deconvolution problem. The aim is to reconstruct blood glucose concentration, $bg(t)$, having access only to noisy measurements, $cgm(t)$, of the interstitial glucose concentration collected by a CGM sensor, $ig_S(t)$. Plasma-interstitium glucose transport and calibration errors can be modeled as the filtering through a linear time-varying system with four unknown parameters. A few accurate samples of the unknown input $bg(t)$ are also used in its reconstruction.

"calibration errors," it can be assumed that CGM sensor does not return directly $ig(t)$ but only a distorted version of it, $ig_s(t)$. The second block in Fig. 11.1 models such a distortion through a static, linear, time-varying deformation of $ig(t)$:

$$ig_s(t) = (\alpha ig(t) + \beta + \gamma \cdot \Delta t), \tag{11.2}$$

where $\Delta t = t - t^{cal}$ is the time difference with respect to the last calibration time t^{cal}. Introducing the time-varying offset $\beta + \gamma \cdot t$ in (11.2) allows capturing sensor performance degradation in time due to changes in sensor sensitivity as illustrated in Ref. [16]. Calibration parameters α, β, and γ are piecewise constants, changing when a CGM sensor calibration is performed. A perfectly calibrated sensor is obtained with $\alpha = 1$, $\beta = \gamma = 0$. CGM reading is then a noisy measure of $ig_s(t)$,

$$cgm(t) = ig_s(t) + w(t) \tag{11.3}$$

where $w(t)$ is an additive, possibly nonwhite, random noise.

As highlighted in Fig. 11.1, the cascade of blood-to-interstitium transport model and calibration-error model is linear, and therefore the problem of reconstructing $bg(t)$ from $cgm(t)$ is a deconvolution problem. In the specific, given that model parameters, α, β, γ, and τ, are estimated together with the input $bg(t)$, this is a so-called semi-blind deconvolution problem, where, in addition to the input, also some model parameters are unknown. The problem under study has the peculiarity that a few sparse accurate measurements of the input to be reconstructed are also available. This allows us to constrain the solution of the input estimation problem to lay in the measurement confidence intervals.

Notation

Let us assume that m reference have been collected at the time instants $t_1, \ldots, t_m$ and define $\mathbf{bg} = [bg(t_1), \ldots, bg(t_m)]^T$, the $\mathbb{R}^m$ vector contain all BG references. Analogously, let us assume that n, $m << n$, CGM readings have been collected at the time instants $t_1^{cgm}, \ldots, t_n^{cgm}$ and define $\mathbf{cgm} = \left[cgm\left(t_1^{cgm}\right), \ldots, cgm\left(t_n^{cgm}\right)\right]^T$ as the $\mathbb{R}^n$ vector containing all CGM measurements. Define moreover $\mathbf{ig}(\tau) = \left[ig\left(t_1^{cgm}\right), \ldots, ig\left(t_n^{cgm}\right)\right]^T$ as the $\mathbb{R}^n$ vector containing estimated interstitial glucose concentrations at the same instants, where the argument τ is used to recall the dependency of $\mathbf{ig}$ on the blood-interstitium diffusion constant τ.

Furthermore, let us assume that calibration events, performed at time instants $t_0^{cal}, \ldots, t_K^{cal}$, divide the dataset in $K+1$ data portions, each one containing data collected between two consecutive calibrations, plus the data portion after the last calibration. A (possibly imaginary) calibration is always assumed to be performed at the beginning of the trial, $t_0^{cal} = t_1^{cgm}$. Let us denote with $\mathbf{bg}_k$ the vector containing only the m_k BG data collected in the kth data portion $k = 1, \ldots, K+1$ at the time instants $t_1^k, \ldots, t_{m_k}^k$

$$\mathbf{bg}_k = \left[bg\left(t_1^k\right), \ldots, bg\left(t_{m_k}^k\right)\right]^T \tag{11.4}$$

Analogously, denote with $\mathbf{cgm}_k$ and $\mathbf{ig}_k$ the vectors containing only data related to the kth data portion. Define, moreover, $\Delta\mathbf{t}_k = \left[t_1^k - t_{k-1}^{cal}, \ldots, t_{m_k}^k - t_{k-1}^{cal}\right]^T$ the vector of BG measurements time-offsets with respect to the initial calibration of the current data portion.

Finally define $\mathbf{w}_k = \left[w\left(t_1^k\right), \ldots, w\left(t_{m_k}^k\right)\right]^T$ the noise vector at reference times and $\mathbb{V}\mathrm{ar}(\mathbf{w}_k) = \Sigma_{\mathbf{w}_k}$ its covariance matrix.

For ease of presentation, let us assume moreover that

$$\left\{t_1^{cal}, \ldots, t_K^{cal}\right\} \subseteq \{t_1, \ldots, t_m\} \subseteq \left\{t_1^{cgm}, \ldots, t_n^{cgm}\right\}. \tag{11.5}$$

However, violation of this hypothesis, for instance, due to CGM connectivity loss, can be easily handled in an implementation phase.

Algorithm description

The retrofitting algorithm is a two-step procedure. Fig. 11.2 illustrates the inputs and outputs of each step on the data of a representative subject. To visually assess the quality of the reconstruction, each panel depicts also the test BG references (empty diamonds), i.e., references collected during the trial to which the retrofitting algorithm had no access.

Since both references measurements and CGM time series are possibly affected by outliers or unreliable/wrong data, a data preprocessing step is needed to detect and isolate possibly erroneous data points either for automatic discharge or to request human intervention. The preprocessing data step is described in Appendix.

Step A: retrospective Bayesian CGM recalibration

Step A aims to compensate for systematic under/overestimation of CGM time series with respect to reference BG values due to calibration errors (uncertainty in calibration process and drift in sensor sensitivity). The parameters of the calibration model presented in section Problem formulation are retrospectively identified through a

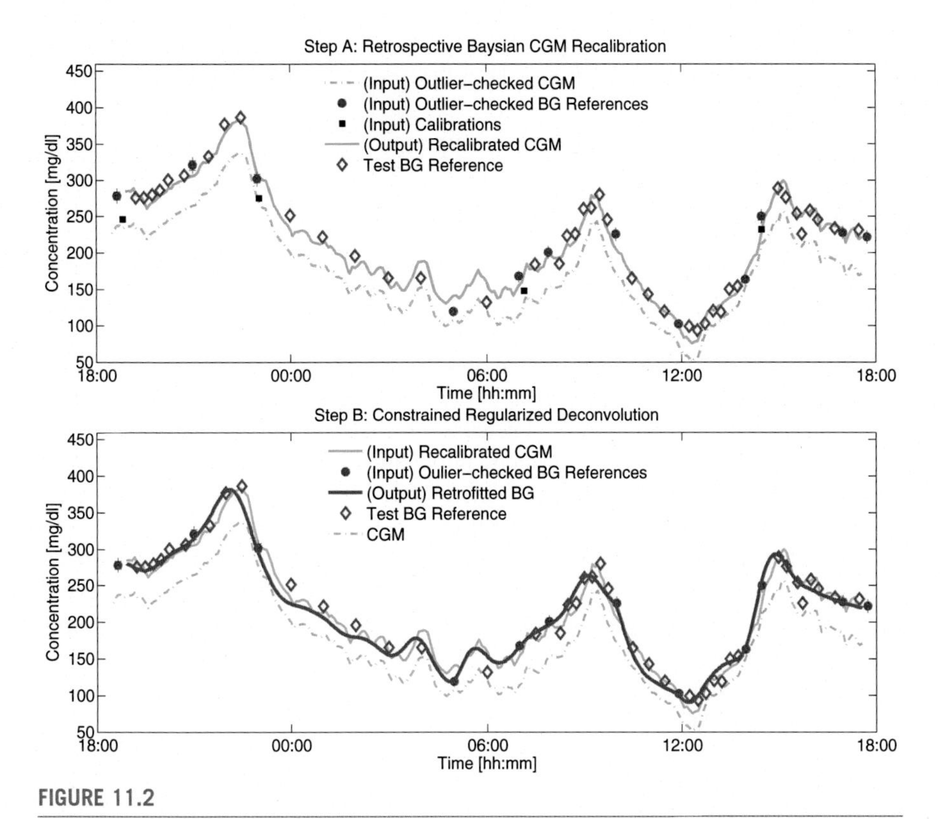

FIGURE 11.2

Example of the two-step method applied on a representative dataset. Top panel: retrospective Bayesian CGM recalibration (Step A) enhances accuracy of the CGM. Bottom panel: constrained regularized deconvolution (Step B) simultaneously compensates for the distortion introduced by the blood-to-interstitium glucose transport and reduces measurement noise exploiting the physiological prior on BG smoothness. To assess the quality of the reconstruction, both panels also show the test BG references (empty diamonds), to which the retrofitting algorithm had no access.

Bayesian estimation procedure. Subsequently, calibration errors are compensated on the basis of these estimates (recalibration). As inputs, Step A takes calibration data together with outliers-checked CGM and BGs data coming from the preprocessing step. As output, it returns the recalibrated CGM trace.

More precisely, for each data portion among two consecutive calibrations (plus the possible data portion after the last calibration), Step A performs the estimation of the vector

$$\theta = [\alpha, \beta, \gamma, \tau]^T \tag{11.6}$$

containing the unknown parameters of the calibration error and glucose diffusion models.

For each data portion $k = 1, \ldots, K + 1$, a Bayesian estimate of the calibration-error parameters is obtained solving the following optimization problem:

$$\left[\widehat{\alpha}_k, \widehat{\beta}_k, \widehat{\gamma}_k, \widehat{\tau}_k\right] = \widehat{\theta}_k = \underset{\theta}{\operatorname{argmin}} J(\theta) \tag{11.7}$$

where

$$\begin{aligned} J(\theta) = & (\mathbf{cgm}_k - \alpha \mathbf{ig}_k(\tau) - \beta - \gamma \Delta \mathbf{t}_k)^T \Sigma_{\mathbf{w}_k}^{-1} . (\mathbf{cgm}_k - \alpha \mathbf{ig}_k(\tau) - \beta - \gamma \Delta \mathbf{t}_k) \\ & + (\theta - \mu_\theta)^T \sum\nolimits_\theta^{-1} (\theta - \mu_\theta) \end{aligned} \tag{11.8}$$

where μ_θ represents the expected value of θ and Σ_θ its covariance matrix, discussed in the remark below. The first addend in (11.8) penalizes errors in the fitting of the observed data while the second one penalizes deviation from a population prior.

This Bayesian formulation, besides allowing the use of prior knowledge on calibration-error model, also prevents identifiability problems when few BG references are available between two calibrations.

When the optimization problem is solved, a number of tests can be performed to detect presence of abnormal dynamics in the CGM signal that do not appear also in the references. These tests include the final cost function value, estimated parameter value, and their coefficient-of-variation. An effective rule of thumb is to exclude data portion where the estimated $\widehat{\alpha}_k$ was smaller, that is, $\widehat{\alpha}_k < 0.3$.

Once the parameters of calibration-error model have been estimated for each data portion, calibration error can be compensated performing a data-portion-wise recalibration by inversion of (11.2), i.e., $\forall k = 1, \ldots, K + 1$ and $\forall t$ in the kth data portion ($t_{k-1}^{cal} \leq t < t_k^{cal}$):

$$cgm_{\text{recal}}(t) = \frac{1}{\widehat{\alpha}_k} \left(cgm(t) - \widehat{\beta} - \widehat{\gamma} \left(t - t_{k-1}^{cal} \right) \right). \tag{11.9}$$

The output of Step A, $cgm_{\text{recal}}(t)$, is a retrospectively calibrated CGM time series, which is more accurate than the original CGM profile (i.e., closer to the reference BG data), as illustrated in Fig. 11.2, upper panel.

Remark: Let us discuss the choice of Σ_θ and μ_θ. In Ref. [16], population distribution of the parameters α, β, γ, and τ has been estimated for the Dexcom SEVEN PLUS sensor (Dexcom Inc., San Diego, CA, USA). The joint second-order description obtained from these distribution provides Σ_θ and μ_θ. For other commercial sensors, it can be assumed that the sensor is on average calibrated ($\alpha = 1$, $\beta = \gamma = 0$) and τ_θ can be set to the population value, for instance, $\tau_\theta = 15$ [min] for Navigator (Abbott Diabetes Care, Alameda, CA) [17], i.e.,

$$\mu_\theta = \mathbb{E} \begin{bmatrix} \alpha \\ \beta \\ \gamma \\ \tau \end{bmatrix} = \begin{bmatrix} 1 \\ 0\ [\text{mg/dL}] \\ 0\ [\text{mg/dL/min}] \\ 15\ [\text{min}] \end{bmatrix}. \tag{11.10}$$

For what concerns Σ_θ, the choice

$$\Sigma_\theta = \gamma_\theta^2 \begin{bmatrix} 0.25 & 12.5 & 0 & 0 \\ 12.5 & 100 & 0 & 0 \\ 0 & 0 & 10 & 0 \\ 0 & 0 & 0 & 3 \end{bmatrix}^2 \tag{11.11}$$

has proven effective, where the constant γ_θ can be learned on a validation dataset or fixed to $\gamma_\theta = 0.1$. With regard to the additive, possibly nonwhite, random noise $w(t)$, at difference with [16] where a further decomposition and analysis of the spectral properties of $w(t)$ for the Dexcom SEVEN PLUS sensor has been proposed, here we simply capture the essential error intersamples correlation with an autoregressive (AR) process of order 1:

$$w(t+1) = aw(t) + e(t) \tag{11.12}$$

with $a = 0.87$ and $e(t)$ a white noise with $\mathbb{V}\text{ar}(e(t)) = 26.6$ [mg^2/dL^2], $\forall t$. The same statistical description can be used for other commercial sensors.

Step B: constrained regularized deconvolution

Step B simultaneously compensates for the distortion introduced by the blood-to-interstitium glucose transportation and reduces measurement noise affecting CGM. In fact, employing regularization [18], it prevents noise amplification due to ill-conditioning [19] and implements a noise filtering leveraging on the physiological prior-knowledge on the BG profile smoothness. Adding the constraints allows using the available accurate BG reference information.

Step B takes as input the recalibrated CGM (produced by Step A) and the outliers-checked BG reference data (produced by the preprocessing step) and returns as output a reconstructed quasi-continuous BG profile, retrofitted BG from now on.

To formulate the deconvolution problem, let us introduce $\mathbf{bg^r}(t)$, the $\mathbb{R}^M$ the vector of all M blood glucose concentrations to be reconstructed. $\mathbf{bg^r}(t)$ is the vectorial representation of the quasi-continuous reconstructed profile, uniformly sampled with an arbitrary small sampling time T_{BG}, from time t_{start} to time t_{end}:

$$\mathbf{bg^r} = [bg(t_{\text{start}}), bg(t_{\text{start}} + T_{BG}), \ldots, bg(t_{\text{end}})]. \tag{11.13}$$

For sake of clarity, it is assumed from now on that $T_{BG} = 1$ [min] and $t_{\text{start}} = 1$ [min], so that the BG profile to be estimated is a quasi-continuous signal and the blood glucose concentration at time t_i, $bg(t_i)$, is stored in the ith entry of the vector $\mathbf{bg^r}$.

The superscript "r" on the $\mathbb{R}^M$ vector $\mathbf{bg^r}$ differentiates the vector to be estimated from the $\mathbb{R}^m$ vector BG containing reference measurements available, $m << M$. In the ideal case of perfect reference measurement,

$$\mathbf{bg} = C \cdot \mathbf{bg^r}, \tag{11.14}$$

where C is the $m \times M$ selection matrix obtained removing from the $M \times M$ identity matrix all the rows except the $t_1, \ldots, t_m$-th ones, i.e., all the entries of C are zeros, except for the ones in position (i, t_i), equal to 1. Although BG references are highly accurate there is always a confidence interval around the measurement where the true concentration can lay, so that the above Eq. (11.13) becomes

$$\mathbf{l}_{ci}(\mathbf{bg}) \leq C \cdot \mathbf{bg}^{\mathbf{r}} \leq \mathbf{u}_{ci}(\mathbf{bg}) \tag{11.15}$$

where $\mathbf{l}_{ci}(\mathbf{bg}) \cdot (\mathbf{u}_{ci}(\mathbf{bg}))$ is the $\mathbb{R}^m$ vectors of lower (upper) limits delimiting measurements Confidence Intervals and the inequality is meant element-wise. As an example, YSI is guaranteed to introduce an error of at most 2%, therefore (11.14) reduces to

$$0.98 \cdot \mathbf{bg} \leq C \cdot \mathbf{bg}^{\mathbf{r}} \leq 1.02 \cdot \mathbf{bg} \tag{11.16}$$

Note that (11.14) easily allows addressing of different accuracies of different reference measurement instruments such as YSI, Hemocue, or traditional capillary finger prick measurements (SMBG). Moreover, it allows taking into account different accuracies of the same instruments in different measurement regions. The constraint expressed in Eq. (11.14) will be explicitly taken into account in the estimation of the blood glucose concentration with a constrained regularized deconvolution in the form of a constrained Tikhonov regularization problem.

To perform the deconvolution in the classical linear time-invariant system framework, τ in (11.1) is fixed to the average of the τ_i previously estimated for each data portion:

$$\tau = \frac{1}{K+1} \sum_{k=1}^{K+1} \widehat{\tau}_k \tag{11.17}$$

Then it holds

$$\mathbf{cgm}_{\text{recal}}(t) = G \cdot \mathbf{bg}^{\mathbf{r}} + \mathbf{w}_{\text{recal}}, \tag{11.18}$$

where G is the $n \times M$ rectangular submatrix of the $M \times M$ transfer matrix $G_{M \times M}(\tau)$, obtained retaining only the $t_1^{\text{cgm}}, \ldots, t_n^{\text{cgm}}$th columns of $G_{M \times M}(\tau)$. The noise $\mathbf{w}_{\text{recal}}$ description can be obtained by noting that, from (11.8), it follows $w_{\text{recal}}(t) = \frac{1}{\hat{\alpha}_k} w(t)$, so that the covariance matrix of $\mathbf{w}_{\text{recal}}$ can be well approximated with a block diagonal matrix,

$$\Sigma = \text{BlockDiag}\left(\frac{1}{\widehat{\alpha}_1^2} \Sigma_w, \ldots, \frac{1}{\widehat{\alpha}_{K+1}^2} \Sigma_w \right). \tag{11.19}$$

To formulate the constrained Tikhonov regularization problem, let us introduce the Toeplitz matrix F:

$$F = \begin{bmatrix} 1 & & & 0 \\ -1 & 1 & & \\ & \ddots & \ddots & \\ 0 & & -1 & 1 \end{bmatrix} \tag{11.20}$$

Finally, the constrained Tikhonov regularization problem has the form:

$$\widehat{\mathbf{bg}}^r = \underset{\mathbf{l}_{ci}(\mathbf{bg}) \le C \cdot \mathbf{bg}^r \le \mathbf{u}_{ci}(\mathbf{bg})}{\operatorname{argmin}} J_r(\mathbf{bg}) \tag{11.21}$$

where

$$J_r(\mathbf{bg^r}) = (\mathbf{cgm}_{\text{recal}} - G\mathbf{bg^r})^T \Sigma^{-1} (\mathbf{cgm}_{\text{recal}} - G\mathbf{bg}^r) + \gamma_{\text{reg}} \% (\mathbf{bg^r})^T \left(F^T F\right)^d \mathbf{bg^r} \tag{11.22}$$

First addend in (11.22) penalizes inadequate data description, while the second addend takes into account the physiological prior knowledge of blood glucose smoothness, where d is fixed, $d = 2$. γ_{reg} is the regularization parameter that trades off data fit and smoothness of the resulting profile.

The current implementation employed a fixed γ_{reg}, manually tuned to achieve satisfactory performances on a validation dataset. The implementation of an algorithm that allows learning of γ_{reg} from the data is deferred to future works.

Retrofitting outpatient study data

In this section, we show that the retrofitting algorithm can be used to enhance precision and accuracy in CGM data collected during outpatient clinical studies such as [11–14], i.e., in a setup offering a relatively large number of highly accurate references to retrospectively enhance the CGM.

An in-depth discussion of this setup can be found in Ref. [6].

We start from a dataset, called "original dataset," collected in an inpatient study and offering frequent reference BG measurements. References BG are then divided into training-set references, available to the retrofitting algorithm, and test-set references. The training set will be called "outpatient-like dataset."

Original dataset

The data used in this section were collected during a large multicenter inpatient clinical trial[20], conducted within the EU-funded project AP@home [21]. The trial aimed to compare two different closed-loop algorithms against the standard open-loop therapy and involved 47 patients in six European centers. Each patient underwent three admissions, lasting about 24 h and employing three different therapies, i.e., open-loop (OL) and two different closed-loop algorithms (CL). Frequent BGs were collected throughout the admission, every hour during the night and at least every 30 min during the day, resulting in the availability of ~55 BG references/day. BG references were measured with YSI2300 STAT Plus analyzer (YSI, Lynchford House, Franborough, United Kingdom) and the CGM sensor was the Dexcom SEVEN PLUS CGM sensor (Dexcom Inc., San Diego, CA, USA). More details on the trial can be found in Ref. [20].

Outpatient-like dataset

In an outpatient setting, BG references are collected much less frequently than inpatient studies. Hence, in order to emulate the few outpatient reference data here we discarded about 80% of the references in the original dataset and retained only those that would have been collected in our outpatient protocols [11–14]. Specifically, we retained:

- BG references collected at calibration times;
- BG references prescribed prior to the admission and the discharge of the patient;
- BG check in response to a CGM hypo-alarm;
- BG references collected at meal time, to compute meal-bolus;
- BG references performed about 2 h after the meal.

For each admission, in the outpatient-like dataset, we retained $12(11-13)$ samples (reported as median and (25th –75th percentile) range), which correspond to $21.4\%(20.0\%-23.6\%)$ of the samples collected during the trial.

An example of the original and the outpatient-like datasets is illustrated in Fig. 11.2. The continuous blue trace represents the CGM readings, and red dots denote the reference BG values that would have been collected in an outpatient study. Gray diamonds are the YSI measurements of the original dataset that are not retained in the outpatient-like dataset.

Accuracy outcomes metrics and statistical analysis

Original CGM and retrofitted CGM values are matched with the gold standard measurement performed at same time, i.e., the test-set YSI measurements.

The main metrics used to assess the accuracy of the signals are the absolute difference (AD) and the relative absolute difference (ARD). These metrics are computed for each data pair. Then, the overall mean (standard deviation) is reported for normally distributed metrics, and population median (25th–75th) percentile is reported for nonnormally distributed metrics. Normality is assessed by Lilliefors test.

To evaluate interpatient variability of the accuracy, we computed for each patient mean AD (MAD) and mean ARD (MARD). Then, the population mean (standard deviation) is reported for normally distributed metrics, and population median (25th–75th) percentile is reported for nonnormally distributed metrics.

Finally, clinical accuracy is assessed with the Clark Error Grid analysis (CEG) [22] by reporting the percentage of data points falling in zone A and with the percentage of points falling in zone Ar of the rate grid (accurate glucose rate) of the rate error grid of the continuous glucose error grid analysis proposed in Ref. [23].

Statistical analysis

Original and retrofitted CGM performances are compared with a paired t-test for normally distributed metrics and with a Wilcoxon signed-rank test for nonnormally distributed data.

Results

Fig. 11.2, bottom panel, shows the outcome of the retrofitting technique in a representative subject (Patient 1, admission 2). CGM trace is depicted in light blue. Full red dots denote BG measurement assigned to the training set and therefore accessible to the retrofitting method. The output of the retrofitting method (tick red line) has to be compared with the test BG references (empty diamonds). The accuracy improvement with respect to CGM is clear also at visual inspection.

Fig. 11.3, top panel, reports the boxplot of the absolute and relative absolute errors on all the test-set references, regardless of the patient and the admission during which they have been collected. Each gray dot represents a measurement. The mean absolute relative deviation (MARD) of the retrofitting method is 7.8%, and 75% of the relative absolute error is below 10.3% as compared with the use of raw CGM, which has an MARD equal to 15.0%, 75th percentile equal to 18.9%. Of note, the 50th percentile of CGM errors (12.25%) is larger than the 75th percentile of retrofitting errors. Similar considerations hold for absolute deviation. Fig. 11.3, bottom panel, reports the distribution boxplot of the mean absolute deviation (MAD) and MARD of CGM and retrofitted CGM in each patient

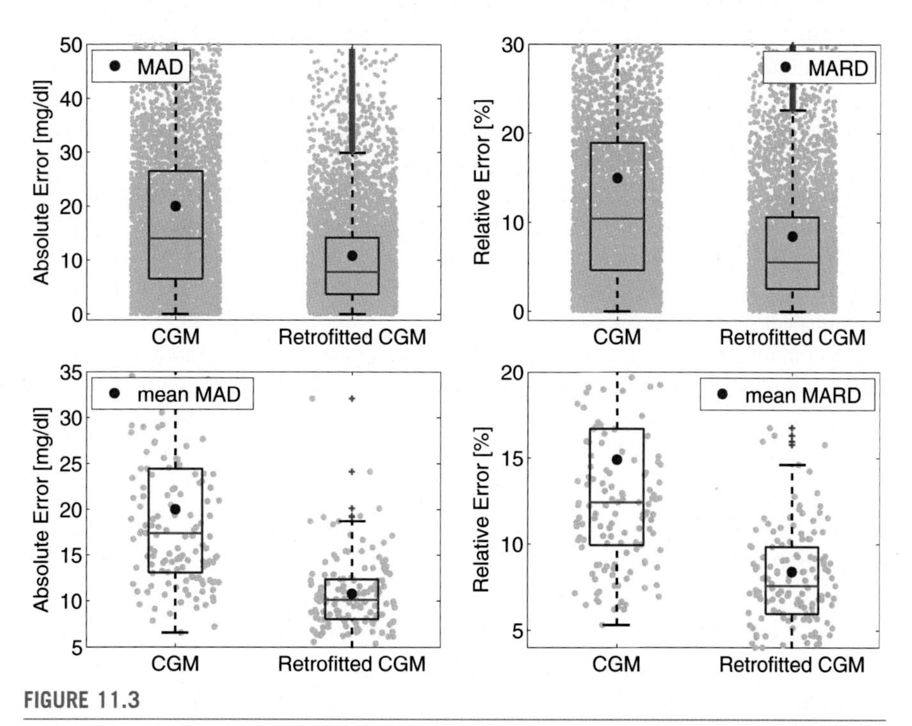

FIGURE 11.3

Evaluation of the retrofitting method on test-set data. Top panel: boxplot of the absolute and relative absolute error (each *gray dot* represents a reference measurement). Bottom panel: boxplot of MAD and MARD population distribution (each *gray dot* represents a patient admission).

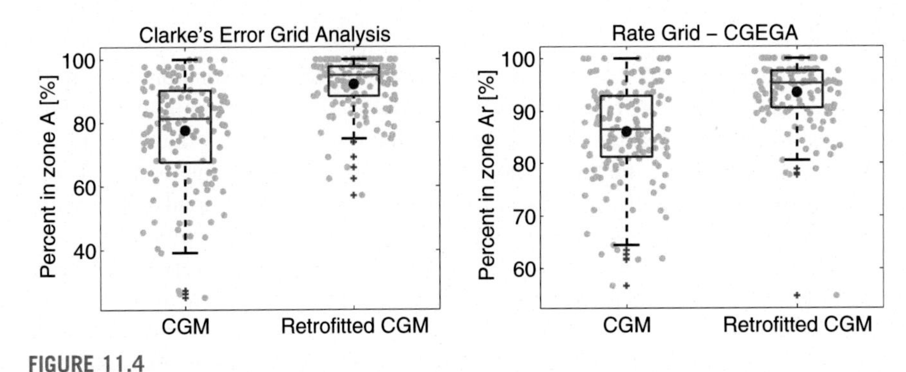

FIGURE 11.4

Evaluation of the retrofitting method on test-set data. Boxplot of percent of points in zone A (Clarke's error grid, left) and zone Ar (rate grid, right). Each *gray dot* represents one patient admission.

in each admission (each gray dot represents a patient admission). Results confirm the improvement provided by the retrofitting method.

Fig. 11.4 reports the percentage of points falling in zone A of Clarke's error grid (accurate measurements) [22]. The percentage was computed for each patient admission and depicted in a boxplot (each gray dot represents a patient admission). Improvement is significant, with the retrofitting achieving more than 90% points in zone A in more than 75% of the patient admissions, while the same percentage in zone A is achieved in less than 25% of the admissions by CGM. Analogously, right panel of Fig. 11.4 shows the percentage of points falling in zone Ar of the rate grid (accurate glucose rate) of the rate error grid of the continuous glucose error grid analysis [23]. Also in this case the percentage was evaluated for each patient admission and depicted in a boxplot. Rate analysis confirms superiority of retrofitted traces with respect to the unprocessed ones.

In conclusion, we showed that retrofitting enhances precision and accuracy of a CGM collected in outpatient-like setups. Collecting CGM data and then retrofitting them is a viable alternative for reducing the frequency of blood glucose sampling without losing temporal resolution.

Retrofitting real-life adjunctive data

In this section, we show that the retrofitting method is effective in improving the accuracy of Dexcom sensor (Dexcom G5) when used in real life as adjunctive treatment to SMBG. This newer sensor reached the 1-digit precision, and it is currently one of the most accurate CGM on the market. The scenario considered in this section is substantially more challenging for the retrofitting algorithm with respect to the one considered in the previous section, since it offers less (~five SMBGs per diurnal session) and less accurate references (SMBG rather than YSI). An in-depth analysis of this setup can be found in Ref. [24].

As for the previous section, we start from datasets offering frequent SMBG and YSI data, called "original dataset." All YSI references and most SMBGs are then discarded to emulate real-life data availability, obtaining what we will denote as "real-life-like dataset." Discarded YSI are used for testing.

Original datasets

The retrofitting algorithm was tested on the data collected in 51 adult subjects, reported in Ref. [25], and in 46 adolescents, 13−17 years old, presented in Ref. [26]. Both group of subjects wore the Dexcom G5 for 7 days and had 12 h in-clinic session on either day 1, 4, or 7. During the admission, accurate BG references were collected with YSI instrument (Yellow Springs, OH) about every 15 min on arterialized venous samples and capillary SMBG about every 30 min using the Bayer USB Contour Next meter.

Remark: We should point out that the data in both Refs. [25,26] were actually collected using the Dexcom G4 equipped with the software 505 (also known as G4AP) and not with the Dexcom G5. Nevertheless, the two models can be considered, for the purpose of this chapter, completely equivalent. Sensing and signal processing technologies are instead identical, and the only difference in the two products is in the data transmission hardware: G4 with software 505 requires an ad-hoc receiver, while G5 allows direct data transmission and processing on the patient's smartphone.

Real-life-like datasets

To mimic a real-life outpatient setting, only $N_{\mathrm{SMBG}} = 5$ references per 12-h sessions are retained. In fact, five SMBG are likely to be collected during the daytime when the CGM sensor is used adjunctively to SMBG: two SMBG measurements for calibration (one at the beginning and one at the end of the session), plus three SMBG checks related to the meals. The five retained SMBG are selected by uniform SMBG sampling, i.e., retaining one SMGB measurement every X available ones, with X suitably chosen to get the desired total number, N_{SMBG}. YSI references are used solely as gold standard to assess the accuracy of CGM and retrofitted CGM.

Accuracy outcomes metrics and statistical analysis

Original CGM and retrofitted CGM values are matched with the YSI measurements performed at same time. Outcome metrics and statistical analysis are analogous to those of the previous section.

Results

Fig. 11.5 reports the boxplots of AD and ARD distribution. By retrofitting, the mean of AD is significantly reduced in both populations: from 16.2 mg/dL to 10.7 mg/dL ($P < .001$) in adults (about 34%) and from 18.1 mg/dL to 11.9 mg/dL ($P < .001$) in

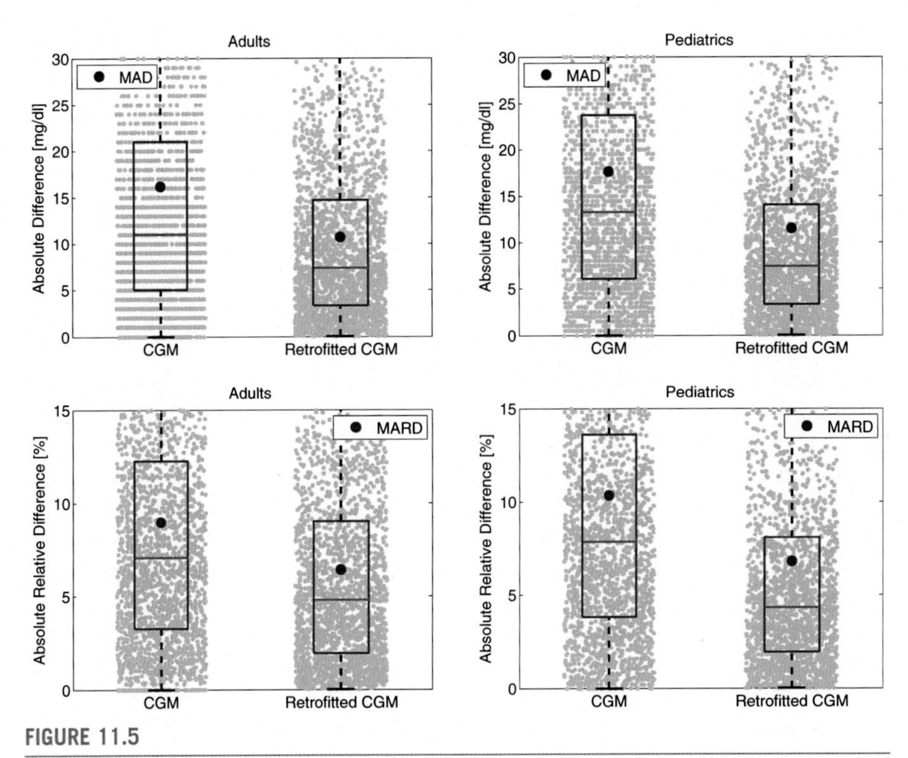

FIGURE 11.5

Distributions of absolute deviation (top panel) and absolute relative deviation (bottom panel) of the original CGM and of the retrofitted CGM obtained with five SMBG per 12-h session. Each *gray dot* represents a reference measurement. The improvement granted by the method is clearly visible.

the adolescents (about 34%). Similarly, the mean of ARD was significantly reduced of about 29% and 34% in the two populations: from 9.0% to 6.4% ($P < .001$) in adults and from 10.7% to 7.1% ($P < .001$) in adolescents.

Fig. 11.6 reports the patient-level analysis. All metrics are computed for each patient and, then, the distribution among 51 adults and 46 adolescents is considered. Overall, MAD and MARD are significantly reduced in both populations by retrofitting. In adults, mean MAD is reduced by 34%, from 16.5 mg/dL to 10.8 mg/dL ($P < .001$), and mean MARD by 28%, from 9% to 6.7 ($P < .001$). Similarly, in adolescents, mean MAD is reduced by 37%, from 19.1 mg/dL to 12.1 mg/dL, and mean MARD by 36%, from 11.3% to 7.2%.

In conclusion, despite the fact that Dexcom G5 is one of the most accurate sensors on the market, in this section we showed that when five SMBG references are available in a 12-h period the retrofitting method can be profitably used offline to push further the accuracy of Dexcom G5 data by retrospective processing. The method allowed decreasing MAD and MARD of roughly 30%, bringing, e.g., the average MARD value from the actual 9.0%–6.4%. Note that this number of

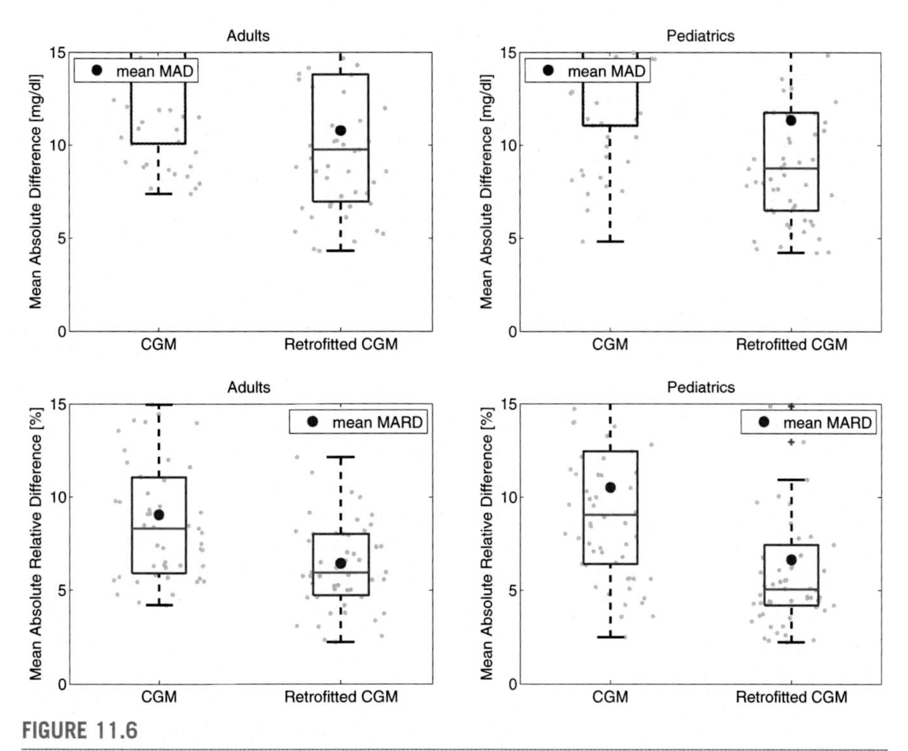

FIGURE 11.6

Boxplot of MAD (top panel) and MARD (bottom panel) population distribution of the original CGM and of the retrofitted CGM obtained with five SMBG per 12-h session (each *gray dot* represents a patient admission). The improvement granted by the method is clearly visible.

references was chosen since it is likely that at least five references in 12 diurnal hours are collected when the CGM sensor is used adjunctively to SMBG, e.g., two SMBG measurements for calibration plus three SMBG checks around meal time per day. Indeed, five [4,7] SMBGs were collected from 07:00 to 19:00 by the patients in our artificial pancreas studies conducted in free-living conditions for 2 months [27] and 1 month [28]. More precisely, in 59.1% of the 12-h periods going from 07:00 to 19:00 the patients collected five or more SMBGs, and only in 23.8% of these 12-h periods less than four SMBG measurements were performed.

Accuracy of retrofitted CGM versus number of references available

In this section, we investigate how the accuracy improvement achieved by retrofitting Dexcom G5 in real-life conditions is affected by the number of SMBG measurements available. In fact, our method leverages on these BG measurements to improve CGM, and as a consequence, an accuracy degradation should be expected as less of them become available.

For this analysis, we employ the same original datasets presented in the previous section, but, at difference with the previous section, we let the number of SMBGs provided to the method vary from $N_{SMBG} = 10$ to $N_{SMBG} = 2$ per 12-h session, spanning both the cases when more or less than NSMBG = 5 are available. For the case $N_{SMBG} = 2$ per 12-h session, the two calibration SMBG measurements were used. More than five SMBGs are expected to be collected by a compliant patient, verifying the CGM reading with an SMBG before taking any therapeutic decision, i.e., at each meal, 2 h after each meal and in case of hypo or hypoglycemia requiring interventions; less than five SMBGs are likely collected by a noncompliant patient or in a 12-h portion including the night. Furthermore, only two SMBGs per day (calibrations) are available in case of nonadjunctive use of CGM, a possible future scenario given by the recent US Food and Drug Administration panel meeting where the panel expressed a positive opinion about the change of the label of Dexcom G5 sensor from adjunctive to nonadjunctive [3].

Fig. 11.7 shows the how the retrofitted CGM accuracy changes when the number of SMBG varies. Upper panels report the mean absolute deviation (mean AD) for the two populations, while the lower panels reports the mean ARD. When many SMBGs are available (8–10 in a 12-h session), the retrofitted trace is nearly as accurate as the SMBG: for $N_{SMBG} = 10$ references per 12-h session, MAD is reduced by about 38% in adults and by 45% in adolescents and similarly MARD is reduced by 30% and 45%, respectively. The benefit of retrofitting decreases gradually as the number of SMBG decreases: for $N_{SMBG} = 5$ references per 12-h session, the improvements in both metrics are around 30%, as previously discussed; when only $N_{SMBG} = 2$ references per 12-h session are available the reduction is lower but still statistically significant: MAD decreases from 16.2 mg/dL to 13.6 mg/dL ($P < .001$) in adults (about 16%) and from 18.1 mg/dL to 15.7 mg/dL ($P < .001$) in adolescents (about 13%); similarly, MARD reduction is from 9.0% to 8.2% ($P < .001$) in adults (about 9%) and from 10.7% to 9.5% ($P < .001$) in adolescents (about 11%).

Conclusions

The retrofitting algorithm, combining a few BG reference samples with frequent, possibly noisy, and biased CGM data, provides an accurate and quasi-continuous BG profile. In this chapter, we provided details on how to implement the algorithm, and we showed several practical applications, e.g., to retrospectively improve CGM data collected in outpatient clinical trial, but also CGM data collected in free-living conditions.

Appendix: data preprocessing

Both reference measurements and CGM time series are possibly affected by outliers or unreliable/wrong data. With regard to BG reference outliers, they are mostly due

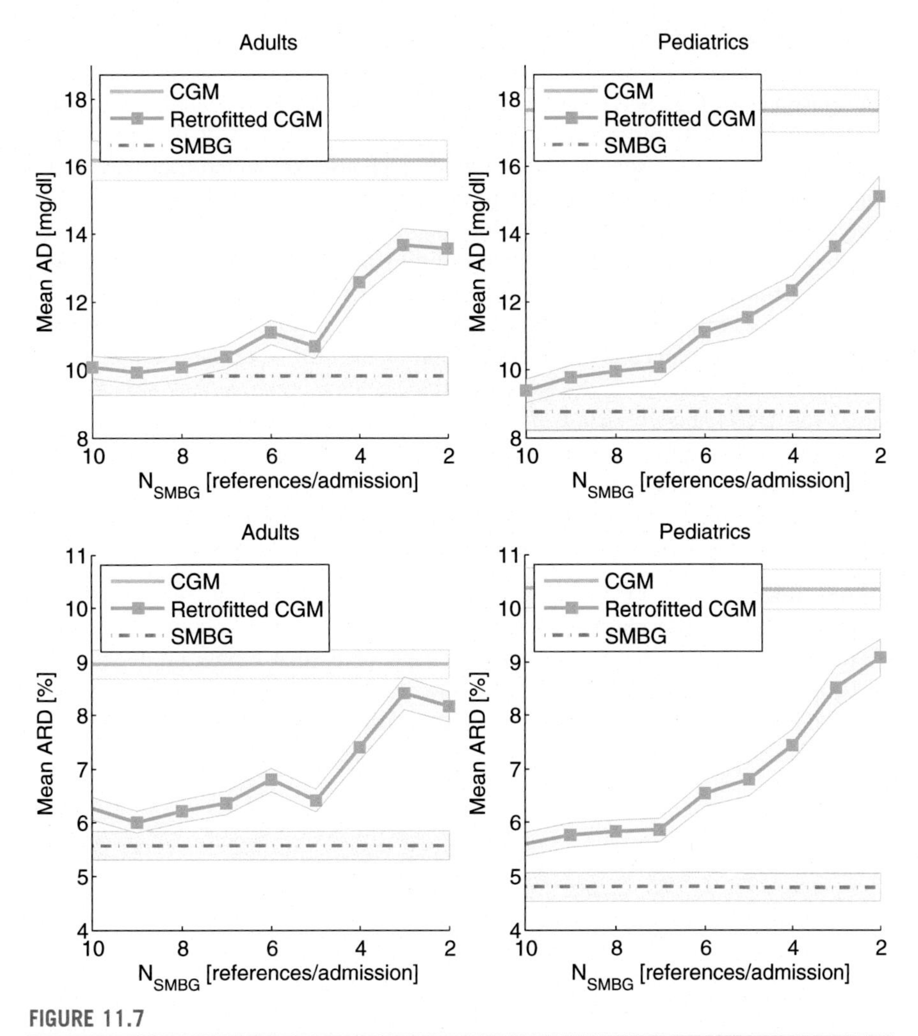

FIGURE 11.7

Accuracy of retrofitted CGM as a function of the number of SMBGs provided to the method. The accuracy of the original CGM and of SMBG is reported as a comparison. Accuracy is assessed by the mean absolute deviation (mean AD) in the upper panels and by the mean absolute relative deviation (mean ARD) in the lower panels. *Shaded areas* represent the 5%–95% confidence intervals on the means.

to human errors, such as erroneous data entry or incorrect blood sampling/processing (an example is reported in Fig. A.1, at 23:30). Moreover, it is not rare that a reference measurement is repeated a second time for double check, shortly after a first measurement, if the latter is considered "abnormal" by the study personnel. This happened in the example of Fig. A.1 at about 06:00. Detection of BG outliers is particularly important for the robustness of the method, as the next steps assume the reference BG as highly accurate.

With regard to CGM data, seldom they suffer isolate readings affected by a significantly larger error with respect to usual one, commonly referred as spikes and illustrated in Fig. A.1, at about 15:00. Moreover, as documented in Ref. [29], CGM may suffer failures related to the biomechanics of the sensor-tissue interface: pressure on the sensor site and foreign body response can create transient effects which generate unphysiological changes of the CGM output, called loss of sensitivity (Fig. A.1, at about 10:30, for 20 min).

The data preprocessing step of the retrofitting method aims to detect and isolate possibly erroneous data points. For most of these points, automatic discharge/correction can be performed, but sometimes human visual inspection and intervention is required. In the second case, the operator running the off-line analysis can manually correct, confirm, or exclude the isolated data. The data preprocessing step takes as inputs row CGM and BG data and returns as output outliers-checked data.

In this work, we limit ourselves to simple fault detection heuristic, that checks data consistency with physiological constraints: for instance, a rate of change larger than 5 mg/(dL·min) or any values outside the range [40–400] mg/dL are considered suspicious and isolated. The physiology-inspired rules are reported in detail below:

- Any reference BG or CGM value in outside the range [40–400] mg/dL is prompted for confirmation to the operator running the retrospective analysis.
- If the Dexcom SEVEN PLUS sensor is used, given that it only produces CGM values in the range [40–400] mg/dL, any eventual value outside this range has to be interpreted as an internal error code and therefore automatically excluded.
- A conservative bound on BG rate of change, r_{BG}, is checked:

$$|r_{BG}(t_{i+1})| = \left|\frac{BG(t_{i+1}) - BG(t_i)}{t_{i+1} - t_i}\right| \geq 5\left[\frac{\text{mg}}{\text{dL}\cdot\text{min}}\right] \quad (11.23)$$

- A similar control is performed on the CGM rate of change, r_{CGM}, but the threshold is loosened not to exclude poorly calibrated data portions:

$$|r_{CGM}(t_{i+1})| = \left|\frac{CGM(t_{i+1}) - CGM(t_i)}{t_{i+1} - t_i}\right| \geq 10\left[\frac{\text{mg}}{\text{dL}\cdot\text{min}}\right] \quad (11.24)$$

- Possibly unreliable reference data (see Fig. 11.2, top panel) are abnormally low (high) with respect to previous and future references. This situation is detected checking the following condition:

$$|r_{BG}(t_{i+1})| \geq 1\left[\frac{\text{mg}}{\text{dL}\cdot\text{min}}\right] \text{ and } |r_{BG}(t_{i+2})| \geq 1\left[\frac{\text{mg}}{\text{dL}\cdot\text{min}}\right] \quad (11.25)$$

- but

$$\text{sign}(r_{BG}(t_{i+1})) \neq \text{sign}(r_{BG}(t_{i+2})) \quad (11.26)$$

- The same heuristics is used to detect CGM spikes.
- If the time interval between two BG references is too close (e.g., absolute difference below 5 min), it is likely that the first reference measurement have been considered "abnormal" by the study personnel and repeated a second time. Therefore the algorithm asks the operator which sample should be trusted.

Data preprocessing could be further improved exploiting more refined fault detection techniques, such as Ref. [30], that rely on CGM and BG reference signal only or exploiting also other signals collected during the trial such as insulin infusion [31] and carbohydrates ingestion [32] or physical activity.

Acknowledgments

This work was supported by the EU project ICT FP7-247138 "Bringing the Artificial Pancreas at Home (AP@home)" (Funding Agency: European Union's Research and Innovation funding programme, FP7 initiative: FP7-ICT-2007-2) and by the Italian SIR project RBSI14JYM2 "Learning Patient-Specific Models for an Adaptive, Fault-Tolerant Artificial Pancreas (Learn4AP)" (Funding agency: MIUR, Italian Ministry of Education, Universities and Research; initiative: SIR—Scientific Independence of young Researchers).

References

[1] Rodbard D. Continuous glucose monitoring: a review of successes, challenges, and opportunities. Diabetes Technology and Therapeutics 2016;18(Suppl. 2):23–213.

[2] Weinzimer S, Miller K, Beck R, Xing D, Fiallo-Scharer R, Gilliam LK, Kollman C, Laffel L, Mauras N, Ruedy K, Tamborlane W, Tsalikian E. Effectiveness of continuous glucose monitoring in a clinical care environment: evidence from the Juvenile Diabetes Research Foundation continuous glucose monitoring (JDRF-CGM) trial. Diabetes Care 2010;33(1):17–22.

[3] Available from: http://www.fda.gov/AdvisoryCommittees/CommitteesMeetingMaterials/MedicalDevices/MedicalDevicesAdvisoryCommittee/ClinicalChemistryandClinicalToxicologyDevicesPanel/ucm511565.htm.

[4] Scheiner G. CGM retrospective data analysis. Diabetes Technology and Therapeutics 2016;18(Suppl. 2):S214–22.

[5] Beck RW, Calhoun P, Kollman C. Use of continuous glucose monitoring as an outcome measure in clinical trials. Diabetes Technology and Therapeutics 2012;14(10):877–82.

[6] Del Favero S, Facchinetti A, Sparacino G, Cobelli C. Retrofitting of continuous glucose monitoring traces allows more accurate assessment of glucose control in outpatient studies. Diabetes Technology and Therapeutics 2015;17(5):355–63.

[7] Schiavon M, Dalla Man C, Kudva YC, Basu A, Cobelli C. Quantitative estimation of insulin sensitivity in type 1 diabetic subjects wearing a sensor-augmented insulin pump. Diabetes Care 2014;37(5):1216–23.

[8] Georga E, Protopappas V, Fotiadis D. "Glucose prediction in type 1 and type 2 diabetic patients using data driven techniques," Knowledge-Oriented Applications in Data Mining. 2011. p. 277–96. Cited by 3.

[9] Toffanin C, Del Favero S, Aiello E, Messori M, Cobelli C, Magni L. Glucose-insulin model identified in free-living conditions for hypoglycaemia prevention. Journal of Process Control 2018;64:27–36.

[10] Del Favero S, Facchinetti A, Sparacino G, Cobelli C. Improving accuracy and precision of glucose sensor profiles: retrospective fitting by constrained deconvolution. IEEE Transactions on Biomedical Engineering 2014;61(4):1044–53.

[11] Cobelli C, Renard E, Kovatchev B, Keith-Hynes P, Ben Brahim N, Place J, Del Favero S, Breton M, Farret A, Bruttomesso D, Dassau E, Zisser H, Doyle III F, Patek S, Avogaro A. Pilot studies of wearable outpatient artificial pancreas in type 1 diabetes. Diabetes Care 2012;35(9):e65–7.

[12] Kovatchev B, Renard E, Cobelli C, Zisser H, Keith-Hynes P, Anderson S, Brown S, Chernavvsky D, Breton M, Farret A, Pelletier M, Place J, Bruttomesso D, Del Favero S, Visentin R, Filippi A, Scotton R, Avogaro A, Doyle III F. Feasibility of outpatient fully integrated closed-loop control: first studies of wearable artificial pancreas. Diabetes Care 2013;36:1851–8.

[13] Del Favero S, Bruttomesso D, Di Palma F, Lanzola G, Visentin R, Filippi A, Scotton R, Toffanin C, Messori M, Scarpellini S, Keith-Hynes P, Kovatchev BP, Devries JH, Renard E, Magni L, Avogaro A, Cobelli C. First use of model predictive control in outpatient wearable artificial pancreas. Diabetes Care 2014;37(5):1212–5.

[14] Del Favero S, Place J, Kropff J, Messori M, Keith-Hynes P, Visentin R, Monaro M, Galasso S, Boscari F, Toffanin C, Di Palma F, Lanzola G, Scarpellini S, Farret A, Kovatchev B, Avogaro A, Bruttomesso D, Magni L, DeVries JH, Cobelli C, Renard E. Multicenter outpatient dinner/overnight reduction of hypoglycemia and increased time of glucose in target with a wearable artificial pancreas using modular model predictive control in adults with type 1 diabetes. Diabetes, Obesity and Metabolism 2015;17(5):468–76.

[15] Rebrin K, Steil GM, van Antwerp WP, Mastrototaro JJ. Subcutaneous glucose predicts plasma glucose independent of insulin: implications for continuous monitoring. American Journal of Physiology 1999;277(3 Pt 1):E561–71.

[16] Facchinetti A, Del Favero S, Sparacino G, Castle JR, Ward WK, Cobelli C. Modeling the glucose sensor error. IEEE Transactions on Biomedical Engineering 2014;61(3): 620–9.

[17] Garg SK, Voelmle M, Gottlieb PA. Time lag characterization of two continuous glucose monitoring systems. Diabetes Research and Clinical Practice 2010;87(3):348–53.

[18] Nicolao GD, Sparacino G, Cobelli C. Nonparametric input estimation in physiological systems: problems, methods, and case studies. Automatica 1997;33(5):851–70. Available from: http://www.sciencedirect.com/science/article/pii/S0005109896002543.

[19] Tikhonov A, Arsenin V. Solutions of ill-posed problems. Washington: Winston/Wiley; 1977.

[20] Luijf YM, DeVries JH, Zwinderman K, Leelarathna L, Nodale M, Caldwell K, Kumareswaran K, Elleri D, Allen JM, Wilinska ME, Evans ML, Hovorka R, Doll W, Ellmerer M, Mader JK, Renard E, Place J, Farret A, Cobelli C, Del Favero S, Dalla Man C, Avogaro A, Bruttomesso D, Filippi A, Scotton R, Magni L, Lanzola G, Di Palma F, Soru P, Toffanin C, De Nicolao G, Arnolds S, Benesch C, Heinemann L. Day and night closed-loop control in adults with type 1 diabetes: a comparison of two closed-loop algorithms driving continuous subcutaneous insulin infusion versus patient self-management. Diabetes Care 2013;36(12):3882–7.

[21] Ap@home eu-fp7 project. website [Accessed 28 April 2013]. Available from: http://www.apathome.eu/.

[22] Clarke WL, Cox D, Gonder-Frederick LA, Carter W, Pohl SL. Evaluating clinical accuracy of systems for self-monitoring of blood glucose. Diabetes Care 1987;10: 622–8.

[23] Kovatchev BP, Gonder-Frederick LA, Cox DJ, Clarke WL. Evaluating the accuracy of continuous glucose-monitoring sensors: continuous glucose-error grid analysis illustrated by TheraSense Freestyle Navigator data. Diabetes Care 2004;27(8):1922–8.

[24] Favero SD, Facchinetti A, Sparacino G, Cobelli C. Retrofitting real-life dexcom g5 data. Diabetes Technology and Therapeutics 2017;19(4):237–45.

[25] Bailey TS, Chang A, Christiansen M. Clinical accuracy of a continuous glucose monitoring system with an advanced algorithm. Journal of Diabetes Science and Technology 2015;9(2):209–14.

[26] Laffel L. Improved accuracy of continuous glucose monitoring systems in pediatric patients with diabetes mellitus: results from two studies. Diabetes Technology and Therapeutics 2016;18(Suppl. 2):S223–33.

[27] Kropff J, Del Favero S, Place J, Toffanin C, Visentin R, Monaro M, Messori M, Di Palma F, Lanzola G, Farret A, Boscari F, Galasso S, Magni P, Avogaro A, Keith-Hynes P, Kovatchev BP, Bruttomesso D, Cobelli C, DeVries JH, Renard E, Magni L. 2 month evening and night closed-loop glucose control in patients with type 1 diabetes under free-living conditions: a randomised crossover trial. Lancet Diabetes and Endocrinology 2015;3(12):939–47.

[28] Renard E, Farret A, Kropff J, Bruttomesso D, Messori M, Place J, Visentin R, Calore R, Toffanin C, Di Palma F, Lanzola G, Magni P, Boscari F, Galasso S, Avogaro A, Keith-Hynes P, Kovatchev B, Del Favero S, Cobelli C, Magni L, DeVries JH. Day-and-Night closed-loop glucose control in patients with type 1 diabetes under free-living conditions: results of a single-arm 1-month experience compared with a previously reported feasibility study of evening and night at home. Diabetes Care 2016;39(7):1151–60.

[29] Helton KL, Ratner BD, Wisniewski NA. Biomechanics of the sensor-tissue interface-effects of motion, pressure, and design on sensor performance and foreign body response-part ii: examples and application. Journal of Diabetes Science and Technology 2011;5(3):647656.

[30] Bequette B. Continuous glucose monitoring: real-time algorithms for calibration, filtering, and alarms. Journal of Diabetes Science and Technology 2010;4(2):404–18.

[31] Facchinetti A, Del Favero S, Sparacino G, Cobelli C. An online failure detection method of the glucose sensor-insulin pump system: improved overnight safety of type-1 diabetic subjects. IEEE Transactions on Biomedical Engineering 2013;60(2): 406–16.

[32] Herrero P, Calm R, Veh J, Armengol J, Georgiou P, Oliver N, Tomazou C. Robust fault detection system for insulin pump therapy using continuous glucose monitoring. Journal of Diabetes Science and Technology 2012;6(5):1131–41.

CHAPTER

Modeling the CGM measurement error

12

Chiara Fabris, PhD [1], Marc D. Breton, PhD [2]

[1]*Assistant Professor, Center for Diabetes Technology, Department of Psychiatry and Neurobehavioral Sciences, University of Virginia, Charlottesville, VA, United States;* [2]*Assistant Professor, Center for Diabetes Technology, Department of Psychiatry and Neurobehavioral Sciences, University of Virginia, Charlottesville, VA, United States*

Introduction

Continuous glucose monitors (CGMs) provide detailed time series of consecutive observations on the underlying process of glucose fluctuations. The feedback of such detailed information to patients with diabetes has been shown to have a positive influence on their glycemic control, including a reduction in glucose variability, time spent in nocturnal hypoglycemia, time spent in hyperglycemia, and levels of glycosylated hemoglobin [1–4]. However, some CGM technology continues to face challenges in terms of sensitivity, stability, calibration, and the physiological time lag between blood glucose (BG) and interstitial glucose (IG) concentration [5–11]. Thus it is frequently concluded that the abundance of information about glucose fluctuations carried by the CGM data stream is to some extent offset by the possibility of sensor errors that exceed in magnitude the errors of the traditional self-monitoring blood glucose (SMBG) devices. Such a conclusion, however, is only partially accurate: while the observed error in an isolated CGM data point is indeed generally larger than the error observed in an SMBG data point, the additional information provided by CGM time series allows the application of error-reduction techniques that are unavailable in SMBG devices. For example, deconvolution and other modeling techniques allow for the mitigation of certain sensor deviations due to blood-to-interstitial time delay [12,13].

The key to CGM error mitigation is a detailed analysis and subsequent mathematical modeling, which allow for the understanding of the sources and magnitude of sensor errors. The analytical approach proposed in this manuscript is the one originally described in Ref. [14] and is based on two principles: (i) CGMs assess BG fluctuations *indirectly*—by measuring the concentration of IG—but are calibrated via SMBG to approximate BG; and (ii) CGM data reflect an underlying process in time and therefore are *time series* consisting of ordered, in-time highly interdependent data points. The first principle stipulates that calibration errors would be responsible for a portion of the sensor deviation from

Glucose Monitoring Devices. https://doi.org/10.1016/B978-0-12-816714-4.00012-0

reference BG [13]. Thus in accuracy studies, the first step of the analysis should be the investigation of calibration errors via simulated recalibration, using all available reference data points. Furthermore, because CGMs operate in the interstitial compartment, which is presumably related to the blood via diffusion across the capillary wall, the second step should entail modeling the sensor deviations from BG describing this diffusion process. Models of blood-to-interstitial glucose transport have been proposed and are reasonably well accepted by the scientific community as an approximation of the possible physiological time lag between BG and IG concentration [15–17]. The second principle stipulates that the temporal structure of CGM data is important and should be taken into account by the analysis of CGM errors. In particular, established accuracy measures, such as mean absolute/relative difference, present an incomplete picture of sensor accuracy because these measures judge the proximity between sensor and reference BG at isolated points in time, without taking into account the temporal structure of the data. In other words, a random reshuffling of the sensor-reference data pairs in time will not change these accuracy estimates. Thus, to account for the *dynamics of sensor errors*, higher-order temporal properties need to be investigated. A wide array of modeling techniques is offered by time-series methods, such as autoregression, autocorrelation, and spectral analysis. In this chapter, we review the use of an autoregressive moving average (ARMA) model to account for the time dependence of consecutive sensor errors.

Finally, the detailed understanding and modeling of sensor errors allow the next step: their computer simulation. This, in turn, allows the development of a simulated "sensor," which is useful for in silico testing of diabetes treatment strategies, such as open- or closed-loop control, under the realistic conditions of imperfect CGM.

Methods

To decompose the sensor errors, we used techniques from linear regression, kernel density estimation, derivative estimation, and time-series analysis, each allowing us to access specific characteristics of the sensor/BG discrepancy. We also provided examples of each analysis using data provided by Abbott Diabetes Care (Alameda, CA).

Datasets

The data used as an example in this chapter come from two different datasets provided by Abbott Diabetes Care:

1. The first dataset is a home-use dataset containing sensor readings from the FreeStyle Navigator taken every 10 min in 136 subjects, for an average of 40 days (e.g., eight sensors, 5-day insertions). The dataset contains 1062 sensors, totaling approximately 4000 days of recording, with 40,745 irregularly

spaced, reference SMBG data points. After elimination of missing data segments and nonfunctioning sensors, the final dataset was composed of 20,660 reference/sensor data pairs.

2. The second dataset is smaller but more controlled, in that BG was measured every 15 min in clinical settings, using YSI instruments (coefficient of variation = 2%). It contains navigator sensor readings taken every minute in 28 patients, two sensors per patient at different sites, that is, 56 sensors. After elimination of missing data and nonfunctioning sensors, the final dataset was composed of approximately 7000 data pairs.

A posteriori recalibration

Ae previously mentioned, CGMs yield estimates that are the product of (at least) two consecutive steps: (1) deduction of BG values from IG-related electrical current recorded by the sensor and (2) blood-to-interstitial glucose transport. To assess the impact of calibration errors (i.e., step (1)), we performed a posteriori recalibration of the data for each sensor leveraging the method generally used by sensor producers, that is, linear/quadratic fitting with time-delay compensation. The major difference between a posteriori and real-time calibration is the availability of all reference BG points; therefore, for a fixed calibration function (the relation between sensor current and BG), a posteriori calibration is optimal in minimizing the sensor readings-reference glucose discrepancy. In this study, a posteriori calibration was performed by linear regression, matching the interpolated sensor readings (if the timing of the reference fell between two readings of the sensor) to the reference measurements. The sum of squares was assessed only at the points of reference measurement. The result of the linear regression was considered to be the recalibrated sensor trace.

Reference-sensor density and delay estimation

Density estimation. Once the sensors were recalibrated, we used kernel density estimation to approximate the distribution of the sensor readings for different glucose references. Each sensor/reference pair was associated with a Gaussian kernel (see Eq. 12.1) centered on the pair and of predetermined width. More details on the selection of the width and kernel function can be found in Ref. [18]. The density was then computed as the weighted sum of all kernels:

$$D(s,r) = \frac{1}{2\pi\sigma^2 N} \sum_{i=1}^{N} e^{-\frac{\left((s-s_i)^2 + (r-r_i)^2\right)}{2\sigma^2}} \tag{12.1}$$

where N is the number of pairs, s_i is the sensor reading of pair i, r_i is the reference measure of pair i, and σ is the kernel width. This estimation of the density is slightly biased by the fact that negative glucose values do not occur; the bias is reduced by choosing σ to be less than 25% of the smallest reference value.

Therefore the mean sensor reading ($\overline{s}$) and the mean error ($\overline{\varepsilon}$) for each reference BG were computed using Eq. (12.2):

$$\overline{s}(r) = \int_{s=0}^{600} s \cdot D(s,r)\ ds, \quad \overline{\varepsilon}(r) = \overline{s}(r) - r \tag{12.2}$$

Delay estimation. The premise behind our methodology for delay estimation is that if the CGM measurements are delayed, then the sensor error will be dependent on the rate of change of glucose. Therefore, by computing $D(s,r)$ for different ranges of glucose rate of change (bins) and studying the differences between bins, we can study the delay. To do so, we needed to estimate the density in particular bins using Eq. (12.3):

$$D_j(s,r) = \frac{1}{2\pi\sigma^2 N_j} \sum_{i=1}^{N_j} I_j(\partial s_i) e^{-\frac{\left((s-s_i)^2+(r-r_i)^2\right)}{2\sigma^2}}, \quad I_j(\partial s_i) = \begin{cases} 1 & \text{if } \partial s_i \text{ is in bin } j \\ 0 & \text{otherwise} \end{cases} \tag{12.3}$$

To compute a robust rate of change, we used sliding linear regression for consecutive 20-min windows.

Estimation and modeling of the sensor error distribution

As discussed earlier, CGMs measure glucose in the interstitial fluid while being used to assess BG, which creates a delay. Therefore, the distribution of sensor errors could not be computed directly by using the difference between reference and sensors at the same time points. As the physiological delay described earlier is not constant across subjects or within a subject over a long period of time, we synchronized sensor and reference BG using a first-order diffusion model in the calibration equation (Eq. 12.4):

$$CGM = \alpha \cdot G_I + \beta\ , \quad \dot{G}_I = -\frac{1}{\tau}(G_I - G_B) \tag{12.4}$$

where G_I and G_B are the interstitial and the blood glucose concentrations, respectively, τ is the diffusion time constant, and α and β are the calibration parameters. Spanning the possible values of τ, we applied the same linear regression technique as before to estimate α and β. The final solution is the set (τ, α, β), which produces the smallest sum of squares. We do not claim that this procedure, derived from the study by Steil et al. [17], models perfectly the transport of glucose from the blood to interstitium nor the functional relationship between electrical current in the sensor and glucose values. However, in the absence of an identifiable model of such a process, we followed the parsimony principle in choosing the simplest available one, that is, a linear transformation added to a first order, gain 1, diffusion.

Once sensor and reference data were synchronized, we computed the differences between sensor and reference estimates, and the first four central moments of their empirical distribution: mean, variance, skewness, and kurtosis.

Finally, based on the Johnson family of distributions [19], we approximated the probability distribution of the sensor error, using a transformation of the normal density [20].

Estimation and modeling of sensor error time dependency

Classical time-series techniques were applied to the recalibrated and synchronized sensor signal to determine the time dependence of sensor errors: the autocorrelation function and the partial autocorrelation functions.

The autocorrelation function (ACF) is fairly straightforward: it is computed as the correlation of an error at time t, with the errors at time $t+h$, where $h = nT$, n is an integer, and T is a fixed time interval (generally, T is set to the time difference between reference measures). Under weak stationary conditions (the mean and variance of the error do not depend on time), the ACF is only dependent on the lag (h) and not on t, and it can be computed using Eq. (12.5):

$$\gamma(h) = \frac{n}{n-h} \frac{\sum_{i=1}^{n-h} (\varepsilon_i - \overline{\varepsilon})(\varepsilon_{i+h} - \overline{\varepsilon})}{\sum_{i=1}^{n} (\varepsilon_i - \overline{\varepsilon})^2} \tag{12.5}$$

The partial autocorrelation function (PACF) can be best described as the correlation between errors at time t and $t+h$, $h = nT$, excluding information transmitted though $t+T$, $t+2T$, $t+3T$, ..., $t+(n-1)T$. It is similar to the concept of the best linear predictor and is commonly computed using the Durbin–Levinson algorithm [21]. For more details on PACF, please refer to Brockwell et al. [22].

Results

Using dataset 1, we studied the sensor response at different reference glucose levels by estimating the probability distribution of the reference/sensor pairs (recalibrated but not synchronized). The distribution is presented in Fig. 12.1, where blue depicts a very low probability and red a very high probability of occurrence. We observed that sensors tend to read low at high reference values (the reference/sensor pair tends to fall below the diagonal when the reference is above 200 mg/dL) and high at low reference levels (the reference/sensor pair tends to fall above the diagonal when the reference is below 110 mg/dL). In addition, the spread of reference/sensor pairs is positively correlated with the reference level: the distribution is flatter at high glucose levels compared to low glucose levels.

Effect of rate of change on sensor error and delay estimation

As presented in the introduction, it is widely believed that there is a delay between BG and IG. To verify this claim, we applied the same technique described in the

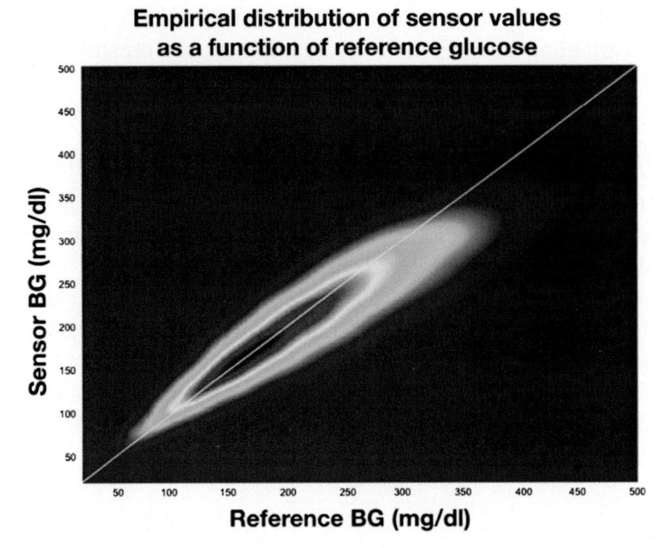

FIGURE 12.1

Distribution of the reference glucose/sensor readings for dataset 1.

previous section but we clustered the reference/sensor pairs by glucose rate of change: eight bins of rate of change were selected, the number of pairs in each bin was not constant but was always greater than 1000, and the number of pairs was fairly symmetric around a 0 rate of change, that is, there were roughly as many pairs between −1.5 and −1 mg/dL/min as between 1 and 1.5 mg/dL/min. This distribution of pairs by rate of change corresponds to the usually accepted distribution of the rate of change in the field [23] therefore indicating an absence of bias. The distributions for each bin are presented in Fig. 12.2.

Observing the distributions in Fig. 12.2, particularly the most likely reference/sensor area (red zone in each distribution), we saw that: (i) at a negative rate of change, the sensors tends to read high (red zone above the diagonal); (ii) at a positive rate, the sensors tend to read low (red zone below the diagonal); and (iii) the extent to which the sensor systematically reads high or low is correlated to the amplitude of the rate of change (e.g., the red zone is further above the 45 degrees line in the top left distribution than the bottom left distribution). To verify the last observation, we computed the average reference/sensor discrepancy as a function of the reference glucose for each glucose rate of change jth bin using the distribution $D_j(s, r)$. The results of this analysis are presented in Fig. 12.3A. We conclude that there exists a correlation between the rate of change and average discrepancy, regardless of reference BG levels. Finally, computing the average reference/sensor discrepancy for a specific rate of change across reference values, we compared these averages with the average rate of change in each bin. The results of this analysis are presented in Fig. 12.3B. The average discrepancy is linearly related to the rate of change ($R^2 = 0.995$), and the slope of this linear relation (without offset) gives an estimate of the delay, which in this dataset is 17 min.

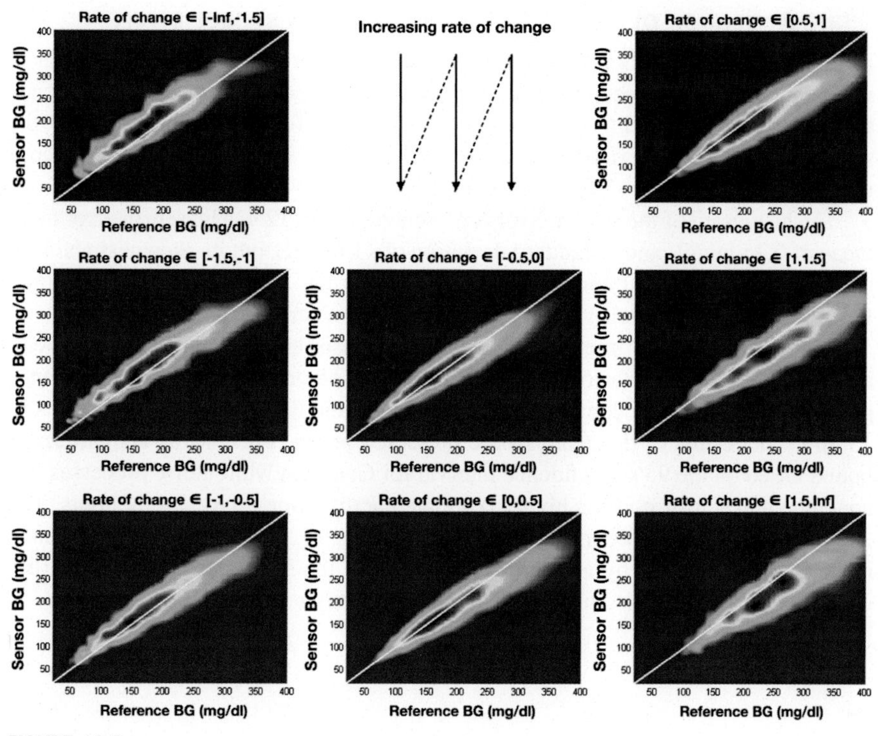

FIGURE 12.2

Empirical distribution of the reference/sensor pairs divided in bins based on the glucose rate of change. BG is expressed in mg/dL.

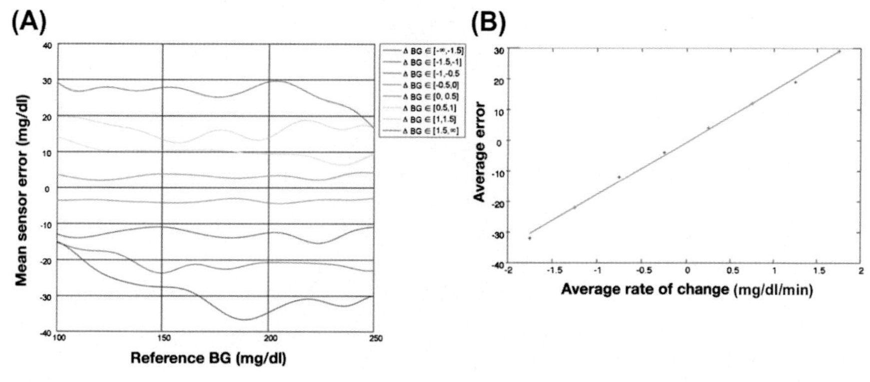

FIGURE 12.3

(A) Effect of the glucose rate of change on the mean sensor/reference discrepancy. (B) Estimation of the sensor delay using the average reference/sensor discrepancy at different levels of glucose rate of change.

Characterization of recalibrated synchronized sensor errors

In this section, we study the third component of sensor difference from reference, which remains after recalibration and synchronization of sensor and reference data. We refer to this difference as the sensor error. To study more precisely the sensor error and its time dependence, we used dataset 2, which contains YSI glucose measurements taken at 15-min intervals.

The histogram of the sensor error is presented in Fig. 12.4. The sensor error has a mean of 0.76 mg/dL and a standard deviation of 11 mg/dL, but its skewness and kurtosis show that the distribution of the error is not normal. To estimate its distribution, we used the Johnson family and obtained the parameters in Table 12.1.

To characterize the time dependency of the sensor error we computed the ACF and PACF of the sensor error as described in the Methods section. As the reference glucose measures were equally spaced at 15 min, the lags are integer multiples of 15 min. The results of this analysis are presented in Fig. 12.5. Using a significance bound based on the 95% confidence interval for Gaussian white noise processes, we conclude that sensor noise is highly correlated across time up to several hours. The lag of autocorrelation depends on each sensor (Fig. 12.5A), but, in general, sensor noise is best predicted by a linear combination of the sensor noise 15 min earlier and a random white noise term (PACF cutting off after lag 1 in Fig. 12.5B).

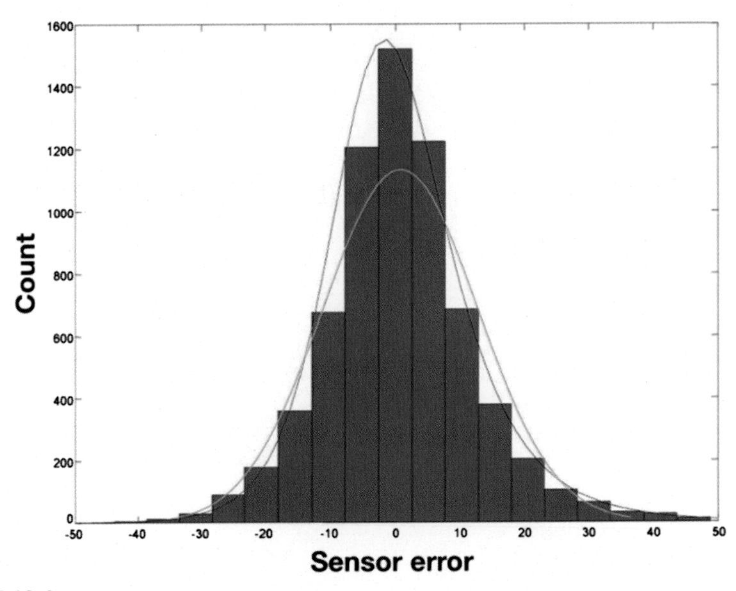

FIGURE 12.4

Histogram of sensor error with the fitted normal distribution (green) and the fitted Johnson distribution (red).

Table 12.1 Johnson parameters of the recalibrated and synchronized sensor error distribution.

Parameters	Sensor error value
Family type	Unbounded system
λ	15.96
ξ	−5.471
δ	1.6898
γ	−0.5444

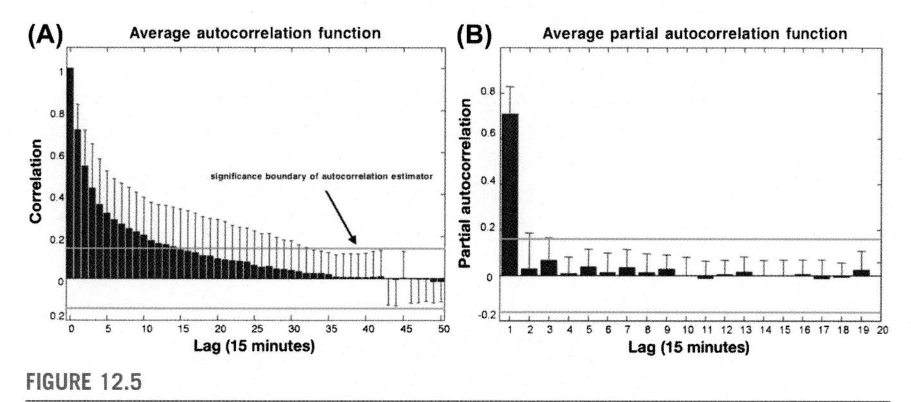

FIGURE 12.5

(A) Autocorrelation function of the sensor error with lag interval of 15 min. (B) Partial autocorrelation function of the sensor error with lag interval of 15 min.

Modeling of subcutaneous sensors

We found that the discrepancy between sensor and reference glucose differ from random noise by having substantial time-lag dependence and other nonindependent identically distributed (iid) characteristics (i.e., the error is independent of previous errors and drawn from the same time-independent probability distribution). The components of the discrepancy are therefore modeled as follows:

(i) Blood-to-interstitium glucose transport described by Eq. (12.6):

$$\begin{cases} \dfrac{\partial \mathrm{IG}}{\partial t} = -\dfrac{1}{\tau}(\mathrm{IG} - \mathrm{BG}) \\ \dfrac{\partial \mathrm{G}}{\partial t} = -\dfrac{1}{\tau_N}(\mathrm{G} - \mathrm{IG}) \end{cases} \tag{12.6}$$

Here IG is the interstitial and BG is the plasma glucose concentration; τ represents the time lag between the two fluids.

(ii) Sensor lag: the time of glucose transport from the interstitium to the sensor needle in Eq. (12.6). Considering that these are two sequential first-order diffusion models, we modeled them with one diffusion equation where the time lag is the resultant single diffusion process representing both the physiological lag and the sensor lag. Empirical estimation gives a time lag of 5 min (which produces a delay of approximately 15 min).

(iii) The noise of the sensor is nonwhite, non-Gaussian. We, therefore, used an ARMA process for its modeling. Based on the analysis of the empirical partial autocorrelation function in dataset 2, we restricted this model to a simple autoregressive model of order 1 (see Eq. 12.7):

$$\begin{cases} e_1 = v_1 \\ e_n = 0.7(e_{n-1} + v_n) \end{cases}, \quad v_n \sim \Phi(0,1) \text{ iid}, \quad \varepsilon_n = \xi + \lambda \cdot \sinh\left(\frac{e_n - \gamma}{\delta}\right) \tag{12.7}$$

This is due to the apparent nonsignificance of any PACF coefficient for lags greater than 1. The sensor noise is ε_n, which is driven by the normally distributed time series e_n. The parameters ξ, λ, δ, and γ are the Johnson system (unbounded system) parameters corresponding to the empirical noise distributions, as shown in Table 12.1. Validation of these model choices and fits is presented in Fig. 12.6.

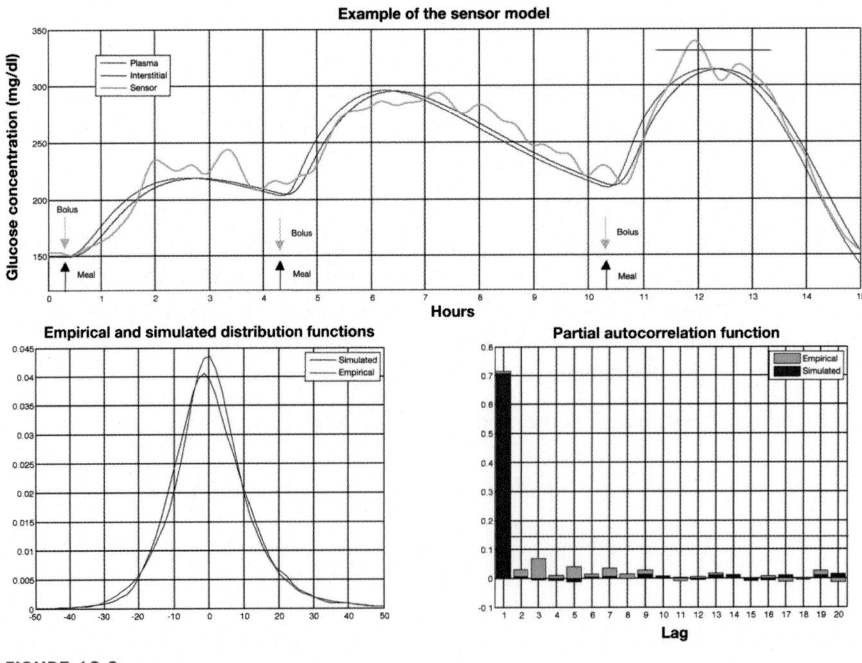

FIGURE 12.6

Simulated sensor trace and validation against empirical distribution and PACF.

These equations were included in a glucose homeostasis simulator [24], implemented in Matlab Simulink. An example is presented in Fig. 12.6, along with validation of the modeling by comparing the sensor error distributions (empirical and from simulation), as well as the partial autocorrelation functions. No difference between empirical and simulated characteristics was apparent. Moreover, a χ^2 test showed that no significant difference exists between the observed and the simulated distribution of sensor errors and the distribution of errors of the FreeStyle Navigator ($P > .46$).

Conclusions

In this chapter, we presented a sequence of analytical techniques beginning with the decomposition of CGM sensor errors into errors due to calibration, blood-to-interstitial glucose transport, and additive noise, leading to computer simulation and building of an in silico sensor based on diffusion and time series modeling. We should emphasize that the term sensor "error," although well accepted, may not be entirely accurate because part of the "error" is due to physiological processes that result in natural deviations of sensor readings from reference BG levels.

Regardless of terminology, sensor deviations from reference BG levels are generally explained by the specifics of CGM technology—the measurement of glucose takes place in the interstitial compartment, which is associated with, but different from the blood. However, CGMs are calibrated against BG, and their accuracy is assessed by comparing IG readings to reference BG. This alone creates two sources of CGM errors, or deviations: calibration and the BG-to-IG gradient.

During analysis, the errors due to calibration can be separated by recalibrating the sensor trace using all available reference BG readings, which is the first step of sensor accuracy analysis. The differences between IG and BG are most evident during rapid BG excursions: sensor errors tend to be negative (high CGM readings) when the BG rate of change is negative and positive (low CGM readings) when the BG rate of change is positive, which is potentially explained by the time lag between the two compartments. This second component of sensor error due to the BG-to-IG gradient can be evaluated by a diffusion model—a commonly accepted technique explaining the possible physiological time lag between blood and IG concentration [15–17]. After recalibration and accounting for the BG-to-IG gradient, the residual sensor noise appears to be nonwhite (non-Gaussian) and the consecutive sensor errors are highly interdependent. Thus the presented analysis of residual (after calibration and diffusion) sensor errors is based on time-series approaches using ARMA noise to account for the interdependence between consecutive sensor errors. As an example, we presented the autocorrelation and the distribution of the errors of the FreeStyle Navigator.

Based on the presented sequence of mathematical models, the computer simulation of sensor errors becomes possible. Such an in silico sensor includes both generic deviations due to physiology and sensor-specific errors due to particular sensor engineering. The generic deviations can be approximated reasonably well by a diffusion model. The sensor-specific errors appear to be generally limited to the order of the ARMA model and the shape of the sensor error distribution. Thus, both types of errors can be described by a few model parameters and included in a simulation. A specific type of error is omitted by this methodology: high-frequency errors (period of 1–15 min) cannot be modeled with our techniques. At this time, no data are available to extract the characteristic of the error at that fine a sample frequency (we would need reference values every minute in a large dataset). Nonetheless, the development of a simulated sensor becomes feasible and became a component of the larger metabolic system simulator developed at the University of Virginia and the University of Padova. This simulator was accepted by the Food and Drug Administration for preclinical testing of closed-loop glucose control strategies. Such in silico testing is now employed by the Juvenile Diabetes Research Foundation Artificial Pancreas Consortium, provides insights on the effectiveness of control algorithms during realistic conditions, and allows the direct transition from in silico testing to clinical trials.

References

[1] Klonoff DC. Continuous glucose monitoring: roadmap for 21st century diabetes therapy. Diabetes Care 2005;28(5):1231–9.

[2] Garg K, Zisser H, Schwartz S, Bailey T, Kaplan R, Ellis S, Jovanovic L. Improvement in glycemic excursions with a transcutaneous, real-time continuous glucose sensor. Diabetes Care 2006;29(1):44–50.

[3] Deiss D, Bolinder J, Riveline J, Battelino T, Bosi E, Tubiana-Rufi N, Kerr D, Phillip M. Improved glycemic control in poorly controlled patients with type 1 diabetes using real-time continuous glucose monitoring. Diabetes Care 2006;29(12):2730–2.

[4] Kovatchev B, Clarke W. Continuous glucose monitoring (CGM) reduces risks for hypo- and hyperglycemia and glucose variability in diabetes. Diabetes 2007;56(Suppl. 1):A23.

[5] Gerritsen M, Jansen JA, Lutterman JA. Performance of subcutaneously implanted glucose sensors for continuous monitoring. The Netherlands Journal of Medicine 1999;54(4):167–79.

[6] Gross TM, Bode BW, Einhorn D, Kayne DM, Reed JH, White NH, Mastrototaro JJ. Performance evaluation of the MiniMed continuous glucose monitoring system during patient home use. Diabetes Technology and Therapeutics 2000;2(1):49–56.

[7] Cheyne EH, Cavan DA, Kerr D. Performance of a continuous glucose monitoring system during controlled hypoglycaemia in healthy volunteers. Diabetes Technology and Therapeutics 2002;4(5):607–13.

[8] Kovatchev BP, Gonder-Frederick LA, Cox DJ, Clarke WL. Evaluating the accuracy of continuous glucose-monitoring sensors: continuous glucose error-grid analysis illustrated by TheraSense FreeStyle Navigator data. Diabetes Care 2004;27(8):1922–8.

[9] Clarke WL, Anderson S, Farhy L, Breton M, Gonder-Frederick L, Cox D, Kovatchev B. Evaluating the clinical accuracy of two continuous glucose sensors using continuous glucose-error grid analysis. Diabetes Care 2005;28(10):2412–7.

[10] Clarke WL, Kovatchev BP. Continuous glucose sensors—continuing questions about clinical accuracy. Journal of Diabetes Science and Technology 2007;1:164–70.
[11] Zisser H, Shwartz S, Rather R, Wise J, Bailey T. Accuracy of a seven-day continuous glucose sensor compared to YSI blood glucose values. Abstract S03. In: Proceedings of the 27th workshop of the study group on artificial insulin delivery, pancreas and islet transplantation (AIDPIT) of the European association for the study of diabetes (EASD) and 2nd European diabetes technology and transplantation meeting (EuDTT). Innsbruck- Igls, Austria: AIDPIT; January 27–29, 2008. p. 90. Available from: http://www.aidpit.org.
[12] Facchinetti A, Sparacino G, Cobelli C. Reconstruction of glucose in plasma from interstitial fluid continuous glucose monitoring data: role of sensor calibration. Journal of Diabetes Science and Technology 2007;1(5):617–23.
[13] King C, Anderson SM, Breton M, Clarke WL, Kovatchev BP. Modeling of calibration effectiveness and blood-to-interstitial glucose dynamics as potential confounders of the accuracy of continuous glucose sensors during hyperinsulinemic clamp. Journal of Diabetes Science and Technology 2007;1(3):317–22.
[14] Breton M, Kovatchev BP. Analysis, modeling, and simulation of the accuracy of conitnuous glucose sensors. Journal of Diabetes Science and Technology 2008;2(5): 853–62.
[15] Kulcu E, Tamada JA, Reach G, Potts RO, Lesho MJ. Physiological differences between interstitial glucose and blood glucose measured in human subjects. Diabetes Care 2003; 26(8):2405–9.
[16] Boyne MS, Silver DM, Kaplan J, Saudek CD. Timing of changes in interstitial and venous blood glucose measured with a continuous subcutaneous glucose sensor. Diabetes 2003;52(11):2790–4.
[17] Steil GM, Rebrin K, Hariri F, Jinagonda S, Tadros S, Darwin C, Saad MF. Interstitial fluid glucose dynamics during insulin induced hypoglycaemia. Diabetologia 2005; 48(9):1833–40.
[18] Hastie T, Tibshirani R, Friedman J. The elements of statistical learning: data mining, inference, and prediction. In: Springer series in statistics. 1st ed. New York: Springer; 2001.
[19] Johnson NL. Systems of frequency curves generated by methods of translation. Biometrika 1949;36(1):149–76.
[20] Breton MD, DeVore MD, Brown DE. A tool for systematically comparing the power of tests for normality. Journal of Statistical Computation and Simulation July 2007. Available from: http://www.informaworld.com/smpp/content~content=a780645596?words=breton%7cdevore&hash=3456547477.
[21] Durbin J. The fitting of time-series models. Review of the International Statistical Institute 1960;28(3):233–44.
[22] Brockwell PJ, Davis RA. Time series: theory and methods. In: Springer series in statistics. 2nd ed. London: Springer; 1998.
[23] Kovatchev BP, Clarke WL, Breton M, Brayman K, McCall A. Quantifying temporal glucose variability in diabetes via continuous glucose monitoring. mathematical methods and clinical application. Diabetes Technology and Therapeutics 2005;7(6): 849–62.
[24] Dalla Man C, Raimondo DM, Rizza RA, Cobelli CGIM. Simulation software of meal glucose—insulin model. Journal of Diabetes Science and Technology 2007;1(3): 323–30.

SECTION

Clinical use of monitoring data

3

CHAPTER

Low glucose suspend systems

13

Viral N. Shah, MD [1], Amanda Rewers, MD [2], Satish Garg, MD [3]

[1]*Assistant Professor of Pediatrics and Medicine, Barbara Davis Center for Diabetes Adult Clinic, University of Colorado Anschutz Medical Center, Aurora, CO, United States;* [2]*Research Assistant, Barbara Davis Center for Diabetes Adult Clinic, Aurora, CO, United States;* [3]*Professor of Pediatrics and Medicine, Barbara Davis Center for Diabetes Adult Clinic, University of Colorado Anschutz Medical Center, Aurora, CO, United States*

Introduction

Diabetes is a chronic health disease that affects millions of people worldwide. According to the International Diabetes Federation (IDF), 451 million people in the world were expected to have diabetes in 2017; that number is projected to increase to 693 million by 2045 [1]. Recent data from Centers for Disease Control and Prevention estimated that 23 million Americans had diagnosed diabetes, 7.2. Americans had undiagnosed diabetes, and 84.1 million had prediabetes in the year 2015 [2]. Type 1 diabetes (T1D) is an autoimmune form of diabetes, characterized by inadequate insulin production due to near-total β-cell destruction [3]. Patients with T1D require life-long insulin therapy to achieve glycemic control to prevent long-term acute and chronic diabetes complications [4]. Type 2 diabetes (T2D) is characterized by inadequate insulin action, initial insulin resistance, later followed by a decrease in insulin secretion requiring treatment with insulin [5]. Most patients with T2D ultimately need insulin therapy 20–30 years after diagnosis [4,5]. Thus insulin is a cornerstone of diabetes management. Insulin use in patients with diabetes is expected to increase from 516·1 million vials per year in 2018 to 633·7 million vials per year in 2030 [6].

In patients with T1D and T2D, intensive diabetes treatment has been shown to reduce long-term micro- and macrovascular complications [7,8]. However, intensive therapy is associated with an increased risk for hypoglycemia [9]. Hypoglycemia is stated by patients, caregivers, and providers as the most feared complication of intensive insulin therapy.

Incidence rates of hypoglycemia associated with intensive insulin therapy span from 115 to 320 events per 100 patients with T1D [10]. Recent data from the T1D Exchange Registry reported a 6% prevalence of severe hypoglycemia defined as loss of consciousness or seizure among subset of 2561 patients with T1D who completed the participant questionnaire [11]. Severe hypoglycemia was more than threefold higher in the intensively treated group of DCCT than in conventionally

Glucose Monitoring Devices. https://doi.org/10.1016/B978-0-12-816714-4.00013-2

treated participants [8]. However, during the long-term observational phase of EDIC, where the HbA1c levels were similar between groups, severe hypoglycemia was not different between groups [12]. Moreover, recent data show a comparable prevalence of severe hypoglycemia across different HbA1c ranges in both adolescents and adults, including the elderly [13,14]. Hypoglycemia is also common among patients with T2D treated with insulin. In a 2013 study, 97,648 emergency visits per year were ascribed to insulin-induced hypoglycemia, and the incidence of hypoglycemic events was twice as high in individuals older than 80 years old, resulting in a projected cost of $640 million in healthcare expenses [15]. Other complications of hypoglycemia include an increased risk of cardiovascular events and cognitive deficits in older adults with diabetes [16,17]. Thus hypoglycemia is a major obstacle in achieving optimal glycemic control in insulin-treated patients with diabetes.

Hypoglycemic events have decreased with the introduction of insulin analogs such as insulin glargine U100 and insulin detemir compared to previous generations of intermediate-acting insulins such as neutral protamine Hagedorn and longer-acting ultralente insulins (now discontinued worldwide) [18]. Moreover, newer longer-acting insulin analogs such as insulin glargine U300 and insulin degludec have been shown to reduce the rate of symptomatic hypoglycemia by 20%–30% compared to insulin glargine U100 in patients with T1D and insulin-treated T2D [19–21]. However, severe hypoglycemia, particularly nocturnal, continues to pose a risk with insulin therapy. Insulin pump therapy and the use of continuous glucose monitors (CGM) have been shown to improve glycemic control and reduce the risk of hypoglycemia [22,23]. However, sensor-augmented pump therapy does not eliminate nocturnal hypoglycemia completely [24]. Moreover, most youths with T1D do not meet the glycemic A1c goal of below 7.5% (for children) or 7% (for adults) as defined by the American Diabetes Association and International Society for Pediatric and Adolescent Diabetes [25]. The open-loop insulin delivery system (use of insulin pump and/or CGM without automation) is subject to human error and can cause unexpected hypo- and/or hyperglycemic events. Therefore closing the loop by integrating insulin pumps and CGMs using necessary algorithms offers promise in the reduction of overall and nocturnal hypoglycemia, meal-related hyperglycemia, and wide glucose excursions as measured by different indices of glycemic variability.

Other chapters in this book have provided detailed information on various closed-loop systems. This chapter focuses only on the low glucose suspend feature of the closed-loop system.

Low glucose suspend system

For patients with T1D, more than half of hypoglycemic events occur at night. In younger individuals with T1D, more than 75% of hypoglycemic seizures and 6% of deaths can be attributed to nocturnal hypoglycemia in children [26]. In both children with T1D and their parents, fear of hypoglycemia is associated with poor quality of life [27]. Additionally, hypoglycemia begets hypoglycemia [28].

Therefore, reducing hypoglycemia should be the first goal of the closed-loop system and may be accomplished by suspending insulin delivery when a predetermined low glucose level is reached [29].

The low glucose suspend (LGS, alternatively called threshold suspend (TS) in the United States) feature suspends insulin delivery when the glucose sensor value nears a programmed lower threshold (unless the patient overrides the alarm/alert) with the aim of reducing the severity of hypoglycemic events. Pump suspension lasts for 2 h in absence of user intervention, but the pump can be manually restarted at any time. The concept of the LGS system is illustrated in Fig. 13.1.

The crucial issue in creating an LGS system was the length of insulin suspension that would decrease the severity of hypoglycemia while not risking serious hyperglycemia or ketoacidosis in the patient. Earlier studies performed in the 1980s assessed glucose response and ketone production for up to 9 h after interruption of insulin delivery from insulin pumps in nine patients with T1D [30]. In the first hour after insulin delivery interruption, no change was observed in plasma glucose or 3-hydroxybutyrate levels. However, after the first hour of interrupted insulin delivery, there was a continuous increase in plasma glucose as well as levels of ketones. A significant increase in plasma glucose was observed 2 h after insulin delivery interruption, while a significant increase in ketones was observed 3 h after insulin delivery interruption [30]. The findings of this early experiment were

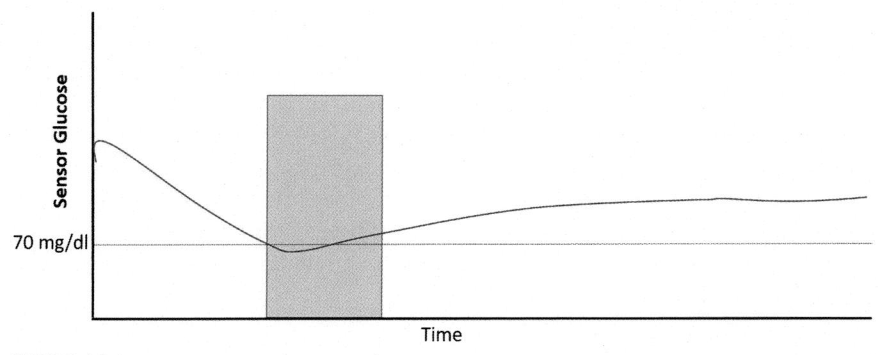

FIGURE 13.1

Illustration shows the concept of low glucose suspend (LGS, also known as Threshold Suspend in the US) system. When sensor glucose (redline) is reached to a preset threshold value as shown in dashed line (preset threshold approved for Medtronic Veo in the Europe was 40-90 mg/dl and for 530 G in the US was 60-90 mg/dl), the insulin pump suspends insulin delivery (gray bar) for up to 2 hours. During LGS activation, patient receives first alarm when insulin is suspended and second alarm 2 minutes after if user does not respond. If patient acknowledges and overrides alarm, insulin pump resumes insulin delivery. If patient does not acknowledge alarm, insulin pump suspends insulin delivery for 120 minutes. After 120 minutes, irrespective of sensor glucose level, insulin pump resumes insulin delivery. However, insulin pump will suspend insulin delivery again after 6 hours if cycle is not interrupted and sensor glucose is still below the threshold value.

replicated by other researchers demonstrating the safety of insulin interruptions for at least 2 h without increasing significantly hyperglycemia or ketone levels [31,32]. Moreover, studies have reported that metabolic abnormalities due to insulin delivery interruptions can be easily corrected by resuming insulin delivery [32]. These early studies on insulin delivery interruptions used human regular insulin, which has a longer duration of action compared to rapid-acting insulin analogs. It is also important to note that these early studies did not assess the role of insulin suspension in the presence of hypoglycemia or impending hypoglycemia, where longer duration of insulin delivery may not be associated with significant hyperglycemia or ketoacidosis.

With the availability of rapid-acting insulin analogs such as insulin lispro in August of 1996, which has a faster onset and shorter duration of action compared to human regular insulin, it became necessary to replicate the earliest experiment to document the safe duration of insulin interruptions. To compare the safety between human regular insulin and insulin lispro after insulin pump interruption, insulin delivery was stopped at night (3:00 a.m.) for up to 6 h in 18 patients with T1D [33]. The investigators also evaluated the differences in the time required by the two insulins to correct metabolic abnormalities after insulin pump resumption in the second phase of the same experiment [33]. They found no significant differences between the human regular insulin group and the insulin lispro group in the reduction in plasma insulin levels or rise in the concentrations of plasma glucose and 3-hydroxybutyrate. However, in those treated with lispro compared to regular human insulin, the increase in insulin levels and decrease in plasma glucose were more rapid [33]. This study reassured the safety of insulin lispro interruption for at least 2 h, but it also provided evidence that insulin lispro in pumps had better outcomes when correcting metabolic abnormalities compared to human regular insulin for insulin-requiring patients with diabetes. Similar results were reported by other investigators, suggesting that the use of rapid-acting insulin analogs might be a better option in treating mildly decompensated patients with T1D [34]. These studies paved the way toward the development of the LGS system for the prevention of severity of hypoglycemia through insulin delivery interruptions in patients with T1D using insulin pumps.

Clinical studies with LGS system

The first LGS system, the Paradigm Veo Insulin Pump, was developed by Medtronic (Northridge, CA) and became commercially available in Europe in June 2009. The Paradigm Veo can automatically suspend basal insulin delivery for up to 2 h when the sensor glucose value is at the predetermined threshold. The threshold for insulin pump suspension in the Paradigm Veo can be set anywhere between 40 and 90 mg/dL depending on patient and provider preference. Clinical studies with LGS system has been summarized in Table 13.1 and detailed here in this chapter.

The safety and efficacy of the Paradigm Veo were assessed in 31 patients with T1D from six centers in the United Kingdom [35]. During the first 2 weeks of the

Table 1 Summary of clinical studies of low glucose suspend system in patients with type 1 diabetes.

First author, Year (Ref)	Number of subjects, total (cases/ controls)	Mean age (SD), years	Men/ Females	Study Design	Study duration	Country	Pump	Results
Choudhary et al. (35)	31	41.9 (10.6)	10/21	Prospective cohort	5 weeks- 2 weeks of LGS off and 3 weeks of LGS on	Six centers from U.K.	Paradigm Veo®	• There were 166 LGS episodes • 66% of daytime LGS episodes were terminated within 10 min, and 20 episodes lasted the maximum 2 hr • ↓ NH duration (46.2 vs 1.8 min/day) in the LGS group in those at the highest quartile of hypoglycemia at baseline
Danne et al. (36)	21	10.8 (3.8)	NA	Prospective cohort	6 weeks	Three centers from Germany	Paradigm Veo®	• Of 1,298 LGS alerts, 853 were shorter than 5 minutes and 24% had a duration of 2 hr. • Number of hypoglycemic excursions and time spent in hypoglycemia ↓ without change in mean blood glucose
Garg et al. (39)	50	34.3 (12.4)	28/22	Randomized crossover study, LGS on and off sessions were separated by washout period of 3-10 days	NA	USA	Medtronic pump with TS feature (530 G)	• ↓ hypoglycemia duration during LGS-on compared to LGS-off sessions by ~ 30 minutes

(continued)

Table 1 Summary of clinical studies of low glucose suspend system in patients with type 1 diabetes.—*cont'd*

First author, Year (Ref)	Number of subjects, total (cases/ controls)	Mean age (SD), years	Men/ Females	Study Design	Study duration	Country	Pump	Results
Bergenstal et al. (40)	247 (121/ 126)	41.6 (12.8) in TS group, 44.8 (13.8) control group	96/151	Randomized controlled trial	3 months	USA	Medtronic pump with TS feature (530 G)	• The changes in A1c values were similar in the two groups. • ↓ mean AUC for NH events by 37.5% in the TS group compared to control group • ↓ NH events by 31.8% in the • TS group than in the control group
Ly et al. (37)	95 (46/49)	18.6 (11.8)	47/48	Randomized controlled trial	6 months	Australia	Paradigm Veo®	• ↓ moderate-to-severe hypoglycemic events by 36% in the LGS group compared to control group. • No change in A1c in either group
Ly et al. (38)	24	17.4 (9.3)	NA	Secondary analysis of data from the primary study	6 months	Australia	Paradigm Veo®	• Total of 3128 LGS events with median duration of 11.2 minutes and 126 full 2-hr LGS events • The mean sensor glucose at the end of the 2-hr suspend period was 99± 6 mg/dL

(continued)

Table 1 Summary of clinical studies of low glucose suspend system in patients with type 1 diabetes.—*cont'd*

First author, Year (Ref)	Number of subjects, total (cases/controls)	Mean age (SD), years	Men/Females	Study Design	Study duration	Country	Pump	Results
Elbarbary et al. (42)	60	15.6 (2.7)	19/41	Observational	NA	Egypt	Paradigm Veo®	• During Ramadan fast, most LGS events occurred between 4-7 PM with mean duration of LGS event was 26.5 minutes • ↓ AUC for hypoglycemia during LGS-on compared to LGS-off periods
Agrawal et al. (45)	935	NA	NA	Real-word data from Europe	>3 months of LGS use	Europe	Paradigm Veo®	• LGS threshold was most commonly set between 50-60 mg/dl • Median duration of LGS events was 9.8 minutes • 11% of LGS events lasted for about 2 hr where event started at mean sensor glucose of 58.8 mg/dl and ended with mean glucose of 102.2 mg/dl
Gomez et al. (46)	111	42 (29-58)	50/61	Single institution-based, prospective follow-up of T1D patients	Mean follow-up 47 months	Colombia	Paradigm Veo®	• Over mean duration of 47 months, total insulin dose was reduced by 0.22 units/kg and A1c reduced by 1.7%

(*continued*)

Table 1 Summary of clinical studies of low glucose suspend system in patients with type 1 diabetes.—*cont'd*

First author, Year (Ref)	Number of subjects, total (cases/ controls)	Mean age (SD), years	Men/ Females	Study Design	Study duration	Country	Pump	Results
				started on LGS system				• ↓ severe hypoglycemic events after LGS start
Agrawal et al. (47)	20,973	NA	NA	Real-world data from USA	NA	USA	530 G	• TS feature was enabled on 82% of patient-days • ↓ sensor glucose <50 mg/dl by 69% during TS-on days compared to TS-off days • The median sensor glucose at the start of TS was 60 mg/dl and at the end of TS was 87 mg/dl

NH; nocturnal hypoglycemia; NA; data not available or not applicable, AUC; area under curve, ↓- reduced, LGS; low glucose suspend, A1c; glycated hemoglobin, TS; threshold suspend, hr; hour, <; less than, >; greater than.

study, participants only used CGM with the LGS off and then the LGS feature was activated for the following 3 weeks. The authors reported a significant reduction in the duration of nocturnal hypoglycemia with the use of the LGS feature in study participants who were in the highest quartile of nocturnal hypoglycemia at baseline (median 46.2 vs. 1.8 min/day, $P = .02$). There was no difference in mean sensor glucose between the LGS off or LGS on periods. All subjects in that clinical trial reported that the LGS was useful and 93% reported feeling safer at night. The findings suggested that using an LGS system was safe in reducing the severity of nocturnal hypoglycemia without an increase in rebound severe hyperglycemia in patients with T1D [35]. Similar results have been reported in a study from Germany with data from 21 pediatric patients with T1D using the LGS system. The area under the curve for sensor glucose <70 mg/dL was reduced from 0.76 to 0.53 and there was a significant reduction in the duration of hypoglycemia by 43% without any serious adverse events (ketoacidosis) among these participants with T1D [36]. In a clinical trial from Australia, 95 patients with T1D (mean age 18.6 years) using insulin pumps were randomized to receive either a CGM with the LGS feature ($n = 46$) or no CGM ($n = 49$) [37]. After 6 months of treatment, the moderate-to-severe hypoglycemic event rates decreased from 28 to 16 in the pump-only group versus 175 to 35 in the low-glucose suspension group [37]. Despite a significant reduction in moderate-to-severe hypoglycemia among LGS users with T1D, there was no change in glycated hemoglobin (HbA1c) in the LGS group compared to insulin pump users. The same group evaluated glucose profiles following 2 h of full LGS suspensions in 24 patients with T1D who had hypoglycemia unawareness as defined by a Clarke score of four or greater [38]. There was an improvement in HbA1c and hypoglycemia awareness scores over 6 months among LGS users with T1D. There were 126 full 2-h suspend events among these 24 T1D patients with hypoglycemia unawareness and, as expected, most suspensions happened at night before 3:00 a.m. All the LGS activations occurred when sensor glucose approached 60 mg/dL and the mean sensor glucose at the end of the 2-h suspend period was 99 ± 6 mg/dL [38]. These data confirmed the effectiveness of LGS systems in reducing the severity and duration of hypoglycemia in patients with T1D with impaired hypoglycemia awareness.

Despite the Medtronic Paradigm Veo being safe and effective in preventing the duration and severity of hypoglycemia, and commercially available in Europe and other parts of the world since June 2009, it was not approved by the US Food and Drug Administration (FDA) until September 2013. The first pivotal study in the United States, the ASPIRE In-Clinic (Automation to Simulate Pancreatic Insulin Response) was conducted to establish the safety of LGS systems in patients with T1D [39]. In this study, 55 subjects with T1D aged 17–58 years were randomized in a cross-over study. Subjects participated in an exercise-induced hypoglycemia session with the LGS function on or off. LGS was set to suspend insulin delivery for 2 h once a sensor glucose value of 70 mg/dL or below was detected. There was a significant reduction in hypoglycemia duration (by 30 min) in LGS-on sessions compared to LGS-off sessions (138.5 ± 76.68 vs. 170.7 ± 75.91 min,

$P = .006$) [39]. These clinical trial data led the FDA approval of the LGS system (known as 530G) in the United States in 2013. The predetermined threshold for insulin pump suspension approved by the US FDA was 60–90 mg/dL depending on the patient and/or provider preference. It is also important to know that this is the first study suggesting that hypoglycemia is an independent risk factor for future hypoglycemic episodes in patients with T1D [28]; thus, the cross-over study design may not be the most appropriate study design for devices that are being evaluated for hypoglycemia reduction because of the possible spillover effect of hypoglycemia occurred during the first half of the cross-over period. This pivotal study was followed by a large, multicenter, randomized trial (ASPIRE In-Home) in patients with T1D [40]. The ASPIRE In-Home randomly assigned 247 patients with T1D (age range 16–70 years) to receive either the LGS feature or sensor-augmented insulin pump therapy for 3 months. The mean area under the curve for nocturnal hypoglycemic events was 37.5% lower in the LGS group compared to the control group. Similarly, nocturnal hypoglycemic events occurred at 31.8% less frequently in the LGS group than in the control group [40].

Despite a number of clinical trials establishing the safety of the LGS system in reducing the severity of hypoglycemia in patients with T1D, there was a potential concern related to suspension of insulin delivery for 2 h due to a failing sensor when blood glucose levels were not actually low, which may result in hyperglycemia and/or ketoacidosis. A small study of 17 participants with T1D showed no clinically significant differences in blood ketone levels between suspend nights compared to nonsuspend nights—even in the absence of low glucose—demonstrating the safety of LGS systems in real-life situations even when sensor glucose may not be accurate [41].

A prospective study from Egypt looked at the effect of the LGS system on the frequency of hypoglycemia in adolescents with T1D who desired to fast during Ramadan [42]. Ramadan is a sacred month of the year in Islamic culture where people refrain from eating, drinking, and using oral medications from predawn to sunset. Diabetes management during Ramadan could be challenging and difficult, especially in people with T1D or T2D requiring insulin therapy [43]. The risk of hypoglycemia increases during the Ramadan fast, and it is the most disliked complication as its treatment signifies the intake of carbohydrates, resulting in premature fast breaking, which produces a sense of guilt and failure in faithful individuals. Therefore, using technology that can reduce the risk of hypoglycemia and maintain glycemic control is of utmost importance during this period. In the study, the safety and efficacy of LGS systems during Ramadan was evaluated among 25 adolescents with T1D who used LGS features compared to 35 adolescents with T1D who used the sensor but turned off the LGS feature [42]. Compared with the group with the LGS feature off, the use of the LGS feature significantly reduced the number of both hypo- and hyperglycemic excursions [42]. There were about four LGS events per patient per day recorded, the mean duration of LGS events was 26.5 min, and most LGS events occurred between 4 and 7 p.m., a few hours before the end of fasting.

Real-life evidence with TS system

A properly designed clinical trial unquestionably remains a gold standard for developing scientific evidence about the safety and efficacy of a medical device/product while improving our understanding of the biological mechanisms involved in its therapeutic action. However, clinical trials are often conducted with specific populations and in specialized environments that differ from the real-life experiences of clinical or home settings. Therefore, real-world evidence may be complementary to randomized controlled clinical trials [44].

There are very few studies published based on real-world evidence regarding LGS systems and most of these studies were conducted and published by Medtronic. In a retrospective data analysis involving 7 months of real-world use of the system by 935 patients with diabetes from Europe, it is shown that LGS-mediated insulin pump suspensions occurred approximately 0.55 times per patient per day (27,216 LGS events in total) [45]. However, the median duration of suspension was only 9 min, suggesting that most of these low glucose alerts were terminated by patients either due to nonreliance on sensor glucose or to the treatment of hypoglycemia based on alerts. Among the 27,216 LGS events, 11% of those LGS events lasted for >115 min (which mainly happened during the night) where mean sensor glucose was 58.8 mg/dL at LGS activation and rose to 102.2 mg/dL by the end of the LGS session. In a subset of patients ($n = 278$) who used the LGS system for more than 3 months, the LGS use significantly reduced the number of sensor glucose readings below 50 mg/dL and greater than 300 mg/dL [45]. This real-life study complemented the findings from the randomized controlled trial in documenting the efficacy of the LGS system in reducing the severity of hypoglycemia, mainly nocturnal hypoglycemia, without increasing hyperglycemia in patients with T1D. Another hospital-based cohort study from Colombia analyzed T1D patients who started on an LGS system between August 2009 and August 2014 and were followed for a mean of 47 months [46]. HbA1c levels were reduced from a baseline mean value of 8.8%–7.1% at follow-up. The incidence of severe hypoglycemia also decreased significantly among patients with T1D using LGS systems (10.8% at baseline vs. 2.7% at the end of the study) [46]. Most patients with T1D were initiated on LGS in the study due to a history of severe hypoglycemia. Therefore, this study showed an even greater reduction in the incidence of severe hypoglycemia. In another real-world study from the United System, sensor glucose data from 20,973 patients with diabetes who enabled the LGS feature at their discretion and uploaded pump and sensor data to CareLink between October 15, 2013 and July 21, 2014 were analyzed to evaluate safety and efficacy of the LGS system [47]. Patient-days in which the LGS feature was enabled had 69% fewer sensor glucose values ≤50 mg/dL compared with patient-days in which the LGS was not enabled. The reduction in hypoglycemia seen on LGS-enabled days was more pronounced during nighttime than during daytime hours, which is consistent with other real-world studies [47].

Overall, the findings of real-world experience studies were similar to what was reported in randomized controlled trials. However, the results of real-world studies need to be interpreted with caution because real-world studies are biased in terms of patient selection. It is likely that patients with T1D who are likely to benefit from the LGS system have been started on this system and, therefore, it is likely to exaggerate its benefits. In addition, the real-word studies lack the information on the change in HbA1c or difficulties in using the system. Moreover, there is no real-life data on reduction in healthcare utilization with the use of the LGS system.

Cost-effectiveness

Cost-effectiveness analysis estimates the health costs and gains of alternative interventions, and it is a significant factor for prioritizing the distribution of resources by the payer as well as policy decisions by government agencies [48]. Few studies have reported on the cost-effectiveness of LGS systems.

The incremental cost-effectiveness ratio (ICER) based on the primary randomized study conducted in Australia [49] reported that the ICER per severe hypoglycemic event avoided was $18,257 for all patients and $14,944 for those aged 12 years and older. Additionally, the cost per quality-adjusted life-year gained was $40,803 for patients aged 12 years and older over 6 months. Based on this study, sensor-augmented insulin pump therapy with the LGS feature is a cost-effective alternative to standard insulin pump therapy with self-monitoring of blood glucose in patients with T1D who have hypoglycemia unawareness [49]. Similar cost-effective analysis from the ASPIRE In-Home clinical trial suggested a savings of $117 per patient, per year [50]. Studies from Hungary, France, and the United Kingdom also found that sensor-augmented pump therapy with the LGS feature is cost-effective compared to insulin pump therapy, especially for patients with T1D who are at high risk for hypoglycemia [51–53].

In conclusion, most studies have found the LGS system to be cost-effective compared to insulin pump therapy alone. The economic models are prone to problems and errors [54]. Therefore, the results of these studies should be interpreted cautiously. First, it is important to note that cost-effectiveness ratios based on economic modeling are merely estimates, which are generally founded on assumptions and may document the potential value of one or more interventions. Second, calculating ICER depends on the reference study used in the model and, consequently, it may yield different results contingent on the study used for the analysis. Third, cost-effectiveness analysis differs between countries, and the analysis may change during the years depending on the economy of the country at any point in time.

The limitations of the low glucose suspend system

Any closed-loop system including the LGS system is prone to errors that can be due to problems with the insulin pumps, infusion set, and sensor inaccuracies [55]. The main issue with the Medtronic LGS system was sensor inaccuracy that resulted in the activation of LGS due to sensor reading a low value when blood glucose was not in the hypoglycemic range or even was elevated. From many clinical trials and real-life studies, it is evident that the mean LGS suspension period was less than 30 min, mainly during the day, suggesting that often patients would terminate LGS events either due to sensor glucose errors or because they acted by consuming carbohydrates before checking capillary blood glucose. Another limitation of LGS systems is the inability to prevent hypoglycemia. In the LGS pump, the suspension is triggered by the absolute prevailing glucose level, but there may be advantages to suspending insulin infusion based on predicted hypoglycemia. Studies under short-term hospital conditions have already shown that algorithms that activate insulin suspend when hypoglycemia is predicted (predicted low glucose suspend system, PLGS) can significantly reduce both daytime and nocturnal hypoglycemic events. The details on the safety and effectiveness of PLGS are discussed in the following chapter of this book. The efficacy of the LGS system should also be considered in the context of currently available insulin pharmacokinetics. It takes at least 30 min after suspension of basal insulin delivery before plasma insulin levels will decline enough to allow blood glucose to increase, suggesting that although the suspension of basal insulin alone will often not be enough to prevent hypoglycemia from occurring, it will usually be sufficient to reduce the duration and depth of hypoglycemia. If hypoglycemia is caused by an excess of bolus insulin, suspending the basal insulin will be less effective, which may explain why there is less of a reduction in day-time hypoglycemia when the LGS feature is used. The LGS systems (Medtronic 530G and Paradigm Veo) suspend insulin delivery for 2 h once the LGS is activated and patients can choose not to override. Therefore, 2 h of insulin interruption may result in hyperglycemia especially when LGS activation may have happened while actual capillary glucose may not be low enough. However, a small interventional study described earlier was reassuring that 2 h of insulin interruption, even with normal glucose levels, does not result in clinically significant ketonemia [33–35].

In summary, marketed LGS systems helped patients with T1D reduce the severity and duration of hypoglycemia, and this result was confirmed in both randomized controlled clinical trials and real-life studies.

Future direction

As discussed earlier, the LGS system does not completely prevent hypoglycemia as it suspends insulin delivery only when sensor glucose reaches a predetermined glucose threshold. The next development toward closing the loop was suspending insulin based on PLGS. Studies have shown that the PLGS system significantly reduced the hypoglycemic episodes compared to the LGS system in patients with T1D [56–59]. The PLGS system led to the development of the hybrid closed-loop system (670G System) where the insulin pump adjusts basal insulin delivery every 5 min based on sensor glucose and suspends or delivers insulin based on predicted sensor glucose values [60,61]. However, patient intervention is still required to enter estimated carbohydrates in meals (to deliver boluses) and to possibly deliver corrections. Future research is ongoing to close the loop without any patient interference. The details on the PLGS system, hybrid closed-loop system, and future closed-loop systems can be found in the next chapters of this book.

References

[1] Cho NH, Shaw JE, Karuranga S, Huang Y, da Rocha Fernandes JD, Ohlrogge AW, Malanda B. IDF Diabetes Atlas: global estimates of diabetes prevalence for 2017 and projections for 2045. Diabetes Research and Clinical Practice April 2018;138:271–81.

[2] Centers for Disease Control and Prevention. New CDC report: More than 100 million Americans have diabetes or prediabetes. Available from: https://www.cdc.gov/media/releases/2017/p0718-diabetes-report.html. [Accessed 18 December 2018].

[3] Atkinson MA, Eisenbarth GS, Michels AW. Type 1 diabetes. Lancet 2014;383:69–82.

[4] Shah VN, Moser E, Blau A, Dhingra M, Garg SK. Future of basal insulin. Diabetes Technology and Therapeutics September 2013;15(9):727–32.

[5] Unnikrishnan R, Shah VN, Mohan V. Challenges in diagnosis and management of diabetes in the young. Clinical Diabetes and Endocrinology November 10, 2016;2: 18. https://doi.org/10.1186/s40842-016-0036-6.

[6] Basu S, Yudkin JS, Kehlenbrink S, Davies JI, Wild SH, Lipska KJ, Sussman JB, Beran D. Estimation of global insulin use for type 2 diabetes, 2018-30: a microsimulation analysis. The Lancet Diabetes and Endocrinology November 20, 2018. https://doi.org/10.1016/S2213-8587(18)30303-6 [Epub ahead of print].

[7] The Diabetes Control and Complications Trial Research Group. The effect of intensive treatment of diabetes on the development and progression of long-term complications in insulin-dependent diabetes mellitus. New England Journal of Medicine 1993;329: 977–86.

[8] Holman RR, Paul SK, Bethel MA, Matthews DR, Neil HA. 10-year follow-up of intensive glucose control in type 2 diabetes. New England Journal of Medicine 2008;359: 1577–89.

[9] The DCCT Research Group. Epidemiology of severe hypoglycemia in the diabetes control and complications trial. The American Journal of Medicine 1991;90:450–9.

[10] Seaquist ER, Anderson J, Childs B, Cryer P, Dagogo-Jack S, Fish L, Heller SR, Rodriguez H, Rosenzweig J, Vigersky R. American diabetes association; endocrine society: hypoglycemia and diabetes: a report of a workgroup of the American diabetes association and the endocrine society. The Journal of Clinical Endocrinology and Metabolism 2013;98:1845–59.

[11] Current state of type 1 diabetes treatment in the U.S.: updated data from the T1D exchange clinic registry. Diabetes Care 2015;38:971–8.

[12] Gubitosi-Klug RA, Braffett BH, White NH, et al. Tamborlane and the diabetes control and complications trial (DCCT)/epidemiology of diabetes interventions and complications (EDIC) research group: risk of severe hypoglycemia in type 1 diabetes over 30 years of follow-up in the DCCT/EDIC study. Diabetes Care 2017;40:1010–6.

[13] Weinstock RS, Xing D, Maahs DM, et al. Severe hypoglycemia and diabetic ketoacidosis in adults with type 1 diabetes: results from the T1D exchange clinic registry. The Journal of Clinical Endocrinology and Metabolism 2013;98:3411–9.

[14] Munshi MN, Slyne C, Segal AR, et al. Liberating A1C goals in older adults may not protect against the risk of hypoglycemia. Journal of Diabetic Complications 2017;31: 1197–9.

[15] Geller AI, Shehab N, Lovegrove MC, Kegler SR, Weidenbach KN, Ryan GJ, Budnitz DS. National estimates of insulin-related hypoglycemia and errors leading to emergency department visits and hospitalizations. JAMA Internal Medicine 2014; 174:678–86.

[16] Chow E, Bernjak A, Williams S, et al. Risk of cardiac arrhythmias during hypoglycemia in patients with type 2 diabetes and cardiovascular risk. Diabetes 2014;63:1738–47.

[17] Whitmer RA, Karter AJ, Yaffe K, Quesenberry Jr CP, Selby JV. Hypoglycemic episodes and risk of dementia in older patients with type 2 diabetes mellitus. Journal of the American Academy of Dermatology 2009;301:1565–72.

[18] Garg SK, Moser EG. How basal insulin analogs have changed diabetes care. Diabetes Technology and Therapeutics 2011;13(Suppl. 1). S-1–S-4.

[19] Díez-Fernández A, Cavero-Redondo I, Moreno-Fernández J, Pozuelo-Carrascosa DP, Garrido-Miguel M, Martínez-Vizcaíno V. Effectiveness of insulin glargine U-300 versus insulin glargine U-100 on nocturnal hypoglycemia and glycemic control in type 1 and type 2 diabetes: a systematic review and meta-analysis. Acta Diabetologica December 3, 2018. https://doi.org/10.1007/s00592-018-1258-0.

[20] Lane W, Bailey TS, Gerety G, Gumprecht J, Philis-Tsimikas A, Hansen CT, Nielsen TSS, Warren M, Group Information, SWITCH 1. Effect of insulin degludec vs insulin glargine U100 on hypoglycemia in patients with type 1 diabetes: the SWITCH 1 randomized clinical trial. Journal of the American Academy of Dermatology July 4, 2017;318(1):33–44.

[21] Wysham C, Bhargava A, Chaykin L, de la Rosa R, Handelsman Y, Troelsen LN, Kvist K, Norwood P. Effect of insulin degludec vs insulin glargine U100 on hypoglycemia in patients with type 2 diabetes: the SWITCH 2 randomized clinical trial. Journal of the American Academy of Dermatology July 4, 2017;318(1):45–56.

[22] Karges B, Schwandt A, Heidtmann B, Kordonouri O, Binder E, Schierloh U, Boettcher C, Kapellen T, Rosenbauer J, Holl RW. Association of insulin pump therapy vs insulin injection therapy with severe hypoglycemia, ketoacidosis, and glycemic control among children, adolescents, and young adults with type 1 diabetes. Journal of the American Academy of Dermatology October 10, 2017;318(14):1358–66.

[23] Adolfsson P, Rentoul D, Klinkenbijl B, Parkin CG. Hypoglycaemia remains the key obstacle to optimal glycaemic control – continuous glucose monitoring is the solution. European Endocrinology September 2018;14(2):50–6.

[24] Beck RW, Riddlesworth TD, Ruedy KJ, Kollman C, Ahmann AJ, Bergenstal RM, Bhargava A, Bode BW, Haller S, Kruger DF, McGill JB, Polonsky W, Price D, Toschi E, DIAMOND Study Group. Effect of initiating use of an insulin pump in adults with type 1 diabetes using multiple daily insulin injections and continuous glucose monitoring (DIAMOND): a multicentre, randomised controlled trial. The Lancet Diabetes and Endocrinology September 2017;5(9):700–8.

[25] Foster N, Beck RW, Miller KM, Clements MA, Rickels MR, et al. State of type 1 diabetes management and outcomes from the T1D exchange in 2016–2018. Diabetes Technology and Therapeutics 2019;21(2):66–72.

[26] Cryer PE. Severe hypoglycemia predicts mortality in diabetes. Diabetes Care 2012;35: 1814–6.

[27] Shi L, Shao H, Zhao Y, Thomas NA. Is hypoglycemia fear independently associated with health-related quality of life? Health and Quality of Life Outcomes 2014;12: 167. https://doi.org/10.1186/s12955-014-0167-3.

[28] Garg SK, Brazg RL, Bailey TS, Buckingham BA, Slover RH, Klonoff DC, Shin J, Welsh JB, Kaufman FR. Hypoglycemia begets hypoglycemia: the order effect in the ASPIRE in-clinic study. Diabetes Technology and Therapeutics March 2014;16(3): 125–30.

[29] Kowalski A. Pathway to artificial pancreas systems revisited: moving downstream. Diabetes Care June 2015;38(6):1036–43.

[30] Pickup JC, Viberti GC, Bilous RW, Keen H, Alberti KG, Home PD, Binder C. Safety of continuous subcutaneous insulin infusion: metabolic deterioration and glycaemic autoregulation after deliberate cessation of infusion. Diabetologia March 1982;22(3):175–9.

[31] Krzentowski G, Scheen A, Castillo M, Luyckx AS, Lefèbvre PJ. A 6-hour nocturnal interruption of a continuous subcutaneous insulin infusion: 1. Metabolic and hormonal consequences and scheme for a prompt return to adequate control. Diabetologia May 1983;24(5):314–8.

[32] Scheen AJ, Krzentowski G, Castillo M, Lefèbvre PJ, Luyckx AS. A 6-hour nocturnal interruption of a continuous subcutaneous insulin infusion: 2. Marked attenuation of the metabolic deterioration by somatostatin. Diabetologia May 1983;24(5):319–25.

[33] Attia N, Jones TW, Holcombe J, Tamborlane WV. Comparison of human regular and lispro insulins after interruption of continuous subcutaneous insulin infusion and in the treatment of acutely decompensated IDDM. Diabetes Care May 1998;21(5): 817–21.

[34] Guerci B, Meyer L, Sallé A, Charrié A, Dousset B, Ziegler O, Drouin P. Comparison of metabolic deterioration between insulin analog and regular insulin after a 5-hour interruption of a continuous subcutaneous insulin infusion in type 1 diabetic patients. The Journal of Clinical Endocrinology and Metabolism August 1999;84(8):2673–8.

[35] Choudhary P, Shin J, Wang Y, et al. Insulin pump therapy with automated insulin suspension in response to hypoglycemia: reduction in nocturnal hypoglycemia in those at greatest risk. Diabetes Care 2011;34(9):2023–5.

[36] Danne T, Kordonouri O, Holder M, Haberland H, Golembowski S, Remus K, Bläsig S, Wadien T, Zierow S, Hartmann R, Thomas A. Prevention of hypoglycemia by using low glucose suspend function in sensor-augmented pump therapy. Diabetes Technology and Therapeutics November 2011;13(11):1129–34.

[37] Ly TT, Nicholas JA, Retterath A, Lim EM, Davis EA, Jones TW. Effect of sensor-augmented insulin pump therapy and automated insulin suspension vs standard insulin pump therapy on hypoglycemia in patients with type 1 diabetes: a randomized clinical trial. Journal of the American Academy of Dermatology 2013;310(12):1240–7.

[38] Ly TT, Nicholas JA, Retterath A, Davis EA, Jones TW. Analysis of glucose responses to automated insulin suspension with sensor-augmented pump therapy. Diabetes Care July 2012;35(7):1462–5.

[39] Garg S, Brazg RL, Bailey TS, et al. Reduction in duration of hypoglycemia by automatic suspension of insulin delivery: the in-clinic ASPIRE study. Diabetes Technology and Therapeutics 2012;14:205–9.

[40] Bergenstal RM, Klonoff DC, Garg SK, Bode BW, Meredith M, Slover RH, Ahmann AJ, Welsh JB, Lee SW, Kaufman FR. ASPIRE In-Home Study Group. Threshold-based insulin-pump interruption for reduction of hypoglycemia. New England Journal of Medicine July 18, 2013;369(3):224−32.

[41] Sherr JL, Palau Collazo M, Cengiz E, et al. Safety of nighttime 2-hour suspension of Basal insulin in pump-treated type 1 diabetes even in the absence of low glucose. Diabetes Care 2014;37(3):773−9.

[42] Elbarbary NS. Effectiveness of the low-glucose suspend feature of insulin pump during fasting during Ramadan in type 1 diabetes mellitus. Diabetes/Metabolism Research and Reviews September 2016;32(6):623−33.

[43] Ahmed MH, Husain NE, Elmadhoun WM, Noor SK, Khalil AA, Almobarak AO. Diabetes and Ramadan: a concise and practical update. Journal of Family Medicine and Primary Care 2017 ;6(1):11−8.

[44] Sherman RE, Anderson SA, Dal Pan GJ, Gray GW, Gross T, Hunter NL, LaVange L, Marinac-Dabic D, Marks PW, Robb MA, Shuren J, Temple R, Woodcock J, Yue LQ, Califf RM. Real-world evidence - what is it and what can it tell us? New England Journal of Medicine December 8, 2016;375(23):2293−7.

[45] Agrawal P, Welsh JB, Kannard B, Askari S, Yang Q, Kaufman FR. Usage and effectiveness of the low glucose suspend feature of the Medtronic Paradigm Veo insulin pump. Journal of Diabetes Science and Technology September 1, 2011;5(5):1137−41.

[46] Gómez AM, Marín Carrillo LF, Muñoz Velandia OM, Rondón Sepúlveda MA, Arévalo Correa CM, Mora Garzón E, Cuervo Diaz MC, Henao Carrillo DC. Long-term efficacy and safety of sensor augmented insulin pump therapy with low-glucose suspend feature in patients with type 1 diabetes. Diabetes Technology and Therapeutics February 2017; 19(2):109−14.

[47] Agrawal P, Zhong A, Welsh JB, Shah R, Kaufman FR. Retrospective analysis of the real-world use of the threshold suspend feature of sensor-augmented insulin pumps. Diabetes Technology and Therapeutics May 2015;17(5):316−9.

[48] Russell LB, Gold MR, Siegel JE, Daniels N, Weinstein MC. The role of cost-effectiveness analysis in health and medicine. Panel on cost-effectiveness in health and medicine. Journal of the American Academy of Dermatology October 9, 1996; 276(14):1172−7.

[49] Ly TT, Brnabic AJ, Eggleston A, et al. A cost-effectiveness analysis of sensor-augmented insulin pump therapy and automated insulin suspension versus standard pump therapy for hypoglycemic unaware patients with type 1 diabetes. Value in Health 2014;17:561−9.

[50] Sussman M, Sierra J, Garg S, Bode B, Friedman M, Gill M, Kaufman F, Vigersky R, Menzin J. Economic impact of hypoglycemia among insulin-treated patients with diabetes. Journal of Medical Economics 2016;19(11):1099−106.

[51] Conget I, Martín-Vaquero P, Roze S, Elías I, Pineda C, Álvarez M, Delbaere A, Ampudia-Blasco FJ. Cost-effectiveness analysis of sensor-augmented pump therapy with low glucose-suspend in patients with type 1 diabetes mellitus and high risk of hypoglycemia in Spain. Endocrinology, Diabetes and Nutrition 2018 Aug - Sep;65(7): 380−6.

[52] Roze S, Smith-Palmer J, Valentine W, Payet V, de Portu S, Papo N, Cucherat M, Hanaire H. Cost-effectiveness of sensor-augmented pump therapy with low glucose suspend versus standard insulin pump therapy in two different patient populations with type 1 diabetes in France. Diabetes Technology and Therapeutics February 2016;18(2):75–84.

[53] National Institute for Health and Care Excellence (NICE) evidence overview. Type 1 diabetes: integrated sensor-augmented pump therapy systems for managing blood glucose levels. Available from: https://www.nice.org.uk/guidance/dg21/documents/type-1-diabetes-managing-blood-glucose-levels-the-minimed-paradigm-veo-system-and-the-vibe-and-g4-platinum-cgm-system-committee-papers2. [Accessed 18 December 2018].

[54] Weintraub WS, Cohen DJ. The limits of cost-effectiveness analysis. Circulation: Cardiovascular Quality and Outcomes January 2009;2(1):55–8.

[55] Shah VN, Shoskes A, Tawfik B, Garg SK. Closed-loop system in the management of diabetes: past, present, and future. Diabetes Technology and Therapeutics August 2014;16(8):477–90.

[56] Zhong A, Choudhary P, McMahon C, Agrawal P, Welsh JB, Cordero TL, Kaufman FR. Effectiveness of automated insulin management features of the MiniMed® 640G sensor-augmented insulin pump. Diabetes Technology and Therapeutics October 2016;18(10):657–63.

[57] De Valk HW, Lablanche S, Bosi E, Choudhary P, Silva JD, Castaneda J, Vorrink L, De Portu S, Cohen O. Study of MiniMed 640G insulin pump with SmartGuard in prevention of low glucose events in adults with type 1 diabetes (SMILE): design of a hypoglycemia prevention trial with continuous glucose monitoring data as outcomes. Diabetes Technology and Therapeutics November 2018;20(11):758–66.

[58] Beato-Víbora PI, Gil-Poch E, Galán-Bueno L, Lázaro-Martín L, Arroyo-Díez FJ. The incremental benefits of the predictive low-glucose suspend function compared to the low-glucose suspend function as automation against hypoglycemia in sensor-augmented pump therapy. Journal of Diabetes Science and Technology November 2018;12(6):1241–3.

[59] Forlenza GP, Li Z, Buckingham BA, Pinsker JE, Cengiz E, Wadwa RP, Ekhlaspour L, Church MM, Weinzimer SA, Jost E, Marcal T, Andre C, Carria L, Swanson V, Lum JW, Kollman C, Woodall W, Beck RW. Predictive low-glucose suspend reduces hypoglycemia in adults, adolescents, and children with type 1 diabetes in an at-home randomized crossover study: results of the PROLOG trial. Diabetes Care October 2018;41(10): 2155–61.

[60] Bergenstal RM, Garg S, Weinzimer SA, Buckingham BA, Bode BW, Tamborlane WV, Kaufman FR. Safety of a hybrid closed-loop insulin delivery system in patients with type 1 diabetes. Journal of the American Academy of Dermatology October 4, 2016; 316(13):1407–8.

[61] Garg SK, Weinzimer SA, Tamborlane WV, Buckingham BA, Bode BW, Bailey TS, Brazg RL, Ilany J, Slover RH, Anderson SM, Bergenstal RM, Grosman B, Roy A, Cordero TL, Shin J, Lee SW, Kaufman FR. Glucose outcomes with the in-home use of a hybrid closed-loop insulin delivery system in adolescents and adults with type 1 diabetes. Diabetes Technology and Therapeutics March 2017;19(3):155–63.

CHAPTER

Predictive low glucose suspend systems

14

Gregory P. Forlenza, MD [1], Laya Ekhlaspour, MD [2]

[1]*Assistant Professor, Barbara Davis Center, University of Colorado Denver, Aurora, CO, United States;* [2]*Instructor, Pediatric Endocrinology, Stanford University, Palo Alto, CA, United States*

Introduction

Hypoglycemia and fear of hypoglycemia have long been the major limiters to improved glycemic control via reduced hemoglobin A1c (HbA1c) for patients with type 1 diabetes (T1D). The Diabetes Control and Complications Trial famously highlighted the tradeoff between improved HbA1c and increased incidence of severe hypoglycemia [1–3]. For decades, the assumption among clinicians and patients had been that one must endure a certain amount of hypoglycemia to have adequate glycemic control. Without commercially available and reliable continuous glucose monitoring (CGM) systems, hypoglycemia was generally identified solely by symptoms among those without hypoglycemia unawareness. Other identification and potential prevention was achieved via routine scheduled self-monitoring of blood glucose (SMBG) testing. This approach created burden and anxiety especially during the overnight period when symptoms of hypoglycemia would not be present.

The physiology around hypoglycemia in patients with T1D is a well-studied phenomenon. The normal physiological response to hypoglycemia involves the shutoff of endogenous insulin production followed by counterregulation with the release of glucagon, cortisol, epinephrine, norepinephrine, and growth hormone [4]. In T1D, the counterregulatory response can become altered, likely due to repeated exposure to hypoglycemia [5]. In severe cases, hypoglycemia regulation can become so disrupted that patients experience hypoglycemia unawareness contributing to a viscous cycle of recurrent severe hypoglycemia and decreased hypoglycemia awareness [6]. Hypoglycemia responsiveness is also inherently impaired during sleep as has been documented by decreased epinephrine and increased arousal threshold in some studies [7,8].

In growing children, hypoglycemia—particularly recurrent severe hypoglycemia—can negatively impact neurocognitive development. A negative impact on memory, learning, intelligence, and verbal fluency has been documented in T1D children with recurrent severe hypoglycemia compared to T1D children who do not experience severe hypoglycemia [9]. Similarly, recurrent severe hypoglycemia has been documented to negatively impact spatial memory, particularly in young children [10]. Both electroencephalogram and magnetic resonance imaging have demonstrated

Glucose Monitoring Devices. https://doi.org/10.1016/B978-0-12-816714-4.00014-4

abnormalities in children who have experienced severe hypoglycemia [11]. It can thus be seen that trading hyperglycemia for hypoglycemia is not beneficial. The solution must be achieving normoglycemia.

In addition to its negative physiological implications, hypoglycemia also contributes to fear of hypoglycemia (FOH). FOH tends to develop after an episode of severe hypoglycemia or recurrent severe hypoglycemia and can be viewed as a sort of phobia of hypoglycemia [12–15]. Patients with FOH experience increased distress related to diabetes [16]. In addition, they tend to maintain higher glycemic control in an effort to avoid subsequent severe hypoglycemic episodes [17,18]. FOH thus contributes to both worsened glycemic control as well as increased distress in patients with diabetes and their families.

It is here where incorporation of diabetes technologies to advance an automated care becomes essential. Continuous subcutaneous insulin infusion (CSII) pumps were first developed in the late 1970s and began to see routine patient use in the late 1990s and early 2000s [19]. CSII pumps administer subcutaneous insulin at smaller dosing intervals than those achieved by multiple daily injection (MDI) therapy, deliver variable basal insulin rates throughout the day, and enable engaged patients or caregivers to set temporary basal rates or suspend basal insulin to further fine-tune insulin dosing. The use of insulin pumps has been associated with HbA1c values that are 0.3%–0.5% below those of MDI users [20–22]. Although some studies have demonstrated hypoglycemia reduction with CSII technology, its use has not been consistently associated with reduced hypoglycemia [23–25].

CGM first became commercially available in the early 2000s, and its use has grown exponentially over the past several years [20,26,27]. Early CGM devices suffered from mean absolute relative difference (MARD) values that were significantly worse than commercially available blood glucose meters [26,28,29]. These devices also relied heavily on threshold alarms to alert users about impending hypoglycemia [30]. Although the use of early generation CGMs was associated with hypoglycemia reduction, the alarms contributed significantly to patient fatigue and ultimately discontinuation of the devices [31–33]. Due to an increased arousal threshold overnight, CGM alarms have also not been helpful at eliminating severe nocturnal hypoglycemia. This is well highlighted in a case published by Buckingham where an adolescent with T1D slept through over 2 h of alarms overnight before a hypoglycemic seizure (Fig. 14.1) [34]. The information from CGM systems could thus be seen to contribute to improved glycemic control and hypoglycemia reduction but the burden of utilizing this information via direct self-management may have been overwhelming for most patients.

Automating hypoglycemia reduction via predictive algorithms and predictive low glucose suspend (PLGS) technology allows for hypoglycemia reduction without creating an increased burden on patients. PLGS systems could be considered a form of automated insulin delivery beyond simply suspending insulin below a certain hypoglycemic threshold—as is done in low glucose suspend devices—and before increasing insulin delivery to minimize hyperglycemia—as is done with hybrid closed-loop (or Artificial Pancreas) technology [35]. In the next sections, we will review the development of PLGS systems and the results from their use in various clinical trials and real-world settings.

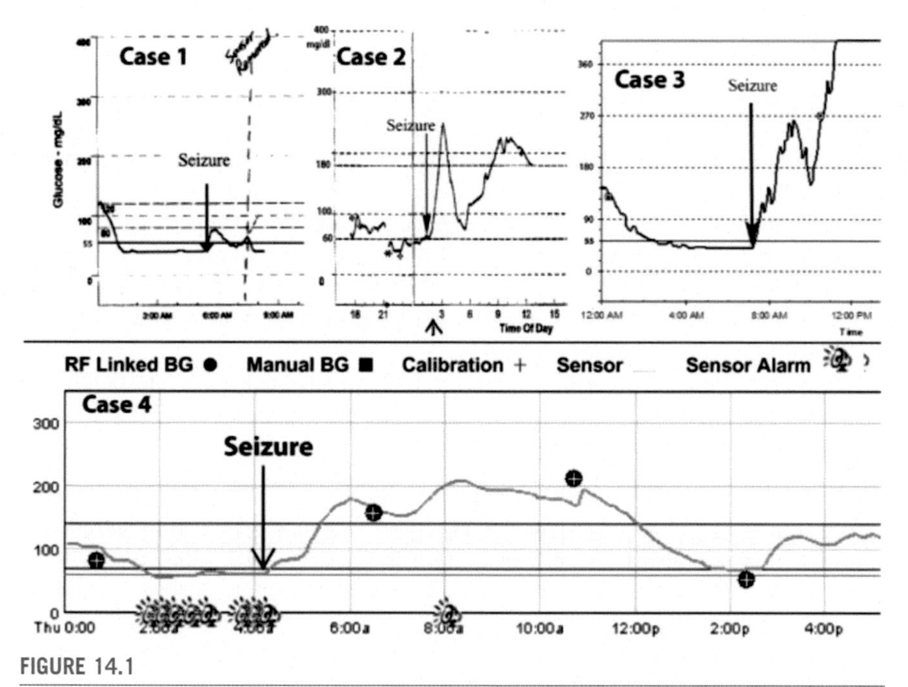

FIGURE 14.1

Case 1: 15-year-old male wearing an original Minimed CGMS. Case 2: 12-year-old female wearing a Minimed CGMS-Gold monitor. Case 3: 16-year-old female wearing the original Minimed CGMS. Case 4: 17-year-old female wearing a MiniMed 722 Paradigm real-time continuous glucose monitor. Alarm "bell" along the time axis at the bottom of the graph indicates alarming (vibratory and then audio). BG, blood glucose [34].

Algorithm development

The use of predictive algorithms to prospectively suspend insulin delivery before hypoglycemia is a very intuitive idea similar to how patients use their own devices in real time. Refinement and selection of appropriate algorithms for this purpose, however, was a process that took almost a decade to progress from theory to clinically approved devices. An early study of PLGS technology published by Buckingham in 2010 demonstrates the complexity of the initial approaches to achieve automated prevention of hypoglycemia [36]. The system tested in this study used a voting scheme between five different algorithms to recommend predictive suspension of insulin [37]. The algorithms deployed included the following:

(1) Linear projection: a projection of future glucose based on linear regression of the past 15 min of CGM values [37];
(2) Kalman filter: a prediction of future glucose based on Kalman filter estimation of glucose and its rate of change; this filter attempts to minimize the effect of sensor noise [37,38];

(3) Hybrid infinite impulse response filter: Uses linear discrete-time signal processing to generate output glucose predictions using previously measured glucose concentrations without the input of insulin infusion; this generates a prediction horizon through recursive application of the filter [37,39];
(4) Statistical prediction: Uses multiple empirical statistical models to generate a probability of hypoglycemia; the methods employed include (a) calibration, to convert CGM signal into an accurate blood glucose (BG), (b) prediction, using training data and recent calibrated BG history to generate predictions and accuracy estimates, (c) hypoglycemia alarming, to transform predictions and accuracy estimates into a probability of hypoglycemia, which is then thresholded into a binary alarm [37,40,41];
(5) Numerical logical algorithm: Feeds a three-point calculated rate of change using backward difference approximation and current CGM value into logical expressions to predict hypoglycemia [37].

The results from these initial trials demonstrated that the use of less voting systems to trigger pump suspension produced better hypoglycemia protection [36]. A subsequent in silico trial utilizing only a refined Kalman filter demonstrated that this simplified approach could reduce hypoglycemia by almost three quarters [42]. Outpatient safety testing of this Kalman filter-based algorithm demonstrated a significant reduction in overnight hypoglycemia, though with a mild increase in mean AM glucose values [43].

Commercial development of PLGS technology has resulted in two different systems: the MiniMed predictive low glucose management (PLGM) system found in the MiniMed 640G and as a feature in the MiniMed 670G, and the Tandem PLGS system available as Basal-IQ [44–46]. These commercial designs (see below for more details) are even more simplistic than those originally proposed by Buckingham as part of the voting system. The PLGM design uses linear regression to estimate future glucose values. The Basal-IQ system uses a linear regression model tuned on the last four sensor glucose values to predict the sensor glucose 30 min into the future.

The development of PLGS technology has thus favored the selection of more simplistic single algorithm designs that suspend basal insulin delivery based on linear prediction models. These systems utilize slightly different rules for the resumption of insulin delivery that may play a role in the postsuspension peak glucose values seen with the different designs. In the next section, we will review clinical trial data on the outpatient and real-world use of PLGS technology.

PLGS clinical studies

In addition to the continuum of developmental studies conducted earlier, there have been numerous clinical efficacy studies conducted on the PLGM and PLGS algorithms. Such studies have been conducted during inpatient, supervised outpatient, and real-world environments. The compilation of these studies is presented below (Table 14.1).

Table 14.1 Clinical studies testing the PLGS algorithms.

The system	N	Study primary outcome
Medtronic Paradigm Veo insulin pump + PLGM [50]	32	The requirement for hypoglycemia treatment. PLGM versus SAPT 86% versus 18% ($P \leq .001$).
Medtronic Paradigm Veo insulin pump + PLGM (predictive low glucose managemenet) software [51]	25	The requirement for hypoglycemia treatment (symptomatic hypoglycemia with plasma glucose <63 mg/dL or plasma glucose <50 mg/dL). PLGM versus SAPT (sensor-augmented pump therapy)32% versus 89%, $P = .003$.
The MiniMed 640G system [52]	40	Duration of pump suspension events: 56.4 ± 9.6 min The mean sensor glucose (SG) nadir:71.8 ± 5.2 mg/dL
MiniMed 640G system [47]	4818 individuals on MiniMed 640G 39,219 individuals on MiniMed 530G 3193 individuals on MiniMed Paradigm Veo system	Effectiveness of the system 640 G: 0.39% ≤ 50 mg/dL, 2.11% between 50 and 70 mg/dL 530 G: 0.7% ≤ 50 mg/dL, 3.4% between 50 and 70 mg/dL Paradigm Veo: 1.0% ≤ 50 mg/dL, 4.1% between 50 and 70 mg/dL
Medtronic MiniMed 640G pump [53]	96	The number of hypoglycemic events below 65 mg/dL. PLGM on versus PLGM off 4.4 ± 4.5 versus 7.4 ± 6.3, $P = .008$.
The MiniMed 640G + SmartGuard [54]	24	The frequency of hypoglycemic episodes SmartGuard versus SAP 0.72 ± 0.36 versus 1.02 ± 0.52 $P = .027$, hypoglycemic intensity (area under the curve) [AUC] and time <70 mg/dL 0.38 ± 0.24 versus 0.76 ± 0.73 $P = .027$
Medtronic MiniMed_ 640G [55]	27	Serious adverse events Hypoglycemia ≤65 mg/dL 60%.
MiniMed 640G [56]	21	Hypoglycemia frequency before versus after (10.4 ± 5.2% vs. 7.6 ± 3.3%, $P = 0.044$)
MiniMed 640G [57]	38	Safety and effectiveness of PLGM system Time<70 (%) PLGS on versus off 0.4 ± 0.4 versus 0.5 ± 0.9 0.4 ± 0.5 $P = .781$ (after 3 months) 0.2 ± 0.7 (after 6 months) 0.5 ± 0.9

Continued

Table 14.1 Clinical studies testing the PLGS algorithms.—*cont'd*

The system	N	Study primary outcome
MiniMed 640G with PLGS [58]	36	Mean sensor glucose and time with hypoglycemia PLGS versus LGS 160 ± 21 mg/dL versus 153 ± 21 mg/dL, $P = .002$. 3.0 ± 2.4 versus 4.1 ± 3.2, $P = .013$
Medtronic MiniMed 640G pump [48]	154	The average percentage of time spent in hypoglycemia (sensor glucose < 63 mg/dL) PLGM versus SAPT 1.5 versus 2.6, $P < .0001$
MiniMed-640G with PLGS [59]	162 (real world)	Hemoglobin A1 C and % SMBG < 70 mg/dL¶ 7.1 ± 0.7 versus 7.2 ± 0.8, $P = .33$. 6 ± 5 versus 10 ± 7, $P = .001$
Medtronic MiniMed 640G pump [60]	68	Time and duration of PLGM-initiated pump suspension. There were 20,183 suspend before low events in 8523 days (2.37 events/day). The mean suspend duration was 55.0 −32.7 min (day 50.0 ± 30.1, night 71.7 ± 35.1; $P < .001$).
Minimed 640G pump [61]	31	The time spent with BG levels below 70 mg/dL with a threshold of 90 versus 70 mg/dL 1.8 versus 2.3%
Statistical prediction algorithm [40]	22	Number of hypoglycemic episodes that were prevented in 30 versus 45 min prediction horizon (60% vs. 80%).
Five predictive alarm algorithms: Modified linear prediction alarm Kalman filtering Adaptive hybrid infinite impulse response (HIIR) filter Statistical prediction Numerical logical algorithm [36]	26	Number of hypoglycemic episodes that were prevented. Hypoglycemia was prevented for 71% of the events (15 events).
Revel CGM device + a hypoglycemia prediction algorithm (Kalman filter-based model) [43]	19	The primary safety outcomes were fasting blood glucose Three algorithms (interventions vs. control) 158 versus 125 151 versus 138 144 versus 133 Efficacy outcome: Percentage of nights with CGM ≤ 70 mg/dL 19% versus 26% 19% versus 33% 16% versus 30%

Table 14.1 Clinical studies testing the PLGS algorithms.—*cont'd*

The system	N	Study primary outcome
Kalman filter [42]	16	The algorithm prevented hypoglycemia in 73% of subjects
MiniMed Paradigm REAL-Time Veo system + a hypoglycemia prediction algorithm (Kalman filter-based model) [62]	45	The proportion of nights in which ≥1 CGM glucose values ≤ 60 mg/dL occurred. Intervention versus control 21 versus 33
MiniMed Paradigm REAL-Time Veo ystem + a hypoglycemia prediction algorithm (Kalman filter-based model) [63]	45	Percent time <70 mg/dL 4.6% in intervention nights versus 4.6% in control nights
Medtronic Veo insulin + a hypoglycemia prediction algorithm (Kalman filter-based model) [64]	127	PLGS phase: The number of blood glucose checks per age group (4−6 years) 75% of nights (7−10 years) 65% (11−14 years) 53% (15−25 years) 33% (26−45 years) 28% Boluses given per age group (4−6 years) 56% of nights (7−10 years) 48% (11−14 years) 33% (15−25 years) 20% (26−45 years) 25% Randomized clinical trial phase Skin reactions: (4−6 years) experienced more frequent and severe skin reactions ($P = .02$) (26−45 years) sensor wear was 26 h longer than the 4−6 years ($P < .001$).
Tandem Diabetes Care t:slim X2 with Basal-IQ Technology [49]	103	CGM-measured percentage of time <70 mg/dL 2.6 (1.4, 4.0) versus 3.2 (1.9, 6.1 (PLGS vs. SAP)
MiniMed 670G "suspend before low" feature [45]	79	Percentage of hypoglycemia ≤55 mg/dL 97.5% (77/79)

Within these trials, several are worth highlighting. The 2016 manuscript by Zhong looked at real-world CareLink data for PLGM and low glucose suspend (LGS) users [47]. This study compared 4818 MiniMed 640G PLGM users to 39,219 MiniMed 530G and 43,193 MiniMed Veo LGS users. They investigated days with advanced features enabled and those with advanced features disabled as well as compared the PLGM and LGS systems. Among the PLGM users, the use of the PLGM feature was associated with a lower percentage of CGM values $\leq$70 mg/dL and a lower percentage of values $\geq$240 mg/dL. For patients switching from an LGS system to a PLGM system, the use of the PLGM system was also associated with a decrease in the number of excursions $\leq$70 mg/dL and $\geq$240 mg/dL. These findings are notable because they reflect a large number of patients, using commercially available systems in the real world without study-level supervision and seeing significant improvement in extreme glycemic exposures. They are also important as they demonstrate a further reduction in extremes when going from an LGS to PLGM system.

A multicenter randomized controlled trial in children and adolescents conducted by Abraham compared the MiniMed 640G PLGM system against sensor-augmented pump therapy (SAPT) in 154 children and adolescents over 6 months of use [48]. This study demonstrated a reduction in hypoglycemia exposure <63 mg/dL for both groups with respect to baseline, though with a greater reduction in PLGM users than SAPT users (−1.4% vs. −0.4%, $P < .0001$). There was no difference in HbA1c values at 6 months between the PLGM and SAPT groups (7.8 ± 0.8% vs. 7.6 ± 1.0%; $P = .35$). This study is important as it has a very strong design as a real-world multicenter randomized controlled trial. In that context, the findings of the significant reduction in hypoglycemia exposure for PLGM compared to a strong control arm of SAPT provides a high level of support for the benefits of this technology.

The pivotal trial for the Tandem PLGS system was a multicenter randomized controlled crossover trial comparing PLGS to SAPT among 103 adults, adolescents and children [49]. This study demonstrated a 31% relative reduction in mean time <70 mg/dL for PLGS compared to SAPT without an increase in percent time >180 mg/dL. The findings from this study also demonstrated excellent system usability with >90% active system time. This study is notable because the randomized crossover design allows for another strong comparison of PLGS versus SAPT and the minimally supervised outpatient setting with broad patient age range allows for generalizability of the study results further supporting the benefits of PLGS technology.

Commercial devices

As it is imperative for clinicians to understand the function and tuning of commercial automated systems, we present the commercially available PLGS systems here using the CARE paradigm previously presented by our group (Table 14.2) [65]. In January 2015, Medtronic announced the international launch of MiniMed 640G. The components of MiniMed 640G (Medtronic, Northridge, California) are the MiniMed 640G insulin pump, Enlite Sensor, GuardianLink 2 transmitter and, SmartGuard feature. The SmartGaurd feature includes a PLGM algorithm that, as described earlier, suspends basal insulin delivery when the sensor glucose value is predicted to reach below a preset low glucose limit within 30 min. Specifically, the system suspends basal insulin when the predicted glucose is at or within 20 mg/dL of the

Table 14.2 Commercial PLGS systems.

	Minimed 640G/670G in open loop	Tandem Basal-IQ
Calculation	PLGS system - Suspends insulin when glucose predicted to be 20 mg/dL above the modifiable threshold of 60 −90 mg/dL in 30 min - Resumes after 30–120 min	PLGS system - Suspends insulin when glucose predicted to be < 80 mg/dL in 30 min - Resumes when glucose rising
Adjustment	Can modify: - All pump settings - Suspend threshold 50–90 mg/dL Cannot modify: N/A	Can modify: - All insulin pump settings Cannot modify: - Suspend threshold
Revert	n/a	n/a
Education	- Consider treating hypoglycemia with less CHO (e.g., 5–10) if the system has not delivered insulin (been suspended) for period of time before low glucose	- Consider treating hypoglycemia with less CHO if the system has not delivered insulin for a period of time before low glucose - System may suspend/resume insulin frequently, leave suspend alerts off for less interruptions
Sensor	Minimed Guardian 3 - Requires 2–4 calibrations for optimal use - 6–7 day sensor life - - Perform fingerstick BG for diabetes management decisions - -Important to calibrate when glucose is stable to prevent calibration errors	Dexcom G6 sensor - Factory calibrated sensor - 10 day sensor life - Can use sensor value for diabetes management if sensor value and arrow are present - Can remotely follow glucose levels with Follow app

modifiable lower threshold of 60–90 mg/dL in the next 30 min. The parameters for the resumption of insulin delivery are as follows: if the sensor glucose value is at least 20 mg/dL (1.1 mmol/L) above the preset low limit and predicted to be at least 40 mg/dL above the low limit in 30 min, and insulin has been suspended for at least 30 min. However, the patient can intervene during the suspend time and override the PLGM feature. The pump suspension does not exceed 2 h, and the system allows the user to receive an alert or not on the insulin resumption.

The t:slim X2 Insulin Pump with Basal-IQ Technology was approved by FDA in June 2018. Based on sensor data, the Basal-IQ algorithm predicts if glucose values fall below 80 mg/dL in the next 30 min or detect if the glucose level is currently below 70 mg/dL. The controller suspends insulin if any of these scenarios occur. The controller then resumes delivery the first time the system receives a CGM glucose reading higher than the previous reading, if glucose is no longer predicted to drop below 80 mg/dL, if no CGM data are available for 10 min, or if the insulin suspension exceeds 120 min in any 150-min period. The algorithm operates silently in the background when it suspends or resumes insulin delivery.

Keys to clinical use

Patient success with PLGS technology is inherently tied to successa maintaining a CGM and insulin infusion site. Thus the first steps to successfully relate to proper skincare. Some patients may have no problems properly maintaining the devices with minimal skin irritation, while for many, particularly pediatric patients, skincare will be the major barrier to success. Our center has previously published comprehensive guidelines on skincare for chronic device use [66]. For patients with allergies to the adhesives, we generally recommend off-label use of nasal steroids applied topically (e.g., fluticasone). We also recommend the use of protective barrier films. For removal, we recommend the use of adhesive removers and a slow low angle folding back of the adhesive tape [66].

Beyond tolerance of the on-body devices, optimization of CGM accuracy is essential to proper PLGS function. Thus, for sensors such as the Medtronic Guardian Sensor 3 which still requires periodic calibration, we recommend calibrating about three times per day, generally when the CGM signal is flat (i.e., trend arrows show less than 1 mg/dL/min of change) and not while the patient is hypoglycemic. Fingerstick reference values should be performed on hands washed with soap and water and then fully dried, and with an accurate glucose meter with a MARD of $<10\%$ [29]. For factory calibrated CGMs such as the Decom G6, we do not recommend additional calibrations as BG meter values may be less accurate than factory-supplied calibrations. We do however advise the patient to not extend sensor life beyond the approved period as this may strongly decrease sensor performance [67].

As insulin suspension to prevent hypoglycemia is intended to replicate normal physiology, we do not view system suspensions as a sign of a problem, but rather a normal aspect of diabetes care. When reviewing device downloads for patients using PLGS technology, it is not uncommon to see 4–8 suspension events per day. If the suspensions are occurring at varied times, we do not generally recommend adjusting pump settings to prevent suspensions from occurring. If, however, suspensions tend to occur after a similar event consistently, such as 2 h after a hyperglycemia correction or 2 h after a certain meal bolus, it may be beneficial to alter pump settings around that event. To avoid rebound hyperglycemia, we also educate patients that following an hour of insulin suspension, often the treatment of hypoglycemia is not required [54].

A more common trend among PLGS users is the request to increase insulin pump dosing after starting this technology. Many patients have reported that with increased confidence in these systems to prevent hypoglycemia, they feel more comfortable being aggressive at minimizing hyperglycemia with their preprogrammed basal rates, insulin-to-carbohydrate ratios, and even correction dosing. These changes should obviously be made cautiously and with the guidance of trained diabetes experts. Such adjustments, however, may help improve time in the target range and reduce HbA1c, as automation reduces hypoglycemia exposure in patients using PLGS technology.

Summary and conclusions

Tight glycemic management is challenging because of an increased risk for hypoglycemia, which can result in seizures, coma or other serious complications. Fear of hypoglycemia is a significant obstacle for optimal glycemic control for both patients with T1D and their families. To optimize blood glucose control, automated insulin delivery systems were introduced. Following the introduction of LGS systems, PLGS systems were successful in the reduction of hypoglycemic episodes.

In this chapter, we reviewed the development of the algorithms that automatically suspend insulin delivery to minimize the number and duration of hypoglycemic episodes. Both Medtronic PLGS system (MiniMed 640G) and Tandem Basal-IQ have been tested in clinical trials that showed a significant reduction in hypoglycemia without a significant increase in HbA1c. Moving forward with the advancement in automated insulin delivery and hybrid closed-loop systems, reduction of nocturnal hypoglycemia, improvement in the overnight glycemic control, and increase in overnight time in the range have been the main advantages for the patients with T1D. The controller in hybrid closed-loop systems adjusts insulin dosing based on data from CGM devices. A meal announcement is still a requirement. The management of postprandial glucose excursions and postexercise hypoglycemia remain challenging. Further studies testing a fully closed-loop system to assess the safety and efficacy of the systems in a free-living environment is necessary.

References

[1] The effect of intensive treatment of diabetes on the development and progression of long-term complications in insulin-dependent diabetes mellitus. The Diabetes Control and Complications trial Research Group. New England Journal of Medicine 1993; 329(14):977–86. https://doi.org/10.1056/NEJM199309303291401. PubMed PMID: 8366922.

[2] Adverse events and their association with treatment regimens in the diabetes control and complications trial. Diabetes Care 1995;18(11):1415–27. Epub 1995/11/01. PubMed PMID: 8722064.

[3] Hypoglycemia in the Diabetes Control and Complications Trial. The Diabetes Control and Complications Trial Research Group. Diabetes 1997;46(2):271–86. Epub 1997/02/01. PubMed PMID: 9000705.

[4] Bolli G, de Feo P, Compagnucci P, Cartechini MG, Angeletti G, Santeusanio F, Brunetti P, Gerich JE. Abnormal glucose counterregulation in insulin-dependent diabetes mellitus. Interaction of anti-insulin antibodies and impaired glucagon and epinephrine secretion. Diabetes 1983;32(2):134–41. Epub 1983/02/01. PubMed PMID: 6337896.

[5] Matyka K, Evans M, Lomas J, Cranston I, Macdonald I, Amiel SA. Altered hierarchy of protective responses against severe hypoglycemia in normal aging in healthy men. Diabetes Care 1997;20(2):135–41. Epub 1997/02/01. PubMed PMID: 9118760.

[6] Cryer PE. Mechanisms of hypoglycemia-associated autonomic failure in diabetes. New England Journal of Medicine 2013;369(4):362–72. https://doi.org/10.1056/NEJMra1215228. Epub 2013/07/26. PubMed PMID: 23883381.

[7] Jones TW, Porter P, Sherwin RS, Davis EA, O'Leary P, Frazer F, Byrne G, Stick S, Tamborlane WV. Decreased epinephrine responses to hypoglycemia during sleep. New England Journal of Medicine 1998;338(23):1657–62. https://doi.org/10.1056/nejm199806043382303. Epub 1998/06/06. PubMed PMID: 9614256.

[8] Ly TT, Gallego PH, Davis EA, Jones TW. Impaired awareness of hypoglycemia in a population-based sample of children and adolescents with type 1 diabetes. Diabetes Care 2009;32(10):1802–6. https://doi.org/10.2337/dc09-0541. Epub 2009/07/10. PubMed PMID: 19587370; PMCID: Pmc2752917.

[9] Blasetti A, Chiuri RM, Tocco AM, Di Giulio C, Mattei PA, Ballone E, Chiarelli F, Verrotti A. The effect of recurrent severe hypoglycemia on cognitive performance in children with type 1 diabetes: a meta-analysis. Journal of Child Neurology 2011; 26(11):1383–91. https://doi.org/10.1177/0883073811406730. Epub 2011/05/17. PubMed PMID: 21572053.

[10] Hershey T, Perantie DC, Warren SL, Zimmerman EC, Sadler M, White NH. Frequency and timing of severe hypoglycemia affects spatial memory in children with type 1 diabetes. Diabetes Care 2005;28(10):2372–7. Epub 2005/09/28. PubMed PMID: 16186265.

[11] Bjorgaas MR. Cerebral effects of severe hypoglycemia in young people with type 1 diabetes. Pediatric Diabetes 2012;13(1):100–7. https://doi.org/10.1111/j.1399-5448.2011.00803.x. Epub 2011/07/28. PubMed PMID: 21790920.

[12] Gonder-Frederick L, Nyer M, Shepard JA, Vajda K, Clarke W. Assessing fear of hypoglycemia in children with type 1 diabetes and their parents. Diabetes Management (London, England) 2011;1(6):627–39. https://doi.org/10.2217/dmt.11.60. Epub 2011/12/20. PubMed PMID: 22180760; PMCID: PMC3237051.

[13] Gonder-Frederick LA, Fisher CD, Ritterband LM, Cox DJ, Hou L, DasGupta AA, Clarke WL. Predictors of fear of hypoglycemia in adolescents with type 1 diabetes and their parents. Pediatric Diabetes 2006;7(4):215–22. https://doi.org/10.1111/j.1399-5448.2006.00182.x. Epub 2006/08/17. PubMed PMID: 16911009.

[14] Driscoll KA, Raymond J, Naranjo D, Patton SR. Fear of hypoglycemia in children and adolescents and their parents with type 1 diabetes. Current Diabetes Reports 2016; 16(8):77. https://doi.org/10.1007/s11892-016-0762-2. Epub 2016/07/03. PubMed PMID: 27370530; PMCID: PMC5371512.

[15] Shepard JA, Vajda K, Nyer M, Clarke W, Gonder-Frederick L. Understanding the construct of fear of hypoglycemia in pediatric type 1 diabetes. Journal of Pediatric Psychology 2014;39(10):1115–25. https://doi.org/10.1093/jpepsy/jsu068. Epub 2014/09/13. PubMed PMID: 25214644; PMCID: PMC4201766.

[16] Patton SR, Dolan LM, Smith LB, Thomas IH, Powers SW. Pediatric parenting stress and its relation to depressive symptoms and fear of hypoglycemia in parents of young children with type 1 diabetes mellitus. Journal of Clinical Psychology in Medical Settings 2011;18(4):345–52. https://doi.org/10.1007/s10880-011-9256-1. Epub 2011/06/07. PubMed PMID: 21643962; PMCID: PMC3199319.

[17] Viaene AS, Van Daele T, Bleys D, Faust K, Massa GG. Fear of hypoglycemia, parenting stress, and metabolic control for children with type 1 diabetes and their parents. Journal of Clinical Psychology in Medical Settings 2017;24(1):74–81. https://doi.org/10.1007/s10880-017-9489-8. Epub 2017/03/11. PubMed PMID: 28280962.

[18] Freckleton E, Sharpe L, Mullan B. The relationship between maternal fear of hypoglycaemia and adherence in children with type-1 diabetes. International Journal of Behavioral Medicine 2014;21(5):804–10. https://doi.org/10.1007/s12529-013-9360-8. Epub 2013/11/06. PubMed PMID: 24190791.

[19] Forlenza GP, Buckingham B, Maahs DM. Progress in diabetes technology: developments in insulin pumps, continuous glucose monitors, and progress towards the artificial pancreas. The Journal of Pediatrics 2016;169:13–20. https://doi.org/10.1016/j.jpeds.2015.10.015. Epub 2015/11/09. PubMed PMID: 26547403.

[20] Foster NC, Beck RW, Miller KM, Clements MA, Rickels MR, DiMeglio LA, Maahs DM, Tamborlane WV, Bergenstal R, Smith E, Olson BA, Garg SK. State of type 1 diabetes management and outcomes from the T1D exchange in 2016–2018. Diabetes Technology and Therapeutics 2019. https://doi.org/10.1089/dia.2018.0384. Epub 2019/01/19. PubMed PMID: 30657336.

[21] Miller KM, Foster NC, Beck RW, Bergenstal RM, DuBose SN, DiMeglio LA, Maahs DM, Tamborlane WV. Current state of type 1 diabetes treatment in the U.S.: updated data from the T1D exchange clinic registry. Diabetes Care 2015;38(6):971–8. https://doi.org/10.2337/dc15-0078. Epub 2015/05/23. PubMed PMID: 25998289.

[22] Sherr JL, Hermann JM, Campbell F, Foster NC, Hofer SE, Allgrove J, Maahs DM, Kapellen TM, Holman N, Tamborlane WV, Holl RW, Beck RW, Warner JT. Use of insulin pump therapy in children and adolescents with type 1 diabetes and its impact on metabolic control: comparison of results from three large, transatlantic paediatric registries. Diabetologia 2016;59(1):87–91. https://doi.org/10.1007/s00125-015-3790-6. Epub 2015/11/08. PubMed PMID: 26546085.

[23] Karges B, Schwandt A, Heidtmann B, Kordonouri O, Binder E, Schierloh U, Boettcher C, Kapellen T, Rosenbauer J, Holl RW. Association of insulin pump therapy vs insulin injection therapy with severe hypoglycemia, ketoacidosis, and glycemic control among children, adolescents, and young adults with type 1 diabetes. JAMA 2017;318(14):1358–66. https://doi.org/10.1001/jama.2017.13994. Epub 2017/10/20. PubMed PMID: 29049584; PMCID: PMC5818842.

[24] Fox LA, Buckloh LM, Smith SD, Wysocki T, Mauras N. A randomized controlled trial of insulin pump therapy in young children with type 1 diabetes. Diabetes Care 2005;28(6):1277–81. Epub 2005/05/28. PubMed PMID: 15920039.

[25] DiMeglio LA, Pottorff TM, Boyd SR, France L, Fineberg N, Eugster EA. A randomized, controlled study of insulin pump therapy in diabetic preschoolers. The Journal of Pediatrics 2004;145(3):380–4. https://doi.org/10.1016/j.jpeds.2004.06.022. Epub 2004/09/03. PubMed PMID: 15343195.

[26] Berget C, Messer LH, Vigers T, Frohnert BI, Pyle L, Wadwa RP, Driscoll KA, Forlenza GP. Six months of hybrid closed loop in the real-world: an evaluation of children and young adults using the 670G system. Epub 2019/12/15 Pediatr Diabetes 2019. https://doi.org/10.1111/pedi.12962. PubMed PMID: 31837064.

[27] DeSalvo DJ, Miller KM, Hermann JM, Maahs DM, Hofer SE, Clements MA, Lilienthal E, Sherr JL, Tauschmann M, Holl RW. Continuous glucose monitoring (CGM) and glycemic control among youth with type 1 diabetes (T1D): international comparison from the T1D exchange and DPV initiative. Pediatric Diabetes 2018. https://doi.org/10.1111/pedi.12711. Epub 2018/06/21. PubMed PMID: 29923262.

[28] Facchinetti A. Continuous glucose monitoring sensors: past, present and future algorithmic challenges. Sensors (Basel, Switzerland) 2016;16(12):2093–104. https://doi.org/10.3390/s16122093. Epub 2016/12/13. PubMed PMID: 27941663; PMCID: PMC5191073.

[29] Ekhlaspour L, Mondesir D, Lautsch N, Balliro C, Hillard M, Magyar K, Radocchia LG, Esmaeili A, Sinha M, Russell SJ. Comparative accuracy of 17 point-of-care glucose meters. Journal of Diabetes Science and Technology 2016. https://doi.org/10.1177/1932296816672237. Epub 2016/10/05. PubMed PMID: 27697848.

[30] Buckingham B, Block J, Burdick J, Kalajian A, Kollman C, Choy M, Wilson DM, Chase P. Response to nocturnal alarms using a real-time glucose sensor. Diabetes Technology and Therapeutics 2005;7(3):440–7. https://doi.org/10.1089/dia.2005.7.440. Epub 2005/06/03. PubMed PMID: 15929675; PMCID: PMC1482828.

[31] Battelino T, Phillip M, Bratina N, Nimri R, Oskarsson P, Bolinder J. Effect of continuous glucose monitoring on hypoglycemia in type 1 diabetes. Diabetes Care 2011; 34(4):795–800. https://doi.org/10.2337/dc10-1989. Epub 2011/02/22. PubMed PMID: 21335621; PMCID: PMC3064030.

[32] Naranjo D, Tanenbaum ML, Iturralde E, Hood KK. Diabetes technology: uptake, outcomes, barriers, and the intersection with distress. Journal of Diabetes Science and Technology 2016;10(4):852–8. https://doi.org/10.1177/1932296816650900. Epub 2016/05/29. PubMed PMID: 27234809; PMCID: PMC4928242.

[33] Tanenbaum ML, Hanes SJ, Miller KM, Naranjo D, Bensen R, Hood KK. Diabetes device use in adults with type 1 diabetes: barriers to uptake and potential intervention targets. Diabetes Care 2017;40(2):181–7. https://doi.org/10.2337/dc16-1536. Epub 2016/12/03. PubMed PMID: 27899489.

[34] Buckingham B, Wilson DM, Lecher T, Hanas R, Kaiserman K, Cameron F. Duration of nocturnal hypoglycemia before seizures. Diabetes Care 2008;31(11):2110–2. https://doi.org/10.2337/dc08-0863. Epub 2008/08/13. PubMed PMID: 18694975; PMCID: PMC2571056.

[35] Kowalski A. Pathway to artificial pancreas systems revisited: moving downstream. Diabetes Care 2015;38(6):1036–43. https://doi.org/10.2337/dc15-0364. Epub 2015/05/23. PubMed PMID: 25998296.

[36] Buckingham B, Chase HP, Dassau E, Cobry E, Clinton P, Gage V, Caswell K, Wilkinson J, Cameron F, Lee H, Bequette BW, Doyle III FJ. Prevention of nocturnal hypoglycemia using predictive alarm algorithms and insulin pump suspension. Diabetes Care 2010;33(5):1013–7. https://doi.org/10.2337/dc09-2303. Epub 2010/03/05. PubMed PMID: 20200307; PMCID: Pmc2858164.

[37] Dassau E, Cameron F, Lee H, Bequette BW, Zisser H, Jovanovic L, Chase HP, Wilson DM, Buckingham BA, Doyle 3rd FJ. Real-Time hypoglycemia prediction suite using continuous glucose monitoring: a safety net for the artificial pancreas. Diabetes Care 2010;33(6):1249–54. https://doi.org/10.2337/dc09-1487. Epub 2010/05/29. PubMed PMID: 20508231; PMCID: PMC2875433.

[38] Palerm CC, Bequette BW. Hypoglycemia detection and prediction using continuous glucose monitoring-a study on hypoglycemic clamp data. Journal of Diabetes Science and Technology 2007;1(5):624–9. https://doi.org/10.1177/193229680700100505. Epub 2007/09/01. PubMed PMID: 19885130; PMCID: PMC2769657.

[39] Oppenheim AV, Schafer RW, Buck JR. Discrete-time signal processing. Upper Saddle River, NJ: Prentice-Hall; 1998.

[40] Buckingham B, Cobry E, Clinton P, Gage V, Caswell K, Kunselman E, Cameron F, Chase HP. Preventing hypoglycemia using predictive alarm algorithms and insulin pump suspension. Diabetes Technology and Therapeutics 2009;11(2):93−7. https://doi.org/10.1089/dia.2008.0032. Epub 2009/10/24. PubMed PMID: 19848575; PMCID: PMC2979338.

[41] Cameron F, Niemeyer G, Gundy-Burlet K, Buckingham B. Statistical hypoglycemia prediction. Journal of Diabetes Science and Technology 2008;2(4):612−21. https://doi.org/10.1177/193229680800200412. Epub 2008/07/01. PubMed PMID: 19885237; PMCID: PMC2769757.

[42] Cameron F, Wilson DM, Buckingham BA, Arzumanyan H, Clinton P, Chase HP, Lum J, Maahs DM, Calhoun PM, Bequette BW. Inpatient studies of a Kalman-filter-based predictive pump shutoff algorithm. Journal of Diabetes Science and Technology 2012;6(5): 1142−7. https://doi.org/10.1177/193229681200600519. Epub 2012/10/16. PubMed PMID: 23063041; PMCID: PMC3570849.

[43] Buckingham BA, Cameron F, Calhoun P, Maahs DM, Wilson DM, Chase HP, Bequette BW, Lum J, Sibayan J, Beck RW, Kollman C. Outpatient safety assessment of an in-home predictive low-glucose suspend system with type 1 diabetes subjects at elevated risk of nocturnal hypoglycemia. Diabetes Technology and Therapeutics 2013;15(8):622−7. https://doi.org/10.1089/dia.2013.0040. Epub 2013/07/26. PubMed PMID: 23883408; PMCID: PMC3746249.

[44] Abraham MB, Nicholas JA, Ly TT, Roby HC, Paramalingam N, Fairchild J, King BR, Ambler GR, Cameron F, Davis EA, Jones TW. Safety and efficacy of the predictive low glucose management system in the prevention of hypoglycaemia: protocol for randomised controlled home trial to evaluate the Suspend before low function. BMJ Open 2016;6(4):e011589. https://doi.org/10.1136/bmjopen-2016-011589. Epub 2016/04/17. PubMed PMID: 27084290; PMCID: PMC4838718.

[45] Wood MA, Shulman DI, Forlenza GP, Bode BW, Pinhas-Hamiel O, Buckingham BA, Kaiserman KB, Liljenquist DR, Bailey TS, Shin J, Huang S, Chen X, Cordero TL, Lee SW, Kaufman FR. In-clinic evaluation of the MiniMed 670G system "suspend before low" feature in children with type 1 diabetes. Diabetes Technology and Therapeutics 2018. https://doi.org/10.1089/dia.2018.0209. Epub 2018/10/10. PubMed PMID: 30299976.

[46] Buckingham BA, Christiansen MP, Forlenza GP, Wadwa RP, Peyser TA, Lee JB, O'Connor J, Dassau E, Huyett LM, Layne JE, Ly TT. Performance of the omnipod personalized model predictive control algorithm with meal bolus challenges in adults with type 1 diabetes. Diabetes Technology and Therapeutics 2018. https://doi.org/10.1089/dia.2018.0138. Epub 2018/08/03. PubMed PMID: 30070928.

[47] Zhong A, Choudhary P, McMahon C, Agrawal P, Welsh JB, Cordero TL, Kaufman FR. Effectiveness of automated insulin management features of the MiniMed((R)) 640G sensor-augmented insulin pump. Diabetes Technology and Therapeutics 2016;18(10): 657−63. https://doi.org/10.1089/dia.2016.0216. Epub 2016/10/18. PubMed PMID: 27672710; PMCID: PMC5111481 were employees of Medtronic during development of this manuscript. P.C. has received research support from Medtronic.

[48] Abraham MB, Nicholas JA, Smith GJ, Fairchild JM, King BR, Ambler GR, Cameron FJ, Davis EA, Jones TW. Reduction in hypoglycemia with the predictive low-glucose management system: a long-term randomized controlled trial in adolescents with type 1 diabetes. Diabetes Care 2018;41(2):303–10. https://doi.org/10.2337/dc17-1604. Epub 2017/12/02. PubMed PMID: 29191844.

[49] Forlenza GP, Li Z, Buckingham BA, Pinsker JE, Cengiz E, Wadwa RP, Ekhlaspour L, Church MM, Weinzimer SA, Jost E, Marcal T, Andre C, Carria L, Swanson V, Lum JW, Kollman C, Woodall W, Beck RW. Predictive low-glucose suspend reduces hypoglycemia in adults, adolescents, and children with type 1 diabetes in an at-home randomized crossover study: results of the PROLOG trial. Diabetes Care 2018;41(10):2155–61. https://doi.org/10.2337/dc18-0771. Epub 2018/08/10. PubMed PMID: 30089663.

[50] Abraham MB, de Bock M, Paramalingam N, O'Grady MJ, Ly TT, George C, Roy A, Spital G, Karula S, Heels K, Gebert R, Fairchild JM, King BR, Ambler GR, Cameron F, Davis EA, Jones TW. Prevention of insulin-induced hypoglycemia in type 1 diabetes with predictive low glucose management system. Diabetes Technology and Therapeutics 2016;18(7):436–43. https://doi.org/10.1089/dia.2015.0364. Epub 2016/05/06. PubMed PMID: 27148807.

[51] Abraham MB, Davey R, O'Grady MJ, Ly TT, Paramalingam N, Fournier PA, Roy A, Grosman B, Kurtz N, Fairchild JM, King BR, Ambler GR, Cameron F, Jones TW, Davis EA. Effectiveness of a predictive algorithm in the prevention of exercise-induced hypoglycemia in type 1 diabetes. Diabetes Technology and Therapeutics 2016;18(9):543–50. https://doi.org/10.1089/dia.2016.0141. Epub 2016/08/10. PubMed PMID: 27505305.

[52] Choudhary P, Olsen BS, Conget I, Welsh JB, Vorrink L, Shin JJ. Hypoglycemia prevention and user acceptance of an insulin pump system with predictive low glucose management. Diabetes Technology and Therapeutics 2016;18(5):288–91. https://doi.org/10.1089/dia.2015.0324. Epub 2016/02/26. PubMed PMID: 26907513; PMCID: PMC4870649.

[53] Battelino T, Nimri R, Dovc K, Phillip M, Bratina N. Prevention of hypoglycemia with predictive low glucose insulin suspension in children with type 1 diabetes: a randomized controlled trial. Diabetes Care 2017;40(6):764–70. https://doi.org/10.2337/dc16-2584. Epub 2017/03/30. PubMed PMID: 28351897.

[54] Biester T, Kordonouri O, Holder M, Remus K, Kieninger-Baum D, Wadien T, Danne T. Let the algorithm do the work": reduction of hypoglycemia using sensor-augmented pump therapy with predictive insulin suspension (SmartGuard) in pediatric type 1 diabetes patients. Diabetes Technology and Therapeutics 2017;19(3):173–82. https://doi.org/10.1089/dia.2016.0349. Epub 2017/01/19. PubMed PMID: 28099035; PMCID: PMC5359639.

[55] Buckingham BA, Bailey TS, Christiansen M, Garg S, Weinzimer S, Bode B, Anderson SM, Brazg R, Ly TT, Kaufman FR. Evaluation of a predictive low-glucose management system in-clinic. Diabetes Technology and Therapeutics 2017;19(5):288–92. https://doi.org/10.1089/dia.2016.0319. Epub 2017/02/22. PubMed PMID: 28221823.

[56] Villafuerte Quispe B, Martin Frias M, Roldan Martin MB, Yelmo Valverde R, Alvarez Gomez MA, Barrio Castellanos R. Effectiveness of MiniMed 640G with SmartGuard(R) System for prevention of hypoglycemia in pediatric patients with type 1 diabetes mellitus. Endocrinología, Diabetes y Nutrición 2017;64(4):198–203. https://doi.org/10.1016/j.endinu.2017.02.008. Epub 2017/04/19. PubMed PMID: 28417874.

[57] Scaramuzza AE, Arnaldi C, Cherubini V, Piccinno E, Rabbone I, Toni S, Tumini S, Candela G, Cipriano P, Ferrito L, Lenzi L, Tinti D, Cohen O, Lombardo F. Use of the predictive low glucose management (PLGM) algorithm in Italian adolescents with type 1 diabetes: CareLink data download in a real-world setting. Acta Diabetologica 2017;54(3):317–9. https://doi.org/10.1007/s00592-016-0927-0. Epub 2016/10/17. PubMed PMID: 27744516.

[58] Beato-Vibora PI, Gil-Poch E, Galan-Bueno L, Lazaro-Martin L, Arroyo-Diez FJ. The incremental benefits of the predictive low-glucose suspend function compared to the low-glucose suspend function as automation against hypoglycemia in sensor-augmented pump therapy. Journal of Diabetes Science and Technology 2018;12(6): 1241–3. https://doi.org/10.1177/1932296818791536. Epub 2018/07/31. PubMed PMID: 30058373; PMCID: PMC6232727.

[59] Beato-Vibora PI, Quiros-Lopez C, Lazaro-Martin L, Martin-Frias M, Barrio-Castellanos R, Gil-Poch E, Arroyo-Diez FJ, Gimenez-Alvarez M. Impact of sensor-augmented pump therapy with predictive low-glucose suspend function on glycemic control and patient satisfaction in adults and children with type 1 diabetes. Diabetes Technology and Therapeutics 2018;20(11):738–43. https://doi.org/10.1089/dia.2018.0199. Epub 2018/09/27. PubMed PMID: 30256132.

[60] Abraham MB, Smith GJ, Nicholas JA, Fairchild JM, King BR, Ambler GR, Cameron FJ, Davis EA, Jones TW, Group PS. Characteristics of automated insulin suspension and glucose responses with the predictive low-glucose management system. Diabetes Technology and Therapeutics 2019;21(1):28–34. https://doi.org/10.1089/dia.2018.0205. Epub 2018/12/27. PubMed PMID: 30585769.

[61] Cherubini V, Gesuita R, Skrami E, Rabbone I, Bonfanti R, Arnaldi C, D'Annunzio G, Frongia A, Lombardo F, Piccinno E, Schiaffini R, Toni S, Tumini S, Tinti D, Cipriano P, Minuto N, Lenzi L, Ferrito L, Ventrici C, Ortolani F, Cohen O, Scaramuzza A. Optimal predictive low glucose management settings during physical exercise in adolescents with type 1 diabetes. Pediatric Diabetes 2019;20(1):107–12. https://doi.org/10.1111/pedi.12792. Epub 2018/11/01. PubMed PMID: 30378759.

[62] Maahs DM, Calhoun P, Buckingham BA, Chase HP, Hramiak I, Lum J, Cameron F, Bequette BW, Aye T, Paul T, Slover R, Wadwa RP, Wilson DM, Kollman C, Beck RW. A randomized trial of a home system to reduce nocturnal hypoglycemia in type 1 diabetes. Diabetes Care 2014;37(7):1885–91. https://doi.org/10.2337/dc13-2159. Epub 2014/05/09. PubMed PMID: 24804697; PMCID: PMC4067393.

[63] Buckingham BA, Raghinaru D, Cameron F, Bequette BW, Chase HP, Maahs DM, Slover R, Wadwa RP, Wilson DM, Ly T, Aye T, Hramiak I, Clarson C, Stein R, Gallego PH, Lum J, Sibayan J, Kollman C, Beck RW. Predictive low-glucose insulin suspension reduces duration of nocturnal hypoglycemia in children without increasing ketosis. Diabetes Care 2015;38(7):1197–204. https://doi.org/10.2337/dc14-3053. Epub 2015/06/08. PubMed PMID: 26049549; PMCID: PMC4477332.

[64] Messer LH, Calhoun P, Buckingham B, Wilson DM, Hramiak I, Ly TT, Driscoll M, Clinton P, Maahs DM. In-home nighttime predictive low glucose suspend experience in children and adults with type 1 diabetes. Pediatric Diabetes 2017;18(5):332–9. https://doi.org/10.1111/pedi.12395. Epub 2016/04/30. PubMed PMID: 27125223; PMCID: PMC5086306.

[65] Messer LH, Forlenza GP, Wadwa RP, Weinzimer SA, Sherr JL, Hood KK, Buckingham BA, Slover RH, Maahs DM. The dawn of automated insulin delivery: a new clinical framework to conceptualize insulin administration. Pediatric Diabetes 2017. https://doi.org/10.1111/pedi.12535. Epub 2017/06/29. PubMed PMID: 28656656.
[66] Messer LH, Berget C, Beatson C, Polsky S, Forlenza GP. Preserving skin integrity with chronic device use in diabetes. Diabetes Technology and Therapeutics 2018;20(S2): S254–64. https://doi.org/10.1089/dia.2018.0080. Epub 2018/06/20. PubMed PMID: 29916740.
[67] Forlenza GP, Kushner T, Messer LH, Wadwa RP, Sankaranarayanan S. Factory-calibrated continuous glucose monitoring: how and why it works, and the dangers of reuse beyond approved duration of wear. Diabetes Technology and Therapeutics 2019. https://doi.org/10.1089/dia.2018.0401. Epub 2019/03/01. PubMed PMID: 30817171.

CHAPTER

Automated closed-loop insulin delivery: system components, performance, and limitations

15

Mudassir Rashid, PhD, BEng [1], Iman Hajizadeh, MSc [2], Sediqeh Samadi, MSc [3], Mert Sevil, MSc [4], Nicole Hobbs, BSc [5], Rachel Brandt, BSc [6], Ali Cinar, PhD [7]

[1] *Senior Research Associate, Department of Chemical and Biological Engineering, Illinois Institute of Technology, Chicago, IL, United States;* [2] *Research Assistant and PhD Student, Chemical and Biological Engineering, Illinois Institute of Technology, Chicago, IL, United States;* [3] *Illinois Institute of Technology, Chemical and Biological Engineering, Chicago, IL, United States;* [4] *Research Assistant and PhD Student, Biomedical Engineering, Illinois Institute of Technology, Chicago, IL, United States;* [5] *Graduate Research Assistant, Department of Biomedical Engineering, Illinois Institute of Technology, Chicago, IL, United States;* [6] *Illinois Institute of Technology, Biomedical Engineering, Chicago, IL, United States;* [7] *Professor, Chemical and Biological Engineering Department, Illinois Institute of Technology, Chicago, IL, United States*

Introduction

Automated insulin delivery provides an advanced technology for the treatment of type 1 diabetes mellitus (T1DM). Despite rapid advancements in insulin analogs and delivery mechanisms, achieving optimal glycemic control and maintaining blood glucose concentrations (BGC) within a safe target range (BGC within 70−180 mg/dL) remains challenging. Moreover, intensive insulin therapy can cause a higher risk of hypoglycemia (BGC $<$ 70 mg/dL), weight gain, and a greater burden of chronic disease self-management [1,2]. In particular, physical activity continues to be a significant impediment to tighter glycemic regulation due to difficult to detect physiologic and metabolic variations such as increased energy expenditure, rapid fluctuations in insulin sensitivity, and possible activation of counterregulatory hormones [3−6]. Advances in technologies that provide intensive insulin therapy and improve glycemic control while minimizing hypoglycemia risk and the burden of disease self-management are needed.

This chapter outlines the components and modules involved in automated insulin delivery systems for regulating BGC. The remainder of the introduction section describes the closed-loop automated insulin delivery systems, provides brief descrip-

Glucose Monitoring Devices. https://doi.org/10.1016/B978-0-12-816714-4.00015-6

tions of the control algorithms and insulin dosage computations involved in the insulin delivery systems, highlights the disturbances affecting glycemic control, and introduces the various paradigms developed to address the challenges to tight glycemic control.

The emergence of continuous glucose monitoring (CGM) sensors, providing real-time measurements of subcutaneous glucose levels, signified a major step toward improved diabetes monitoring and treatment [7]. CGM enables frequent feedback to make corrections and appropriate changes in insulin delivery. CGMs enable users to take preventative measures and make adjustments in insulin therapy based on real-time interstitial glucose readings and alerts for impending hypoglycemia or hyperglycemia excursions [8–11]. Sensor-augmented pump therapy that combines CGMs with continuous subcutaneous insulin infusion (CSII) pumps is shown to improve glycemic control compared with multiple daily injection therapy [11,12] and provides increased functionality that includes personalized bolus calculators, preprogrammed temporary insulin infusion suspension based on preset hypoglycemic thresholds, and automated insulin delivery.

Closed-loop control of glucose concentrations through the pairing of CGM sensors and CSII pumps using control algorithms builds upon the concept of sensor-responsive insulin delivery and is a topic of significant interest [13]. Closing the loop between glucose concentration sensing and insulin infusion through control algorithms, termed automated insulin delivery or artificial pancreas (AP) systems, allows the algorithms to automatically adjust the insulin infusion in real time based on feedback from the CGM sensors [1,14–17]. The core of the AP system is the control algorithm that computes the appropriate amount of insulin to administer to subjects [18,19]. Various control algorithms are developed for AP systems to autonomously manipulate the subcutaneous delivery of insulin on the basis of real-time sensor glucose levels, including proportional-integral-derivative (PID) control, fuzzy logic control, neural networks, and model predictive control (MPC) [9,15,16,20–30]. The classical PID controller manipulates insulin delivery by assessing the deviation of current glucose measurements from the target glucose level (the proportional component), the area under the curve between measured and target glucose levels (the integral component), and the rate of change in the measured glucose level (the derivative component). Fuzzy logic control adjusts the insulin infusion rate based on approximate encoded rules that mathematically express the empirical clinician knowledge acquired by diabetes practitioners. Artificial neural networks approximate nonlinear uncertain systems that are then readily exploited in the synthesis of nonlinear controllers. Among the AP control algorithms, MPC, a control strategy based on optimal control concepts, has become increasingly prevalent because they have theoretically proven closed-loop stability properties, are readily able to handle complex multivariable systems, and can systematically deal with state and input constraints [14,31–33].

MPC algorithms utilize dynamic models of the system in the optimization problem to predict the future evolution of the glucose measurements over a finite-time horizon to determine the optimal future insulin infusion rates with respect to a specified performance index. Furthermore, MPC can explicitly consider the system constraints and multivariable interactions in the optimization problem, and the MPC formulations are not inexorably restricted by the type of model, objective function, or constraints. As predictive controllers compute the optimal control actions on the basis of a model and cost function, the glucose control performance is substantiated by the fidelity of the glucose-insulin models. The MPC formulations employed in many early clinical trials are synthesized using an average linear model relating insulin to glucose concentrations. The use of a general linear model to describe the complex nonlinear glycemic dynamics is compelled by the feasibility and computational complexity of the controller implementation on portable systems with limited computational and power resources. However, subjects exhibit diverse glucose-insulin dynamics over time (intrasubject variability) and in relation to other subjects (intersubject variability) due to diverse time-varying biological and physiological characteristics of people. Adaptive and patient-individualized glucose-insulin models can thus improve the glucose control performance [21,34–37].

An adaptive and customized MPC formulation with tailored glycemic models developed through efficient identification techniques can improve the glucose control performance. Despite these controller enhancements, the closed-loop control performance of glucose regulation is restricted by the exclusive reliance on glycemic measurements and trends. This restrictive observation into the physiological characteristics of people with T1DM limits the control performance, especially when physical activity is confronted, leaving subjects to manually manage glycemic variations during and after exercise. Conducting moderate-intensity aerobic exercise lowers glucose concentrations as sensitivity to insulin and glucose uptake to working muscles increases. Although moderating insulin administration before exercise can alleviate exercise-induced hypoglycemia, it requires prior planning and deliberate organization, which is demanding and disregards spontaneous activities. Neglecting the effects of physical activity in glycemic models and closed-loop insulin control can contribute to a worsening of glycemic control [38]. These limitations can be addressed by explicitly considering physiological measurements from wearable devices, such as heart rate and energy expenditure, that are representative of physical activity. Incorporating these additional variables in glycemic models can extend the capability of the models beyond the usual univariate control architecture for manipulating the infused insulin based solely on the glucose measurements. Developing a multivariable AP (mAP) architecture where physiological variables from wearables can characterize the glycemic effects of physical activity offers immediate benefits for glucose control [39,40]. The improvement in the glucose control performance is reinforced by the alleviation of manual user entries for physical activity and meals as additional physiological variables and real-time estimated parameters can quantify the prandial and exercise effects. The estimation of physiological parameters also ensures improved glycemic control when exercise and meals are unannounced or incorrect estimates are entered into the system.

Despite the advancement in AP systems, a clinically significant number of hypoglycemic events were reported during tests of closed-loop delivery systems [41]. To further reduce the risk of hypoglycemia, dual-hormone closed-loop delivery systems are developed that combine insulin delivery with subcutaneous glucagon delivery [42–45]. Glucagon, a hormone produced in the alpha cells of the pancreas, counters the effects of insulin through promoting the breakdown of glycogen to glucose by the liver, thus stabilizing glucose concentrations and preventing hypoglycemia. Other approaches to reduce the risk of hypoglycemia include suggesting rescue carbohydrates with hypoglycemia, which is predicted by the control algorithms.

The incorporation of additional physiological variables in mAP systems and the estimation of model parameters and meal consumption enhance the available information on the physiologic and metabolic state of the subject. In addition to these innovations, the amount of previously administered insulin that is present in the blood or the subcutaneous space, called as the insulin on board (IOB), must be quantified to prevent overdosing. The IOB is typically determined in infusion pumps through static approximations of the insulin action curves [30,46,47]. The insulin action curves, though convenient, do not account for the time-varying dynamics and kinetics of the metabolic states of individuals due to the diurnal variations in the insulin diffusion, absorption, and utilization. Therefore the insulin decay profiles and action curves used in calculating IOB are not accurate enough over the diverse conditions encountered throughout the day to be reliably used in an mAP system. In contrast, accurate estimates of the insulin concentration in the bloodstream can be obtained using CGM measurements with adaptive observers designed for simultaneous state and parameter estimation. The estimated plasma insulin concentration (PIC) can be subsequently used to design a predictive control algorithm that is dynamically constrained by the estimated PIC and thus explicitly considers the insulin concentration in the bloodstream as part of the optimal control solution [48]. Incorporating PIC constraints in the optimal control problems can prevent insulin stacking that may lead to hypoglycemia.

The remainder of the chapter describes the various facets of the AP systems in detail. First, the popular closed-loop glycemic control algorithms, including PID control and MPC, are discussed. Then, dynamic glucose-insulin modeling methods and adaptive control techniques are presented, followed by a review of state and parameter estimation techniques. Finally, closed-loop glycemic control results are demonstrated using an adaptive MPC approach and possible future research directions are briefly outlined.

Closed-loop glycemic control algorithms

Proportional-integral-derivative control

PID control is widely adopted in various industries to regulate an output variable by manipulating an input variable. PID control computes the control action based on the difference between the reference (or desired) glucose concentration and the measured value of the glucose concentration [9,49,50]. The error between the reference and measured glucose is assessed as a proportional, integral, and derivative term. The proportional term considers the current value of the error, the integral term considers the sum of the errors over a past time window, and the derivative term considers the rate of change in the current error from the previous error. The three terms are multiplied by coefficients that adjust the contributions of the individual terms to the overall amount of insulin dosage to be infused. The three coefficients are the adjustable parameters of the controller that can be tuned by practitioners to render the PID controller more aggressive or more conservative. Although the simplicity and ease of implementation of the PID controller make it an appealing choice for a control algorithm, the unsophisticated structure limits its ability to effectively control systems with multitudes of disturbances, time-varying delays, and temporal dynamics. The PID control algorithm is employed in early versions of AP systems.

Fuzzy logic control

Fuzzy logic and knowledge-based expert systems are developed for closed-loop insulin delivery by manually encoding the experience and knowledge of practitioners as a set of rules or by using learning mechanisms [15,24]. The most common form of the fuzzy logic controller involves constructing the set of logic rules to be evaluated with all the available information at each sampling time. The fuzzy logic system computes the input through the fuzzy rules and the (input and output) membership functions developed on expert knowledge or through the observation of the control actions taken by the practitioner. Challenges in developing and maintaining the set of fuzzy logic rules are limitations, as a survey of experts may lead to a diverse array of possible rules and uncertainties about fuzzy set membership functions. Moreover, the set of rules may become overly expansive, rendering personalization of the controller to individual subjects or adaption of the rules over time a challenge. A large set of rules may also cause conflicting actions to arise among the rules, which will require conflict resolution schemes to draw insulin-dosing decisions.

Model predictive control

MPC is widely adopted in controlled drug-delivery applications as an effective approach to deal with large multivariable constrained control problems. The principal function of MPC is to choose the optimal control actions (i.e., insulin infusion quantitates) by repeatedly solving a constrained optimization problem online that minimizes a performance index (for instance, predicted glucose tracking error from target glycemic set-point) over a finite prediction horizon with predictions obtained using a dynamic glucose-insulin system model [16,35,39,51–53]. Therefore the three main components of MPC are as follows: (i) a *dynamic model* of the glucose-insulin system that is used to predict in open loop the future evolution trajectory of the glycemic dynamics; (ii) a *performance index* such as the quadratic difference between predicted and target glucose values to be minimized over a finite time horizon subject to constraints imposed by the glucose-insulin model, restrictions on the maximum allowable insulin infusion (control inputs), and system states to obtain a trajectory of optimal future insulin infusions at each sampling time; and (iii) *a receding-horizon implementation scheme* that introduces feedback in the control law with new glucose measurements and updated state information at each sampling instance to compensate for disturbances (meals and physical activity) and modeling errors. The reliance on a glucose-insulin model means that the effectiveness of the controller depends highly on the accuracy of the model. Fig. 15.1 illustrates the mechanism of MPC.

Consider a discrete-time glucose-insulin dynamic system as

$$x_{k+1} = f(x_k, u_k)$$
$$y_k = g(x_k, u_k)$$

where x denotes the state of the system, y denotes the output glucose measurements, and u denotes the inputs (infused insulin) and with constraints on the insulin infusion generalized as $h(x_k, u_k) \leq 0$. The MPC formulation that regulates the glucose concentrations involves solving, for each current system state x, the following constrained optimal control problem

$$V_N(x) = \min_{x_0, u_0, \ldots, x_N} J(x, u)$$

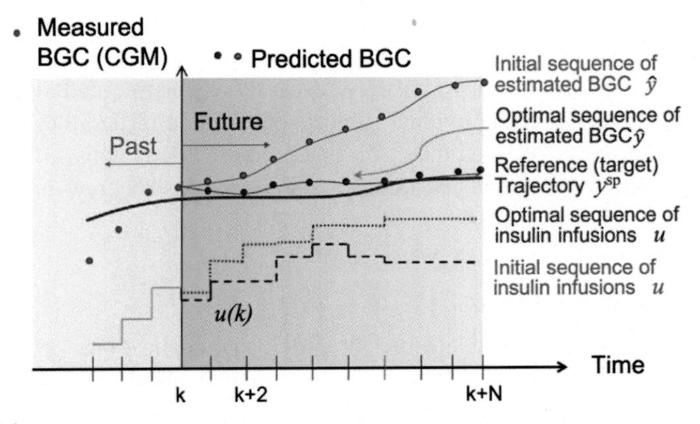

FIGURE 15.1

A diagram of the MPC algorithm with the model used to predict the future sequence of glucose measurements and an optimization approach used to select the best input sequence that minimizes the deviation of the predicted outputs from the reference set-point trajectory.

subject to:

$$
\begin{aligned}
x_0 &= x \\
x_{k+1} &= f(x_k, u_k), \quad k \in \mathbb{I}_{0N-1} \\
y_k &= g(x_k, u_k) \quad k \in \mathbb{I}_{0N} \\
h(x_k, u_k) &\leq 0, \quad k \in \mathbb{I}_{0N} \\
x_k \in \mathbb{X}, &\quad u_k \in \mathbb{U}, \quad k \in \mathbb{I}_{0N}
\end{aligned}
$$

where the objective function $J(x)$ is defined by

$$J(x,u) = \sum_{k=0}^{N} (y_k - y^{sp})^T Q_k (y_k - y^{sp}) + (u_k - u^s)^T R_k (u_k - u^s)$$

with Q_k as a (possibly varying) positive semidefinite weighting matrix penalizing the deviation of the controlled variables from the target set-point y^{sp} and R_k as a (possibly varying) strictly positive definite weighting matrix to penalize the amount of input actions away from a reference input u^s. The set $\mathbb{X}$ denotes that the state variables are constrained as $x^{min} \leq x \leq x^{max}$, with x^{min} and x^{max} as the minimum and maximum values for the state variables. Similarly, the set $\mathbb{U}$ denotes that the input variables are constrained as $u^{min} \leq u \leq u^{max}$, with u^{min} and u^{max} as the minimum and maximum values for the input. Therefore the maximum allowable insulin infusion can be limited based on the estimated insulin on board or the plasma insulin concentration. The optimal insulin infusion sequence $\{u_0, u_1, \ldots, u_N\}$ is termed feasible for a given initial state x if the insulin infusion sequence and the corresponding optimal state sequence $\{x_0, x_1, \ldots, x_N\}$ computed by the glucose-insulin dynamic model satisfy the constraints. The mathematical programming problem is solved at each sampling instance and the first value of the optimal solution (u_0) is implemented to infuse insulin over the current sampling interval. The MPC computation and insulin infusion implementation is repeated at subsequent sampling instances using new glucose measurements and updated state estimates. Extensions to the MPC paradigm include explicit MPC and advanced-step MPC algorithms [54–57]. Explicit MPC involves multiparametric programming, where the state of the system is represented as a vector of parameters so that the optimal solution for all possible realizations of the state vector can be precomputed as explicit functions to render the online decisions as expediated function look-ups and evaluations. Advanced-step MPC uses the prediction of the future state to solve the optimization problem within the sampling time, and applies a sensitivity-based update to compute the manipulated variable online once the new measurement is available.

Zone model predictive control

In contrast to controlling the glucose values to the desired set-point in conventional MPC, zone MPC is developed for systems that lack a specific set-point. The controller objective in zone MPC is to keep the controlled glucose concentrations in a predefined zone, such as a desirable euglycemic zone. An inherent benefit of control to the zone is that the pump actuation and infusion activity will be inherently limited when the glucose levels are within the defined zone. The lack of corrective suggestions when glycemic measurements are within the desired zone has the potential to reduce power consumption on handheld and mobile devices typically running the control algorithms [14,29].

The novelty of the zone MPC algorithm lies in the cost function formulation. Zone MPC, analogous to typical MPC algorithms, predicts the dynamic future trajectory profile of the glycemic values using an explicit model of the glucose-insulin dynamics and a candidate input trajectory to be optimized. In contrast to typical MPC that drives the controlled variable to a specific fixed set-point, the optimization problem in zone MPC attempts to maintain or drive the predicted outputs into a predefined zone described by upper and lower bounds [29,58]. The optimization problem of zone MPC for a given current state x is

$$V_N(x) = \min_{x_0, u_0, \ldots, x_N} J(x, u)$$

subject to:

$$\begin{aligned} x_0 &= x \\ x_{k+1} &= f(x_k, u_k), \quad k \in \mathbb{I}_{0N-1} \\ y_k &= g(x_k, u_k) \quad k \in \mathbb{I}_{0N} \\ h(x_k, u_k) &\leq 0, \quad k \in \mathbb{I}_{0N} \\ x_k \in \mathbb{X}, &\ u_k \in \mathbb{U}, \ k \in \mathbb{I}_{0N} \end{aligned}$$

where the objective function $J(x)$ is defined by

$$J(x, u) = \sum_{k=0}^{N} y_k^{range^T} Q_k y_k^{range} + (u_k - u^s)^T R_k (u_k - u^s)$$

with y^{range} as the superposition of all the predicted glucose outputs exceeding the permitted euglycemic range, and is given by

$$y_k^{range} = \left[y_k^{lo} - y^{lb}, y_k^{hi} - y^{ub} \right]$$

Further, y_k^{lo} and y_k^{hi} denote the aggregation of all predicted controlled outputs below the lower bound y^{lb} and above the upper bound y^{ub}, where y^{lb} and y^{ub} characterize the desired glycemic range. This aggregation is conducted by setting all predicted output values that are within the safe target range $[y^{lb}\ y^{ub}]$ to zero as follows:

$$y_k^{lo} = \begin{cases} y_k & \text{if } y_k < y^{lb} \\ 0 & \text{otherwise} \end{cases}$$

and

$$y_k^{hi} = \begin{cases} y_k & \text{if } y_k > y^{ub} \\ 0 & \text{otherwise} \end{cases}$$

Therefore the glycemic excursions in the undesirable high predicted glycemic values (within the hyperglycemia and severe hyperglycemia range) and the undesirable low predicted glycemic values (within the hypoglycemic or severe hypoglycemic range) are minimized as the zone MPC manipulates the insulin delivery to maximize the time spent in the desired euglycemic zone. The zone MPC algorithm reduces variations in the control input moves and can attenuate abrupt variations in the pump activity in response to noisy glucose measurements.

Adaptive control

The glucose-insulin dynamics vary substantially over time (intrasubject variability), which renders a single time-invariant model of the glycemic dynamics inaccurate for controlled insulin delivery. This realization has motivated adaptive control techniques that accommodate the intrasubject variability by adapting aspects of the control law computation such as the dynamic model or the controller parameters.

Recursive modeling

Adapting the glucose-insulin models employed in the design of model-based predictive controllers to track the time-varying glycemic dynamics is a common feature in adaptive MPC algorithms. The models may be adapted with each new glucose measurement sample or be adapted after a predefined elapsed time period to ensure the validity of the models. Adapting models after a specified time period may require less computation time and can be readily implemented through the reidentification of the model parameters. Updating the model parameters at each sampling instance can better capture the transient dynamics and unknown disturbance effects, thus better tracking the evolving glycemic dynamics.

A number of techniques ranging from recursive subspace-based system identification to nonlinear recursive filtering algorithms are employed to model the glucose measurement and infused insulin data. A commonly employed algorithm for its numerical simplicity and computational tractability is developing autoregressive exogenous input (ARX) models or autoregressive moving average with exogenous input (ARMAX) models with the parameters identified online through recursive least squares [39,59]. The ARX models are linear difference equation models that characterize the relationship between the current output variable and previous values of the output and input variables. The ARX model has the form

$$y_k = -a_1 y_{k-1} - a_2 y_{k-2} - \ldots - a_{n_y} y_{k-n_y} + b_1 u_{k-1} + b_2 u_{k-2} + \ldots + b_{n_u} u_{k-n_u} + \gamma + \varepsilon_k$$

where the output glucose concentration y_k is a linear combination of past output measurements $y_{k-1} \ldots y_{k-n_y}$ and past exogenous input variables $u_{k-1} \ldots u_{k-n_y}$, γ is a constant disturbance, and ε is white Gaussian noise. The input variables typically include the insulin infusion rate and the amount of carbohydrates consumed in meals and snacks. The parameters of the ARX model to be identified include the coefficients and the disturbance term $\theta = \left[a_1 \ldots a_{n_y}\ b_1 \ldots b_{n_u}\ \gamma\right]^T$. The integers n_y and n_u represent the order of the model, specifically the order of lags for the past outputs and inputs. The orders of the lags are also to be determined either by trial and error or explicit enumeration to find the best order of the model. The ARX model can be written as

$$A(q^{-1})y_k = B(q^{-1})u_k + \gamma + \varepsilon_k$$

where q^{-1} is the backward shift operator, that is, $q^{-1}y_k = y_{k-1}$, and $A(q^{-1})$ and $B(q^{-1})$ denote the polynomials

$$A(q^{-1}) = 1 + a_1 q^{-1} + a_2 q^{-2} + \ldots + a_{n_y} q^{-n_y}$$

$$B(q^{-1}) = b_1 q^{-1} + b_2 q^{-2} + \ldots + b_{n_y} q^{-n_y}$$

An attractive feature of the ARX models is that the parameters can be readily estimated using a set of available rich training data involving the solution to optimization problems with theoretically proven and desired convergence and efficiency properties. The practical identification of nonrecursive, or batch, ARX models is especially convenient if the glucose-insulin dynamics are assumed to be invariant as estimates of the model parameters can be obtained analytically by means of the proven and well-known least-squares solution, which minimizes the sum of squares of the one-step-ahead prediction errors for the training data. To update the ARX models using recently collected data, the batch modeling scheme can be retriggered at specific time instances (or after a certain elapsed time) to reidentify a new model using the recently accrued data.

In practice, however, the transient nonlinearity and time-varying nature of the glucose-insulin dynamics necessitate a recursive identification technique that updates the model parameters online as new data become available. Therefore the recursive technique can enable the updated model parameters to handle the evolving system nonlinearity and adapt to the changing conditions. The parameters of the recursively identified ARX models can be updated online using a weighted recursive least-squares solution, which places greater importance on the more recent information and gradually diminishes the contribution of the older data to discount the older information. The discounting of the older data samples is achieved through a forgetting factor λ, $0 < \lambda < 1$, typically chosen to be slightly less than one. The recursive least square (RLS) algorithm is initiated by specifying an initial covariance matrix P_0 and an initial vector of the model coefficients θ_0. Then at each sampling time, given a new output measurement y_k and regressor data sample $\psi_k = \left[y_{k-1} \ldots y_{k-n_y}\ u_{k-1}^T \ldots\ u_{k-n_u}^T\right]^T$, a gain κ_k is computed

$$\kappa_k = \frac{\lambda^{-1} P_{k-1} \psi_k}{1 + \lambda^{-1} \psi_k^T P_{k-1} \psi_k}$$

and the error of the previous set of coefficients is calculated:

$$e_k = y_k - \theta_{k-1}^T \psi_k$$

An innovation term is introduced to update the model coefficients

$$\theta_k = \theta_{k-1} + \kappa_k e_k$$

and the covariance matrix is updated:

$$P_k = \lambda^{-1} P_{k-1} - \lambda^{-1} \kappa_k \psi_k^T P_{k-1}$$

The RLS algorithm results in the data from the ith preceding sampling instance carrying a weight of λ^i relative to the newest sample, thus discounting the older samples over time.

The stability of the identified models is an important consideration for control-relevant models [39,60]. The identified ARX or ARMAX models can be written in a state-space form as

$$\begin{aligned} x_{k+1} &= Ax_k + Bu_k + Ke_k \\ y_k &= Cx_k + Du_k + e_k \end{aligned}$$

with the state x as a memory of past inputs and outputs and the system matrices developed through the model parameters. The updates of the model parameters θ can be constrained based on the physiological meanings of the parameters. The optimization problem can also be formulated to ensure the stability of the identified model as

$$\theta_k = \min_{\theta_k} (\theta_k - \theta_{k-1})^T P_{k-1} (\theta_k - \theta_{k-1}) + e_k^T Q e_k$$

subject to:

$$\begin{aligned} \rho(A) &\leq 1 \\ \theta^{min} \leq \theta_k &\leq \theta^{max} \end{aligned}$$

where $\rho(A)$ is the spectral radius of the state transition matrix for the state-space representation of the ARX/ARMAX model and θ^{min} and θ^{max} are the minimum and maximum of the identified model parameters. The former constraint satisfies the stability condition of the model and the latter satisfies the physiological properties of the system.

Subspace-based state-space system identification

Subspace-based state-space system identification techniques are used to determine the system matrices of a discrete-time state-space model of the form

$$\begin{aligned} x_{k+1} &= Ax_k + Bu_k \\ y_k &= Cx_k + Du_k \end{aligned}$$

where x, y, and u denote the state, input, and output variables, respectively, and A, B, C, and D. are the system matrices to be determined using input–output data. A realization of the system matrices can be obtained by solving complicated optimization problems such as nuclear norm-based structural rank minimization and maximum likelihood estimation through expectation-maximization algorithms.

In contrast to the optimization-based approaches, the fundamental operation in the computationally tractable subspace model identification is a projection, which may emanate from prudent numerical techniques like singular value decomposition (SVD) or even QR factorization.

The subspace-based system identification techniques utilize Hankel matrices constructed from the output measurements and input data [61,62]. To establish these Hankel matrices, define a vector of stacked output measurements as

$$y_{k|i} = \left[y_k^T\, y_{k+1}^T \cdots y_{k+i-1}^T\right]^T$$

where i is a user-specified parameter greater than the observability index or, for simplicity, the system order n. Similarly, define a vector of stacked input variables as $u_{k|i}$. Through the repeated iterative application of the state-space, it is straightforward to verify the expression for the stacked quantities:

$$y_{k|i} = \Gamma_i x_k + \Phi_i u_{k|i}$$

where

$$\Gamma_i = \begin{bmatrix} C \\ CA \\ \vdots \\ CA^{i-1} \end{bmatrix}$$

$$\Phi_i = \begin{bmatrix} D & 0 & \cdots & 0 \\ CB & D & \cdots & 0 \\ \vdots & \vdots & \ddots & \vdots \\ CA^{i-2}B & CA^{i-3}B & \cdots & D \end{bmatrix}$$

Consider the block Hankel matrix for the outputs

$$Y_i = \left[y_{1|i}\, y_{2|i} \cdots\, y_{j|i}\right]$$

and similarly U_i as a block Hankel matrix of inputs.

$$Y_i = \Gamma_i X_i + \Phi_i U_i$$

where

$$X_i = [x_1\, x_2 \ldots\, x_j]$$

The next step is to estimate the extended observability matrix, followed by retrieving the system matrices. The basic underlying idea of many common system identification methods is the orthogonal projection matrix on the null space of U_i as

$$\Pi_{U_i}^{\perp} = I - U_i^T\left(U_i U_i^T\right)^{-1} U_i$$

The projection matrix is multiplied by the black Hankel representation of the system to yield

$$Y_i \Pi_{U_i}^{\perp} = \Gamma_i X_i \Pi_{U_i}^{\perp}$$

where $Y_i \Pi_{U_i}^{\perp}$ is computable using numeral algorithms such as LQ factorization. An efficient implementation of this scheme is the multiinput, multioutput output-error state-space (MOESP) algorithm, where an estimate of Γ_i is obtained through the dominant left singular vectors of $Y_i \Pi_{U_i}^{\perp}$. Moreover, numerous variations of this approach (for instance, multiplying the matrix $Y_i \Pi_{U_i}^{\perp}$ with instrumental variables and/or nonsingular weight matrices before computing the SVD) are proposed to improve the consistency of the estimate. Four major variants of this method are PO-MOESP (past outputs MOESP), N4SID (numerical algorithms for subspace state-space system identification), IVM (instrumental variable method), and CVA (canonical variate analysis) approach, which differ by the choice of weight matrices that pre- and postmultiply the matrix $Y_i \Pi_{U_i}^{\perp} \Psi^T$, where Ψ is the instrumental variable matrix. Once an estimate of Γ_i is determined from the dominant left singular vectors of $W_1 Y_i \Pi_{U_i}^{\perp} \Psi^T W_2$, where $W_1$and W_2 denote the weight matrices, a system realization can be calculated by retrieving estimates of system matrices A and C from Γ_i, while estimates of B and D can be determined by solving a least-squares problem. This approach yields linear, time-invariant state-space models, while other system identification techniques are extended to model time-varying systems.

Recursive system identification

The predictor-based subspace identification approach is able to track a time-varying linear system and can be coupled with a constrained optimization solver to guarantee the stability of the model [61]. Consider a vector autoregression with exogenous variables (VARX) model

$$\widehat{y}_{k+1|k} = \sum_{i-0}^{p} \theta_{k-i}^{u} u_{k-1} \sum_{i-1}^{p} \theta_{k-i}^{y} y_{k-1}$$

where $\widehat{y}_{k+1|k}$ is the predicted output for the kth sampling instance using the past inputs $u_k, \ldots, u_{k-p}$ and outputs $y_{k-1}, \ldots, y_{k-p}$. The VARX model parameter p is the length of the past window of data considered when predicting future outputs. The coefficient matrices θ^u and θ^y are readily estimated through RLS techniques at each sampling time. Furthermore, the stacked vector $y_{k-p,p}$ is defined with respect to the past window of length p as

$$y_{k-p,p} = \left[y_{k-p}^T \, y_{k-p+1}^T \cdots y_{k-1}^T \right]^T$$

The stacked vector $u_{k-p,p}$ is also similarly defined. Furthermore, recognizing that the predicted state $\widehat{x}_k$ is given by

$$\widehat{x}_k = A^p \widehat{x}_{k-p} + L u_{k-p,p} + K y_{k-p,p}$$

where L and K denote the extended controllability matrices $L = \left[A^{p-1}B \ldots AB\ B\right]$ and $K = \left[A^{p-1}K \ldots AK\ K\right]$ and assuming that the state transition matrix is nilpotent with degree p ($A^p = 0$), that is the contribution of the initial state $\widehat{x}_{k-p}$ is negligible for sufficiently large p, the predicted state can be expressed as

$$\widehat{x}_k = Lu_{k-p,p} + Ky_{k-p,p}$$

Premultiplying the predicted state by the observability matrix Γ gives

$$\Gamma\widehat{x}_k = \Gamma Lu_{k-p,p} + \Gamma Ky_{k-p,p}$$

with

$$\Gamma = \begin{bmatrix} C \\ CA \\ \vdots \\ CA^{p-1} \end{bmatrix}$$

The product of the matrices ΓL and ΓK can be constructed from the VARX model coefficient matrices as

$$\Gamma L = \begin{bmatrix} \theta^u_{k-p} & \theta^u_{k-p+1} & \cdots & \theta^u_{k-1} \\ 0 & \theta^u_{k-p} & \cdots & \theta^u_{k-2} \\ \vdots\ \vdots & & \ddots & \vdots \\ 0\ \ 0 & & \cdots & \theta^u_{k-f} \end{bmatrix}$$

and

$$\Gamma K = \begin{bmatrix} \theta^y_{k-p} & \theta^y_{k-p+1} & \cdots & \theta^y_{k-1} \\ 0 & \theta^y_{k-p} & \cdots & \theta^y_{k-2} \\ \vdots\ \vdots & & \ddots & \vdots \\ 0\ \ 0 & & \cdots & \theta^y_{k-f} \end{bmatrix}$$

where f is the user-specified parameter for the future window length.

Therefore after estimating the VARX coefficient matrices, the estimated coefficient matrices θ^u and θ^y can be used to determine all quantities on the right-hand side of the state evolution equation, and an SVD can be used to readily obtain a low-rank approximation of the state sequence. For recursive identification, a selection matrix S of appropriate dimensions can be determined such that the basis of the state estimation is consistent at each sampling time as

$$\widehat{x}_k = S_k W_k \left(\Gamma Lu_{k-p,p} + \Gamma Ky_{k-p,p}\right)$$

where W_k is a predefined weight matrix and the selection matrix S can be recursively updated through the projection approximation subspace tracking method. The estimated state sequence is then employed along with the inputs and measured outputs to estimate the system matrices A_k, B_k, C_k, D_k, and K_k by the solution of recursive least-squares problems. Specifically, after computing an estimate of the state sequence $\widehat{x}_k$, two recursive least-squares problems that ensure the stability of the estimated system are used to determine the state-space matrices, thus yielding the identified model

$$\begin{aligned} x_{k+1} &= Ax_k + Bu_k + K_k e_k \\ y_k &= Cx_k + Du_k + e_k \end{aligned}$$

where K_k is the Kalman gain matrix and the $e_k = y_k - \widehat{y}_k$.

Adaptive generalized predictive control

Model-based predictive controllers, such as MPC and generalized predictive control (GPC), are widely used to control a variety of systems with complex dynamics and long-time delays [39,40]. GPC uses recursively identified time-series models to optimize the cost function.

$$J(u,k) = \sum_{j=n_1}^{n_2} \left(y_{k+j} - y_{k+j}^{sp}\right)^2 + \sum_{j=1}^{n_u} \lambda_j \Delta u_{k+j-1}^2$$

with the predictions for the future outputs made using the recursively updated model.

$$A\left(q^{-1}\right)y_k = B\left(q^{-1}\right)u_k + \gamma + \varepsilon_k$$

where n_1 and n_2 are the finite minimum and maximum cost horizons and n_u is the control horizon; λ is a weighting sequence for the input. Explicit solutions to the GPC control problem exist for certain cases, and additional constraints on the inputs can also be included in the optimization problem. Only the first calculated input u_k is applied to the system and in the following sampling instances, the problem solution is repeated with new model parameters updated using the new measurements made available.

Run-to-run control

Run-to-run (R2R) control schemes have been adapted from batch industrial processes to address the problem of glucose regulation. These techniques have been successfully implemented in industrial practice to control batch processes with improvement in control performance based on learning from recent batches. In batch control, quantitative measures of batch performance from the previous run (such as the final product quality) are used to determine the input profile trajectory (recipe) for the next run. The repeated nature of the batch is thus exploited to correct the future batch based on the performance of the previous batch [63]. To apply R2R control to the glucose regulation problem, the postprandial period or the daily cycle of a subject can be considered as a batch. Similarities between a chemical batch process and a particular time window of a subject with T1DM include: following a particular protocol for meals and diurnal activities (similar to batch recipe) with variations in amounts, compositions, type, and timings of meals; the quality variables of batch process product and the performance measure for the quality of glycemic variations during the batch; and the differences between subjects (intersubject variability), as well as evolving glucose-insulin dynamics and glycemic disturbances similar to variations in raw material feed and disturbances over the course of the batch. Therefore the glucose measurements obtained during a batch can be used to adjust the insulin therapy for the subsequent batch [64].

R2R control algorithms typically update the control law on a time scale of the entire cycle (i.e., one correction allowed at the end of the batch), thus refining the control action over the course of multiple cycles until desirable control performance is obtained. Consider a linear, input–output model of the process for the kth batch run

$$y_k = Au_k + b_k + e_k$$

where k denotes the batch index, y is the system output, u is the input, A is the coefficients matrix for relating the inputs to the outputs, b is the bias/drift coefficient matrix, and e denotes the disturbances. Note that this model represents a batch run, hence the inputs are a trajectory profile applied to the process, and the outputs are the measurements collected over the batch run. The target for the process outputs y is denoted as y^{sp}. If A and b are known, and A is invertible, then the optimal control for the system is

$$u_k = A^{-1}(y^{sp} - b_k)$$

The mismatch between the model and the system and batch-to-batch (run-to-run) variations are common in practice as systems evolve temporally and disturbances affect the batch runs. An iterative scheme called exponentially weighted moving average (EWMA) filter for updating the estimates of the bias coefficient matrix is proposed as

$$b_k = \lambda(y_{k-1} - Au_{k-1}) + (1 - \lambda)b_{k-1}$$

where $0 < \lambda < 1$ is an adjustable tuning parameter for the EWMA filter. Substituting the optimal control law into the EWMA filter for updating the bias coefficient matrix

$$b_k = b_{k-1} + \lambda(y_{k-1} - y^{sp})$$

and hence integrating the optimal control law from the $k-1$ batch run yields

$$u_k = u_{k-1} + \lambda A^{-1}(y^{sp} - y_{k-1})$$

which is the EWMA-type R2R or proportional-type (P-type) R2R control algorithm due to the proportional contribution of performance measure from the previous batch. A more general form of R2R is

$$u_k = \alpha u_{k-1} + r_k$$

where $0 < \alpha < 1$ is the forgetting factor and r_k the updating law that can be determined as

$$r_k = K(y^{sp} - y_{k-1})$$

with K as an appropriate gain matrix. It is readily shown that the eigenvalues of the matrix $[I - AK]$ should be within the unit circle for the algorithm to converge, with the rate of convergence governed by the system matrix A and controller gain K.

Insulin doses can be adjusted using the R2R algorithm by considering the glucose-insulin data from each postprandial period or day as a batch run. Consider the manipulated variable u to be the insulin-to-carbohydrate ratio. The insulin dosage regimen, specifically the insulin-to-carbohydrate ratio, for a batch can be corrected based on a scalar performance measure y that quantifies the glycemic excursion during the particular time window. The desired glycemic response trajectory during the batch is denoted y^{sp}. The performance measure can be the rate of change of glucose measurements in the postprandial period normalized by the carbohydrate amount of the meal or a glycemic index summarizing the degree of glycemic control during the day. The insulin-to-carbohydrate ratio for the next batch u_k is then given by the ratio for the previous batch and the performance measure for the previous batch as

$$u_k = u_{k-1} + K(y^{sp} - y_{k-1})$$

with the gain K as a tuning parameter that determines how aggressive the R2R algorithm is in correcting the insulin-to-carbohydrate ratio based on the previous batch results. The ability to learn from past periods to correct the insulin therapy regimen for the subsequent batch is an attractive proposition for improving glycemic control.

Iterative learning control

Iterative learning control (ILC) is an improvement in run-to-run control that uses more frequent measurements in the form of the error trajectory from the previous batch to update the control signal for the subsequent batch run [63]. Compared to the R2R control algorithm of

$$u_k = u_{k-1} + r_k$$

where u denotes the input trajectory and r is the updating law, the ILC algorithm is of the form

$$u_{i,k} = u_{i,k-1} + r_{i,k}$$

where $u_{i,k}$ is the input trajectory at the ith sampling instant in batch k, $u_{i,k-1}$ is the input trajectory from the ith sampling instant of the previous batch, and $r_{i,k}$ is the updating law. Different choices for the updating law result in different ILC schemes, such as P-type ILC. Defining the tracking error as

$$e_{i,k} = y_i^{sp} - y_{i,k}$$

a simple formulation of ILC is

$$u_{i,k} = u_{i,k-1} + K\left(y_i^{sp} - y_{i,k-1}\right)$$

where K is the learning gain matrix. The objective of ILC is that

$$\lim_{k \to \infty} e_{i,k} = 0$$

Therefore the input trajectory for the current batch is determined based on the input trajectory implemented in the previous batch plus the proportional contribution of tracking error. In certain cases, in practice $e_{i+1,k-1}$ can be considered as an approximate prediction of $e_{i+1,k}$, and therefore the P-type ILC is implemented as

$$u_{i,k} = u_{i,k-1} + K\left(y_i^{sp} - y_{i+1,k-1}\right)$$

Other types of ILC are proposed for improved robustness and formulations that employ information from the past batch beyond the current sampling instance for anticipatory or phase-lead type ILC. Real-time information can be incorporated into the ILC as

$$u_{i,k} = u_{i,k-1} + r_{i,k}^{p} + r_{i,k}^{c}$$

where

$$r_{i,k}^{p} = K^{p}\left(y_i^{sp} - y_{i,k-1}\right)$$

and

$$r_{i,k}^{c} = K^{c}\left(y_i^{sp} - y_{i,k-1}\right)$$

The gains K^p and K^c correspond to the information in the previous and current cycles, respectively. Note that future or anticipatory information can be used to design $r_{i,k}^{p}$, though future information cannot be used to design $r_{i,k}^{c}$ as it is physically unrealizable. Another ILC approach integrates feedback control with ILC as

$$u_{i,k} = u_{i,k}^{ILC} + u_{i,k}^{FB}$$

where $u_{i,k}^{ILC}$ is the input computed by the ILC and $u_{i,k}^{FB}$ is the input computed through a real-time feedback controller such as PID control or MPC [65,66].

MPC may also be integrated with ILC to involve future predictions of the current batch in the control law calculation [67]. Consider an ARX model of the form

$$A\left(q^{-1}\right)\Delta^{K} y_{i,k} = B\left(q^{-1}\right)\Delta^{K} u_{i-d,k} + \varepsilon_{i,k}$$

where a delay of order d is included and the differences between batches are represented as

$$\Delta^{K} y_{i,k} = y_{i,k} - y_{i,k-1}$$

$$\Delta^{K} u_{i,k} = u_{i,k} - u_{i,k-1}$$

and $\varepsilon_{i,k}$ denotes disturbances or uncertainties. Given that ILC can be expressed as

$$u_{i,k} = u_{i,k-1} + r_{i,k}$$

which is equivalent to

$$r_{i,k} = \Delta^K u_{i,k}$$

The objective of the model predictive iterative learning control (MPILC) algorithm is to minimize the term

$$e_{i,k} = y_i^{sp} - y_{i,k}$$

which can be rewritten as. $e_{i,k} = y_i^{sp} - y_{i,k-1} - \Delta^K y_{i,k}$

Hence, y_i^{sp} and $y_{i,k-1}$ are known, while $\Delta^K y_{i,k}$ can be predicted using the aforementioned ARX model. The MPILC optimization problem isthen

$$V_N = \min_{r_{i,k},\ldots,r_{i+N,k}} \sum_{j=0}^{N} q_1 \cdot e_{i+j,k}^2 + q_2 \cdot r_{i+j,k}^2$$

subject to:

$$e_{i+j,k} = y_{i+j}^{sp} - y_{i+j,k-1} - \Delta^K y_{i+j,k}$$

$$r_{i+j,k} = \Delta^K u_{i+j,k} \qquad \textit{for } j \in \mathbb{I}_{0:N}$$

$$A(q^{-1}) \Delta^K y_{i+j,k} = B(q^{-1}) \Delta^K u_{i+j-d,k}$$

where q_1 and q_2 are penalty weights for the deviation of the error and the input moves and N is the prediction/control horizon. Additional constraints on the inputs and controlled outputs may be readily incorporated into the optimization problem.

Adaptive weights through glycemic risk index

Variations in the glycemic measurements can alter the precedence for insulin infusion. The fluctuating importance of insulin infusion can be reflected in the varying weights of the MPC objective function. Manipulating the penalty weights of the objective function based on the glucose measurements can vary the necessity for insulin infusion. The weights of the objective function can be modified through a glycemic risk index (GRI) that relates the glucose measurement to the weight on the glucose tracking error (Fig. 15.2). The glycemic risk index disproportionally increases the penalty weights as the CGM measurements deviate from the desired set-point target, which in this case is considered to be 110 mg/dL. The asymmetry of the GRI is due to the fact that hypoglycemia is associated with severe immediate and short-term adverse effects, such as coma or death, the penalty on the set-point tracking error should be increased rapidly in response to decreasing glucose measurements. Uncontrolled T1DM leads to prolonged hyperglycemia that can cause more long-term ailments like micro- (i.e., neuropathy, retinopathy, and nephropathy) and/or macrovascular (i.e., myocardial infarction and stroke) complications, and premature mortality. Therefore the GRI increases steadily in response to hyperglycemia. The GRI weight can be used to adjust the penalty weight for the

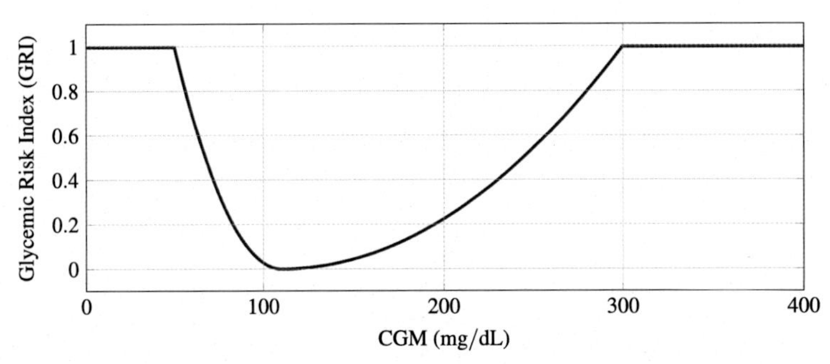

FIGURE 15.2

The glycemic risk index for adapting the penalty weight on the tracking error based on the CGM measurement value.

set-point tracking error $Q_k = Q(GRI_k)$, with GRI_k as the glycemic risk index value at sampling instance k for a given glucose measurement reading. Manipulating the penalty weights can thus inform the MPC of changing priorities and improve glycemic control.

Quantifying plasma insulin concentrations

AP systems require a safety constraint to moderate the potential aggressive control actions (insulin overdosing) and minimize the risk of hypoglycemia for significantly improving the performance of the AP. Estimating the amount of available insulin in the body is challenging because of the inter- and intrasubject variability attributed to physiological differences and metabolic changes throughout the course of the day. Despite the lack of direct measurement, maintaining the BGC in the target range requires AP systems that are cognizant of the quantity of insulin previously administrated, which if not appropriately incorporated into the control algorithm may cause overcorrection for the postprandial rise in BGC. Such excessive dosing in either the bolus or basal insulin administered through CSII pumps can potentially lead to hypoglycemia. Hence, in addition to the current and target BGC, a constraint expressing an approximation of the insulin present in the body, such as the conventional IOB estimates, is needed for insulin-dosing calculations.

The IOB is an estimate of the amount of insulin that is present in the blood and the interstitial fluid cavity. It is typically determined through the approximation of the insulin decay curves, which represent the amount of insulin still remaining in the body due to the prior insulin infusions. Static approximations of the insulin action curves are typically utilized in insulin pumps, with IOB calculations primarily relying on basic insulin decay profiles. Furthermore, significant time-varying delays induced by the absorption and utilization of the subcutaneously administrated insulin as well as diurnal variations in the metabolic state of individuals have significant effects on the IOB. Therefore the insulin action curves for IOB calculations, usually involving static models with basal and bolus insulin as inputs and active insulin as the output, are not accurate enough to be used in an AP control system. Regardless of the sophistication of the IOB calculation, the information obtained from insulin action curves is usually an approximation of the active insulin in the body and is not a direct estimate of the concentration of insulin in the bloodstream.

Accurate estimates of PIC can be obtained by using CGM measurements with adaptive observers designed for simultaneous state and parameter estimation based on reliable glucose-insulin models. The glucose-insulin dynamic model can be written in the form

$$\frac{dx(t)}{dt} = f(x(t), u(t)) + w(t), \quad w \sim \mathcal{N}(0, Q_w)$$

$$y_k = h(x_k) + v_k, \quad v \sim \mathcal{N}(0, R_v)$$

where $x(t)$ denotes the vector of state variables including the PIC as a state variable, $h(x_k)$ denotes the measurement function with y_k as the subcutaneous glucose output measurement, $w(t)$ and v_k represent the process and observation noise vectors, respectively, Q_w and R_v denote the covariance and variance of the process and measurement noise, respectively. To design a state estimator for the augmented system, the augmented system should be observable.

Nonlinear observers, such as extended or unscented Kalman filters (UKF) and moving horizon estimation, can be designed for the estimation of the model states and parameters. The UKF algorithm can handle the nonlinear dynamics of the glucose-insulin model, is robust to measurement noise, and can compensate for deviations and converge to the true value of the augmented states through the Kalman gain correction term added to the estimation. In the UKF algorithm, the unscented transformation (UT) method is employed for calculating the statistics of a random variable that undergoes a nonlinear transformation such as the glucose-insulin dynamics model. The UT characterizes the mean and covariance estimates with a minimal set of sample points called sigma points. Let the set of sigma points at sampling instance k be denoted

$$\mathscr{X}'_{i,k}, \quad i \in \{0, \ldots, 2n\}$$

with each point being associated with a corresponding weight w_i. In the UKF approach, both the sigma points and the weights are determined deterministically via specific criteria and equations. The sigma points are propagated through the nonlinear state dynamics of the glucose-insulin function, which yields propagated states $\mathscr{X}'_{i,k|k-1}$ and the corresponding mean representing the prior estimates $\widehat{X}_{k|k-1|}$ approximated by the weighted average of the transformed points as

$$\widehat{X}'_{k|k-1} = \sum_{i=0}^{2n} w_i \mathscr{X}'_{i,k|k-1}$$

where

$$\sum_{i=0}^{2n} w_i = 1$$

and the covariance of the prior state estimates $P_{x,k|k-1}$ is computed by the weighted outer product of the transformed points as

$$P_{x,k|k-1} = \sum_{i=0}^{2n} w_i \left(\mathscr{X}'_{i,k|k-1} - \widehat{X}'_{k|k-1} \right) \left(\mathscr{X}'_{i,k|k-1} - \widehat{X}'_{k|k-1} \right)^T$$

. The sigma points are similarly propagated through the measurement function as

$$y_{i,k|k-1} = h' \left(\mathscr{X}'_{i,k|k-1} \right)$$

and the estimated prior CGM output $\widehat{y}_{k|k-1}$ is approximated by the weighted average of the transformed points as

$$\widehat{y}_{k|k-1} = \sum_{i=0}^{2n} w_i y_{i,k|k-1}$$

as well as the estimated covariance matrices

$$P_y = \sum_{i=0}^{2n} w_i \left(y_{i,k|k-1} - \widehat{y}_{k|k-1} \right) \left(y_{i,k|k-1} - \widehat{y}_{k|k-1} \right)^T$$

$$P_{xy} = \sum_{i=0}^{2n} w_i \left(\mathscr{X}'_{i,k|k-1} - \widehat{X}'_{k|k-1} \right) \left(y_{i,k|k-1} - \widehat{y}_{k|k-1} \right)^T$$

The Kalman gain K_k and posterior updates for the augmented state estimate $\widehat{X}'_{k|k|}$ as well as the posterior error covariance matrix $P_{x,k|k}$ of the augmented state estimate are given by the standard Kalman update equations

$$K_k = P_{xy} P_y^{-1}$$

$$\widehat{X}'_{k|k} = \widehat{X}'_{k|k-1} + K_k \left(y_k - \widehat{y}_{k|k-1} \right)$$

$$P_{x,k|k} = P_{x,k|k-1} - K_k P_y K_k^T$$

. The UKF algorithm can handle the nonlinear dynamics of the glucose-insulin model, is robust to noise, and has the ability to compensate for deviations and converge to the true value of the state variables. The state variables in the glucose-insulin dynamic models represent a physiological process based on first-principles, and the state variables should be maintained within a physically realizable range. For example, a negative value for the PIC due to measurement noise and system uncertainty is not physically possible. Therefore constraints can be employed in the UKF algorithm to ensure the augmented state estimates correspond with the physical definitions. In addition, maximum rates of change constraints are defined for the parameters estimated simultaneously to the states to avoid sudden changes in the parameter values due to measurement noise or unknown disturbances that may result in inappropriate corrections.

Modulating insulin infusion

Estimates of the PIC can be used to moderate the aggressiveness of the MPC algorithm and dynamically constrain the insulin infusion. The MPC thus explicitly considers the insulin concentration in the bloodstream within the control law computation. A plasma insulin risk index can manipulate the weighting matrix for penalizing the amount of input actuation (aggressiveness of insulin dosing) depending on the estimated PIC, thus suppressing the infusion rate if sufficient insulin is present in the bloodstream (Fig. 15.3). As the plasma insulin concentration increases, the penalty weight on the input action is also simultaneously increased as $R_k = R(PIRI_k)$, with $PIRI_k$ as the risk index at sampling instance k derived from the estimated PIC. Incorporating PIC constraints in the optimal control problems can prevent insulin stacking that may lead to hypoglycemia, which can yield a safe and reliable AP system even in the presence of significant uncertainty in the system.

Closed-loop glycemic control results

In this section, we illustrate the efficacy of an MPC formulation that employs adaptive models recursively identified through subspace-based techniques. The adaptive MPC incorporates variable weights in the objective function through the glycemic and plasma insulin risk indexes. The adaptive MPC is also dynamically

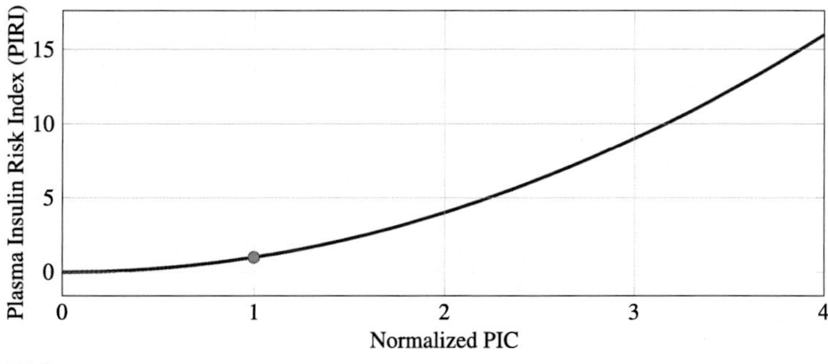

FIGURE 15.3

The plasma insulin risk index for increasing the penalty weight on insulin infusion based on plasma insulin concentration estimates.

constrained through predictions of the PIC obtained over the prediction horizon. The ability of the adaptive MPC is demonstrated using the multivariable Glucose, Insulin, and Physiological variable simulator (mGIPsim), which is based on a modified Hovorka's glucose-insulin dynamic model that takes into account the effects of various physical activities. In addition to the CGM values, mGIPsim generates physiological variable signals reported by noninvasive wearable devices. These physiological variables are used to evaluate the ability of the mAP system and compare its performance to that of the conventional single input AP (sAP) system. Aerobic medium-intensity exercises with treadmill and bicycle are used for testing the mAP system. Twenty virtual subjects are simulated for 3 days with varying times and quantities of meals consumed on each day and different types and times of physical activities (Tables 15.1 and 15.2). The meal and physical activity information are not entered manually to the AP, as the AP controller is designed to regulate the BGC in the presence of significant disturbances such as unannounced meals and exercises. The metabolic equivalent of task (MET) values computed by the simulator are used as physiological signals in the recursive system identification technique. To show the efficacy of using physiological signals in the AP system (the mAP case), the sAP case is also considered where no information of physiological signals (MET values) are used in the AP system. The results based on these two different cases (the mAP and sAP) are compared.

The evaluations of the closed-loop results based on the mAP and sAP are presented in Figs. 15.4–15.7, and the quantitative metrics for comparing the sAP

Table 15.1 Meal scenario for 3-day closed-loop experiment using mGIPsim—the integrated multivariable metabolic and physiologic simulator.

	First day		Second day		Third day	
Meal	**Time**	**Amount (g)**	**Time**	**Amount (g)**	**Time**	**Amount (g)**
Breakfast	7:00	70	8:00	50	7:30	60
Lunch	12:00	60	12:30	80	13:00	70
Dinner	18:00	50	19:00	60	18:30	80
Snack	22:00	30	22:30	25	21:30	20

Table 15.2 Exercise scenario for one-hour duration for 3-day closed-loop experiment using mGIPsim—the integrated multivariable metabolic and physiologic simulator.

Exercise	First day	Second day	Third day
Morning	Treadmill at 10:30	Bicycling at 09:45	Treadmill at 10:00
Afternoon	Bicycling at 16:00	Treadmill at 16:45	Bicycling at 16:15

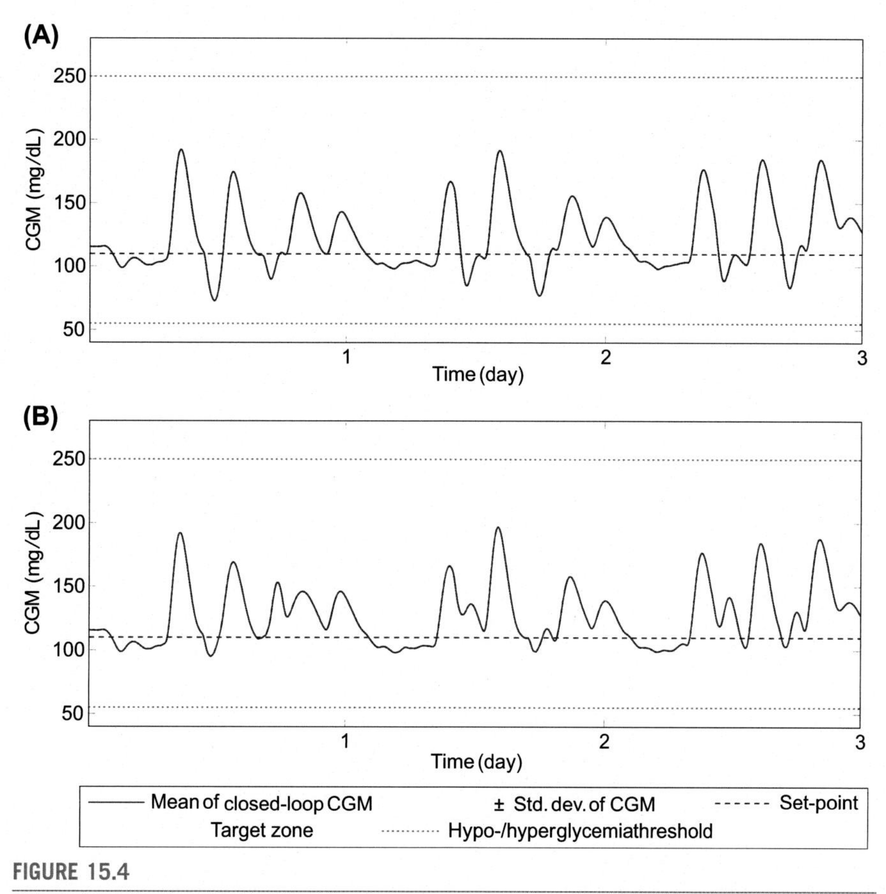

FIGURE 15.4

Closed-loop results in mGIPsim for the CGM values using (A) sAP and (B) mAP.

and mAP systems is presented in Table 15.3. Fig. 15.7 presents a detailed comparison of the sAP and mAP during the one-hour exercise period and subsequent two-hour postexercise recovery period for all subjects. The changes in the CGM during the one-hour exercise session and the two-hour postexercise recovery session are plotted to show the efficacy of using physiological signals in an mAP system.

No hypoglycemia occurs when the mAP is used, while hypoglycemia (BGC < 70 mg/dL) occurs with the sAP system during the exercise period where the CGM values drop significantly. The use of physiological variables in the mAP can benefit AP systems by improving the regulation of the BGC and thus avoid any potential hypoglycemia. For both the sAP and mAP systems, the CGM measurements never drop below 55 mg/dL or rise higher than 250 mg/dL. Overall, the mAP performs better than the sAP system across all subjects in regulating glucose concentrations within a safe target range while accommodating unannounced disturbances such as meals and physical activity.

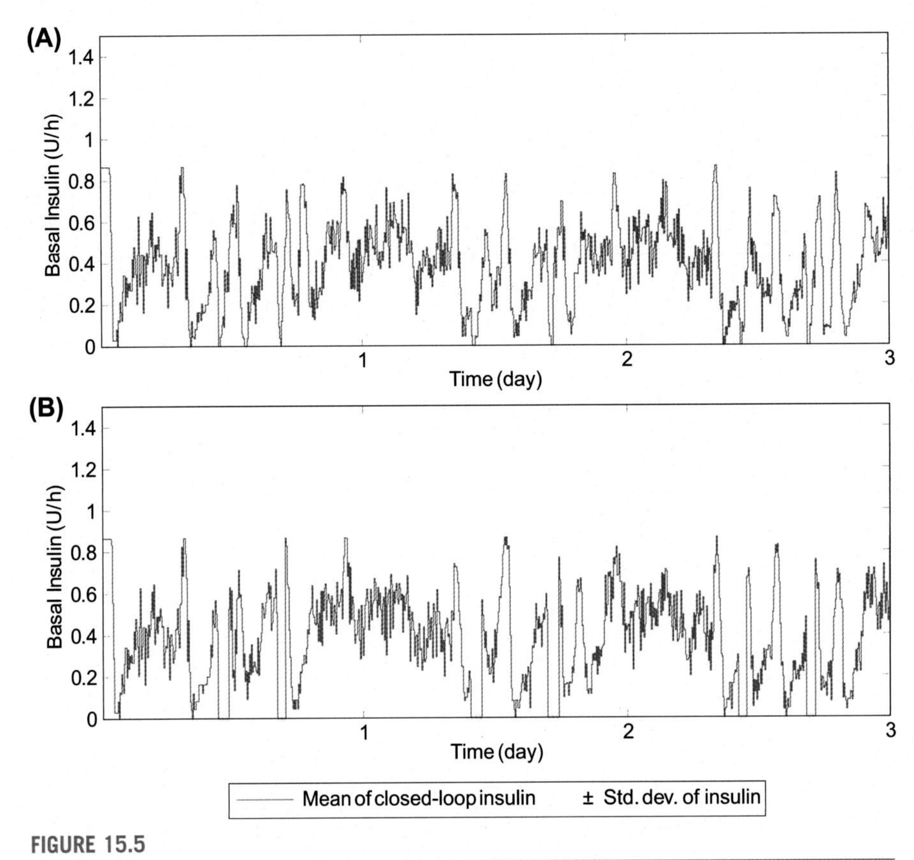

FIGURE 15.5

Closed-loop results in mGIPsim for the basal insulin infusion using (A) sAP and (B) mAP.

Future directions

Meals and physical activity pose significant challenges for glycemic control in people with diabetes. Addressing these challenges in closed-loop automated insulin delivery systems is a priority. Other disturbances and factors affecting euglycemia include the sleep architecture and acute psychological stress. The characteristics of sleep can affect the glycemic dynamics and insulin sensitivity during the subsequent day. Therefore recognizing the overnight sleep characteristics and adjusting the insulin-dosing algorithms accordingly to counteract the dawn phenomenon or the glycemic variations of the subsequent day will improve glycemic control. Psychological stress is associated with the activation of certain counterregulatory hormones, which cause an increase in the BGC. Detecting and responding to episodes of stress is also a possible future direction for AP systems. As shown in Fig. 15.8, these additional modules can form a comprehensive multivariable AP system for safe and effective glucose control.

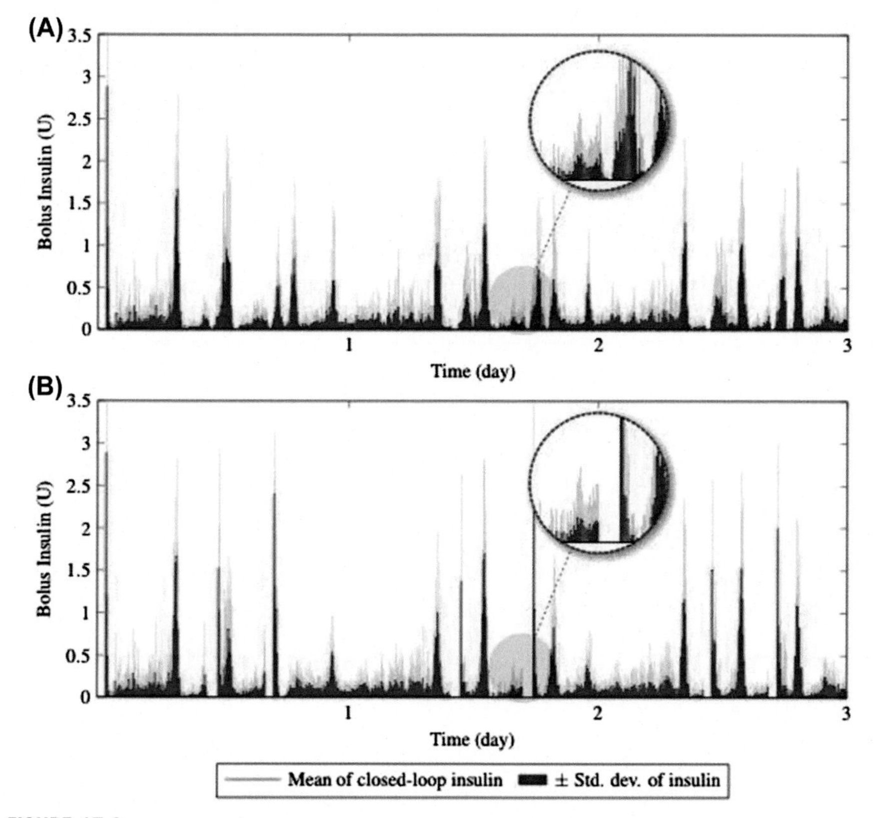

FIGURE 15.6

Closed-loop results in mGIPsim for the bolus insulin infusion using (A) sAP and (B) mAP (the insets zoom into the same exercise periods on the afternoon of the second day).

Conclusions

This chapter reviews and summarizes the state-of-the-art control techniques used in the development of artificial pancreas systems. Recent advancements in adaptive control techniques and predictive control algorithms have the potential to improve glycemic control in people with T1DM.

Acknowledgments

The multivariable artificial pancreas research is supported by the National Institutes of Health (NIH) under grants 1DP3DK101077-01 and 1DP3DK101075-01 and by Juvenile Diabetes Research Foundation International (JDRF) under grants 17-2013-472 and 2-SRA-2017-506-M-B.

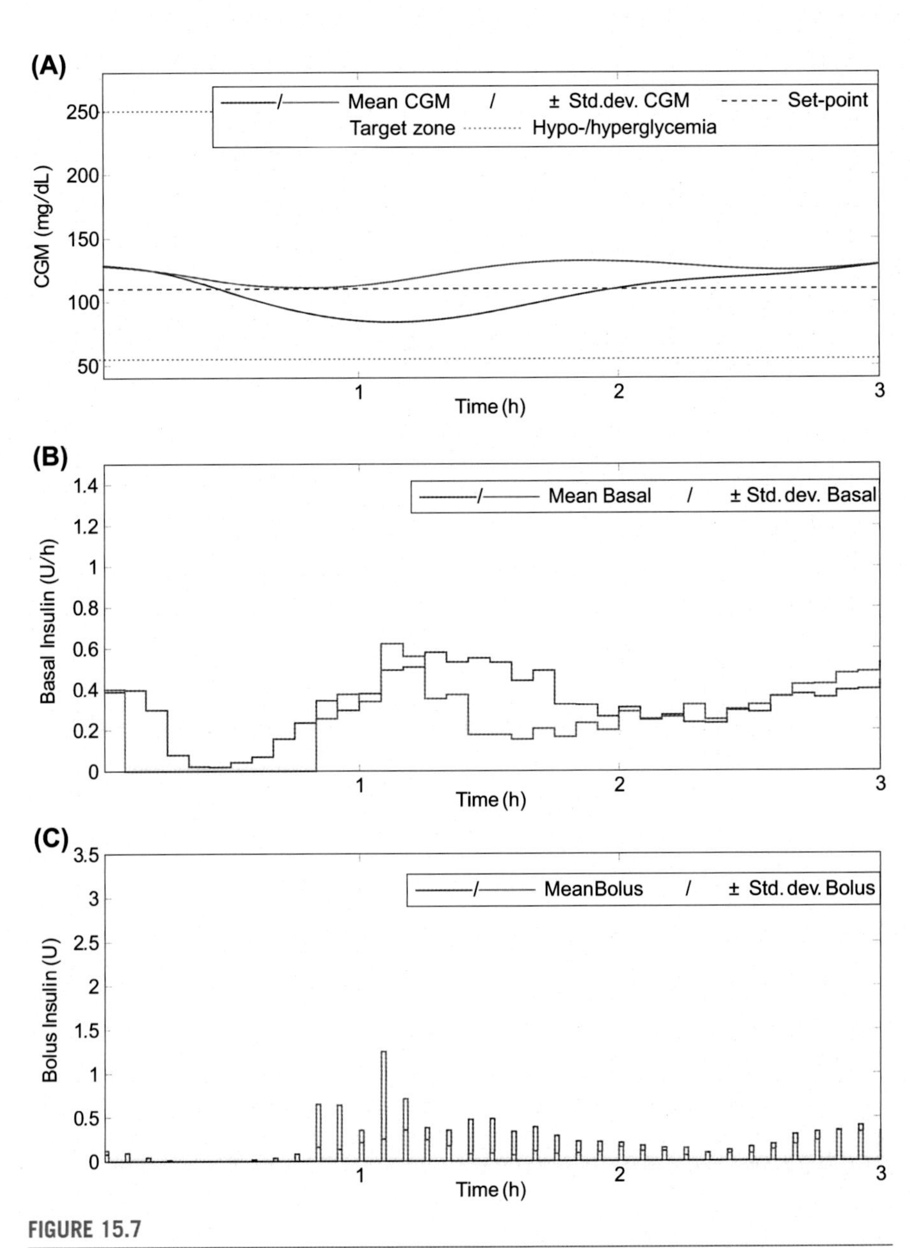

FIGURE 15.7

Closed-loop results for the exercise period (first one-hour) and the postexercise recovery periods (2 h succeeding exercise) for the sAP (blue) and mAP (red) systems.

Table 15.3 Glycemic ranges and selected statistics across all subjects for closed-loop glucose control with sAP and mAP systems during exercise and over the entire experiment (SD, standard deviation; Med, median; Q1, first quartile; Q3, third quartile).

Performance index		Exercise time		Entire experiment	
		sAP	**mAP**	**sAP**	**mAP**
70–180 mg/dL (%)	Mean (Med)	95.04 (96.97)	98.34 (100)	93.25 (92.99)	93.82 (93.10)
	SD (Q1/Q3)	5.77 (92.56/100)	2.87 (97.80/100)	5.14 (90/98.29)	4.82 (89.62/98.80)
180–250 mg/dL (%)	Mean (Med)	0.59 (0.0)	1.66 (0.0)	5.99 (6.90)	6.18 (6.91)
	SD (Q1/Q3)	1.02 (0/1.24)	2.87 (0/2.20)	4.92 (1.05/10)	4.82 (1.19/10.38)
55–70 mg/dL (%)	Mean (Med)	3.91 (1.93)	0.0 (0.0)	0.68 (0.32)	0.0 (0.0)
	SD (Q1/Q3)	4.54 (0/6.96)	0.0 (0/0)	0.79 (0/1.17)	0.0 (0.0/0.0)
>250 mg/dL (%)	Mean (Med)	0.0 (0.0)	0.0 (0.0)	0.0 (0.0)	0.0 (0.0)
	SD (Q1/Q3)	0 (0/0)	0.0 (0/0)	0.0 (0.0)	0.0 (0.0/0.0)
<55 mg/dL (%)	Mean (Med)	0.46 (0.0)	0.0 (0.0)	0.08 (0.0)	0.0 (0.0)
	SD (Q1/Q3)	1.43 (0/0)	0.0 (0/0)	0.24 (0.0/0.0)	0.0 (0.0/0.0)
Mean of CGM (mg/dL)	Mean (Med)	100.76 (100.97)	121.39 (119.59)	122.78 (126.80)	127.28 (131.56)
	SD (Q1/Q3)	7.27 (95.40/104.36)	13.93 (112.22/130.33)	10.10 (112.86/130.55)	11.91 (116.63/135.85)
SD of CGM (mg/dL)	Mean (Med)	21.40 (21.95)	21.04 (21.02)	28.86 (30.16)	27.22 (27.59)
	SD (Q1/Q3)	5.05 (17.91/23.97)	6.10 (16.15/24.20)	5.93 (23.99/33.81)	5.21 (23.15/32.13)
Min (mg/dL)	Mean (Med)	68 (69)	88 (83)	68 (69)	81 (82)
	SD (Q1/Q3)	8.92 (61.94/73.17)	14.30 (77.35/94.71)	8.92 (62/73)	6.49 (75.90/87.15)
Max (mg/dL)	Mean (Med)	166 (168)	170 (176)	199 (199)	202 (203)
	SD (Q1/Q3)	19.71 (154/186)	21.00 (155/185)	20.31 (183/217)	20.52 (185.43/219.40)

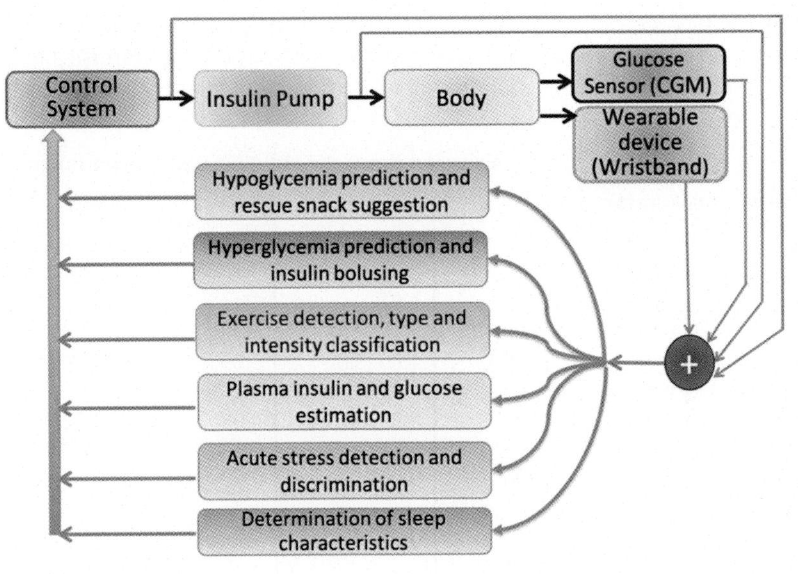

FIGURE 15.8

A diagram of the various components and modules of an mAP system.

References

[1] Thabit H, Hovorka R. Coming of age: the artificial pancreas for type 1 diabetes. Diabetologia 2016;59(9):1795–805.

[2] Reddy M, et al. Metabolic control with the bio-inspired artificial pancreas in adults with type 1 diabetes: a 24-hour randomized controlled crossover study. Journal of Diabetes Science and Technology 2015;10(2):405–13.

[3] Breton MD. Physical activity-the major unaccounted impediment to closed loop control. Journal of Diabetes Science and Technology 2008;2(1):169–74.

[4] Riddell M, Perkins BA. Exercise and glucose metabolism in persons with diabetes mellitus: perspectives on the role for continuous glucose monitoring. Journal of Diabetes Science and Technology 2009;3(4):914–23.

[5] Galassetti P, Riddell MC. Exercise and type 1 diabetes (T1DM). Comparative Physiology 2013;3(3):1309–36.

[6] Zecchin C, et al. Physical activity measured by physical activity monitoring system correlates with glucose trends reconstructed from continuous glucose monitoring. Diabetes Technology and Therapeutics 2013;15(10):836–44.

[7] Dassau E, Bequette BW, Buckingham B a, Doyle III FJ. Detection of a meal using continuous glucose monitoring. Diabetes Care 2008;31(2):295–300.

[8] Clarke WL, Anderson S, Breton M, Patek S, Kashmer L, Kovatchev B. Closed-loop artificial pancreas using subcutaneous glucose sensing and insulin delivery and a model predictive control algorithm: the Virginia experience. Journal of Diabetes Science and Technology 2009;3(5):1031–8.

[9] Steil GM, et al. The effect of insulin feedback on closed loop glucose control. The Journal of Cinical Endocrinology and Metabolism 2011;96(5):1402–8.

[10] Cengiz E, Swan KL, Tamborlane WV, Steil GM, Steffen AT, Weinzimer SA. Is an automatic pump suspension feature safe for children with type 1 diabetes? An exploratory analysis with a closed-loop system. Diabetes Technology and Therapeutics 2009;11(4): 207–10.
[11] Sherr JL, et al. Safety of nighttime 2-hour suspension of basal insulin in pump-treated type 1 diabetes even in the absence of low glucose. Diabetes Care 2014;37(3):773–9.
[12] Agrawal P, Welsh JB, Kannard B, Askari S, Yang Q, Kaufman FR. Usage and effectiveness of the low glucose suspend feature of the Medtronic Paradigm Veo insulin pump. Journal of Diabetes Science and Technology 2011;5(5):1137–41.
[13] Renard E, et al. Day and night closed-loop glucose control in patients with type 1 diabetes under free-living conditions: results of a single-arm 1-month experience compared with a previously reported feasibility study of evening and night at home. Diabetes Care 2016:dc160008.
[14] Harvey RA, et al. Clinical evaluation of an automated artificial pancreas using zone-model predictive control and health monitoring system. Diabetes Technology and Therapeutics 2014;16(6):348–57.
[15] Nimri R, et al. Night glucose control with MD-logic artificial pancreas in home setting: a single blind, randomized crossover trial-interim analysis. Pediatric Diabetes 2014;15: 91–9.
[16] Kropff J, et al. 2 month evening and night closed-loop glucose control in patients with type 1 diabetes under free-living conditions: a randomised crossover trial. Lancet Diabetes and Endocrinology 2015;3(12):939–47.
[17] Dassau E, et al. Multicenter outpatient randomized crossover trial of zone-MPC artificial pancreas in type 1 diabetes: effects of initialization strategies. In: Diabetes, vol. 64; 2015. A59–60.
[18] Doyle FJ, Huyett LM, Lee JB, Zisser HC, Dassau E. Closed-loop artificial pancreas systems: engineering the algorithms. Diabetes Care 2014;37(5):1191–7.
[19] Patek SD, et al. In silico preclinical trials: methodology and engineering guide to closed-loop control in type 1 diabetes mellitus. Journal of Diabetes Science and Technology 2009;3(2):269–82.
[20] Lee JB, Dassau E, Gondhalekar R, Seborg DE, Pinsker JE, Doyle III FJ. Enhanced model predictive control (eMPC) strategy for automated glucose control. Industrial and Engineering Chemistry Research 2016;55(46):11857–68.
[21] Kovatchev B, Tamborlane WV, Cefalu WT, Cobelli C. The artificial pancreas in 2016: a digital treatment ecosystem for diabetes. Diabetes Care 2016;39(7):1123–6.
[22] Toffanin C, Messori M, Di Palma F, De Nicolao G, Cobelli C, Magni L. Artificial pancreas: model predictive control design from clinical experience. Journal of Diabetes Science and Technology 2013;7(6):1470–83.
[23] Renard E, et al. Reduction of hyper-and hypoglycemia during two months with a wearable artificial pancreas from dinner to breakfast in patients with type 1 diabetes. In: Diabetes, vol. 64; 2015. A237–8.
[24] Nimri R, Atlas E, Ajzensztejn M, Miller S, Oron T, Phillip M. Feasibility study of automated overnight closed-loop glucose control under MD-logic artificial pancreas in patients with type 1 diabetes: the DREAM project. Diabetes Technology and Therapeutics 2012;14(8):728–35.
[25] Steil GM, Rebrin K, Darwin C, Hariri F, Saad MF. Feasibility of automating insulin delivery for the treatment of type 1 diabetes. Diabetes 2006;55(12):3344–50.

[26] Steil GM, Clark B, Kanderian S, Rebrin K. Modeling insulin action for development of a closed-loop artificial pancreas. Diabetes Technology and Therapeutics 2005;7(1): 94–108.

[27] Bequette BW. Challenges and recent progress in the development of a closed-loop artificial pancreas. Annual Reviews in Control 2012;36(2):255–66.

[28] Cameron FM, et al. Closed-loop control without meal announcement in type 1 diabetes. Diabetes Technology and Therapeutics 2017;19(9):527–32.

[29] Gondhalekar R, Dassau E, Doyle FJ. Periodic zone-MPC with asymmetric costs for outpatient-ready safety of an artificial pancreas to treat type 1 diabetes. Automatica 2016;71:237–46.

[30] Toffanin C, Zisser H, Doyle FJ, Dassau E. Dynamic insulin on board: incorporation of circadian insulin sensitivity variation. Journal of Diabetes Science and Technology 2013;7(4):928–40.

[31] Lee H, Bequette BW. A closed-loop artificial pancreas based on model predictive control: human-friendly identification and automatic meal disturbance rejection. Biomedical Signal Processing and Control 2009;4(4):347–54.

[32] Hovorka R, et al. Nonlinear model predictive control of glucose concentration in subjects with type 1 diabetes. Physiological Measurement 2004;25(4):905.

[33] Cameron F, Niemeyer G, Bequette BW. Extended multiple model prediction with application to blood glucose regulation. Journal of Process Control 2012;22(8):1422–32.

[34] Trevitt S, Simpson S, Wood A. Artificial pancreas device systems for the closed-loop control of type 1 diabetes what systems are in development? Journal of Diabetes Science and Technology 2016;10(3):714–23.

[35] Del Favero S, et al. Multicenter outpatient dinner/overnight reduction of hypoglycemia and increased time of glucose in target with a wearable artificial pancreas using modular model predictive control in adults with type 1 diabetes. Diabetes, Obesity and Metabolism 2015;17(5):468–76.

[36] Del Favero S, et al. Randomized summer camp crossover trial in 5-to 9-year-old children: outpatient wearable artificial pancreas is feasible and safe. Diabetes Care 2016: dc152815.

[37] Messori M, et al. Individually adaptive artificial pancreas in subjects with type 1 diabetes: a one-month proof-of-concept trial in free-living conditions. Diabetes Technology and Therapeutics 2017;19(10):560–71.

[38] Breton MD, et al. Closed loop control during intense prolonged outdoor exercise in adolescents with type 1 diabetes: the artificial pancreas Ski study. Diabetes Care 2017; 40(12):1644–50.

[39] Turksoy K, Quinn L, Littlejohn E, Cinar A. Multivariable adaptive identification and control for artificial pancreas systems. IEEE Transactions on Biomedical Engineering 2014;61(3):883–91.

[40] Turksoy K, Bayrak ES, Quinn L, Littlejohn E, Cinar A. Multivariable adaptive closed-loop control of an artificial pancreas without meal and activity announcement. Diabetes Technology and Therapeutics 2013;15(5):386–400.

[41] Kovatchev B, Cobelli C. Glucose variability: timing, risk analysis, and relationship to hypoglycemia in diabetes. Diabetes Care 2016;39(4):502–10.

[42] Jacobs PG, et al. Automated control of an adaptive bihormonal, dual-sensor artificial pancreas and evaluation during inpatient studies. IEEE Transactions on Biomedical Engineering 2014;61(10):2569–81.

[43] Taleb N, Haidar A, Messier V, Gingras V, Legault L, Rabasa-Lhoret R. Glucagon in artificial pancreas systems: potential benefits and safety profile of future chronic use. Diabetes, Obesity and Metabolism 2017;19(1):13–23.
[44] Russell SJ, et al. Outpatient glycemic control with a bionic pancreas in type 1 diabetes. The New England Journal of Medicine 2014;371(4):313–25.
[45] Haidar A, Legault L, Messier V, Mitre TM, Leroux C, Rabasa-Lhoret R. Comparison of dual-hormone artificial pancreas, single-hormone artificial pancreas, and conventional insulin pump therapy for glycaemic control in patients with type 1 diabetes: an open-label randomised controlled crossover trial. Lancet Diabetes and Endocrinology 2015;3(1):17–26.
[46] Ellingsen C, et al. Safety constraints in an artificial pancreatic beta cell: an implementation of model predictive control with insulin on board. Journal of Diabetes Science and Technology 2009;3(3):536–44.
[47] Rossetti P, et al. Closed-loop control of postprandial glycemia using an insulin-on-board limitation through continuous action on glucose target. Diabetes Technology and Therapeutics 2017;19(6):355–62.
[48] Hajizadeh I, et al. Adaptive and personalized plasma insulin concentration estimation for artificial pancreas systems. Journal of Diabetes Science and Technology 2018; 12(3):639–49.
[49] Steil GM, Rebrin K. Closed loop system for controlling insulin infusion. Google Patents. 2008.
[50] Steil GM. Algorithms for a closed-loop artificial pancreas: the case for proportional-integral-derivative control. Journal of Diabetes Science and Technology 2013;7(6): 1621–31.
[51] Bequette BW. Algorithms for a closed-loop artificial pancreas: the case for model predictive control. Journal of Diabetes Science and Technology 2013;7(6):1632–43.
[52] Hovorka R, Canonico V, Chassin LJ, Haueter U, Massi-Benedetti M, Orsini Federici M, Pieber TR, Schaller HC, Schaupp L, Vering. Nonlinear model predictive control of glucose concentration in subjects with type 1 diabetes. Physiological Measurement 2005;25(4):905–20.
[53] Cinar A, Turksoy K, Hajizadeh I. Multivariable artificial pancreas method and system. 2016.
[54] Biegler LT, Yang X, Fischer GAG. Advances in sensitivity-based nonlinear model predictive control and dynamic real-time optimization. Journal of Process Control 2015;30: 104–16.
[55] Yu ZJ, Biegler LT. Advanced-step multistage nonlinear model predictive control. IFAC-PapersOnLine 2018;51(20):122–7.
[56] Oberdieck R, Pistikopoulos EN. Explicit hybrid model-predictive control: the exact solution. Automatica 2015;58:152–9.
[57] Rivotti P, Pistikopoulos EN. A dynamic programming based approach for explicit model predictive control of hybrid systems. Computers and Chemical Engineering 2015;72:126–44.
[58] Cao Z, Gondhalekar R, Dassau E, Doyle FJ. Extremum seeking control for personalized zone adaptation in model predictive control for type 1 diabetes. IEEE Transactions on Biomedical Engineering Aug. 2018;65(8):1859–70.
[59] Eren-Oruklu M, Cinar A, Rollins DK, Quinn L. Adaptive system identification for estimating future glucose concentrations and hypoglycemia alarms. Automatica 2012; 48(8):1892–7.

[60] Turksoy K, Bayrak ES, Quinn L, Littlejohn E, Cinar A. Guaranteed stability of recursive multi-input-single-output time series models. In: Proc. 2013 American control conf. IEEE, Washington, DC; 2013. p. 77–82.

[61] Hajizadeh I, et al. Multivariable recursive subspace identification with application to artificial pancreas systems. IFAC-PapersOnLine 2017;50(1):886–91.

[62] Van Overschee P, De Moor B. Subspace identification for linear systems: theory, implementation Methods. Springer; 1996.

[63] Wang Y, Gao F, Doyle FJ. Survey on iterative learning control, repetitive control, and run-to-run control. Journal of Process Control 2009;19(10):1589–600.

[64] Owens C, Zisser H, Jovanovic L, Srinivasan B, Bonvin D, Doyle III FJ. Run-to-run control of blood glucose concentrations for people with type 1 diabetes mellitus. IEEE Transactions on Biomedical Engineering 2006;53(6):996–1005.

[65] Magni L, et al. Run-to-run tuning of model predictive control for type 1 diabetes subjects: in silico trial. Journal of Diabetes Science and Technology 2009;3(5):1091–8.

[66] Palerm CC, Zisser H, Jovanovič L, Doyle FJ. A run-to-run control strategy to adjust basal insulin infusion rates in type 1 diabetes. Journal of Process Control 2008;18(3): 258–65.

[67] Wang Y, Dassau E, Doyle IIIFJ. Closed-loop control of artificial pancreatic $β$ -cell in type 1 diabetes mellitus using model predictive iterative learning control. IEEE Transactions on Biomedical Engineering Feb. 2010;57(2):211–9.

CHAPTER

The dawn of automated insulin delivery: from promise to product

16

Laura M. Nally, MD [1], Jennifer L. Sherr, MD, PhD [2]

[1]*Associate Professor, Pediatric Endocrinology, Yale Children's Diabetes Program, Yale University School of Medicine, New Haven, CT, United States;* [2]*Instructor, Pediatric Endocrinology, Yale Children's Diabetes Program, Yale University School of Medicine, New Haven, CT, United States*

Introduction

Although the discovery of insulin nearly a century ago allowed those diagnosed with type 1 diabetes (T1D) to live with this chronic medical condition, intensive insulin therapy proven in the Diabetes Control and Complications Trial (DCCT) and the follow-up Epidemiology of Diabetes Complications to stave off long-term complications does not come without a price [1,2]. Meticulous self-care in the form of frequent glucose monitoring and adjustment of insulin regimen, while accounting for a multitude of factors, including food, physical activity, stress, and sickness, can be overwhelming [3,4]. Yet, recent technological advancements in the form of continuous subcutaneous insulin infusion (CSII) pumps and continuous glucose monitors (CGM) have made it possible to provide more physiologic insulin delivery. The ultimate goal is to attain physiologically normal glucose levels, thus preventing short- and long-term complications. To achieve restoration of normal function, a combination of therapies will likely be required, composed of both immune modulators as well as β-cell regeneration therapies.

Although cellular therapies are developed, a "mechanical cure" may allow persons with diabetes to achieve targeted glycemia while minimizing the burden placed on both themselves and, in many cases, their families. A mechanical cure would ideally be composed of a fully closed-loop insulin delivery system; however, given the pharmacokinetic properties of currently available rapid-acting insulin analogs, to date, the focus has been placed on hybrid closed-loop systems.

The concept of a closed-loop system was first realized by Kadish and colleagues in 1964. That early system incorporated continuous real-time intravenous glucose readings to adjust intravenous (IV) infusions of insulin, glucose and/or glucagon [5]. Yet, the determination for adjustments of these substrates was made by a clinician monitoring the system constantly, making the practical application of the system nearly impossible. A decade later, the Biostator was introduced, automating the process. An algorithm on a microcomputer used real-time whole blood analyses of glucose levels to automate insulin delivery [6]. However, this system could not be

Glucose Monitoring Devices. https://doi.org/10.1016/B978-0-12-816714-4.00016-8

used in the home setting because it required IV blood sampling and IV insulin delivery. Though, these systems provided proof of the concept that automated insulin delivery was possible. With technological advancements over the past 30 years, including the development of CSII pumps and subcutaneous CGM, outpatient studies employing algorithms to automate insulin delivery have flourished, paving the way for commercial approval of the first hybrid closed-loop system.

Herein, we review the components necessary for closed-loop systems and the testing of these systems in various environments ranging from strictly controlled clinical research centers to transitional environments like hotels and camps, and finally, to the outpatient setting. The discussion culminates with a review of the first commercially available system, real-world data on its use, and how other systems are likely to come to market in the near future.

Continuous subcutaneous insulin infusion therapy: the first building block in developing a closed-loop system

More than 40 years ago, the first reports on portable CSII pumps were published [7,8]. These initial systems demonstrated the feasibility of using CSII to administer insulin by deploying a miniature, battery-driven syringe pump that allowed for both basal insulin delivery and bolus insulin delivery to cover carbohydrate intake [7].

Despite the improvement in glycemic control demonstrated with early pump therapy studies, the penetration of the device into the clinical care of those with diabetes lagged. After the DCCT demonstrated the beneficial effect of intensive insulin therapy and with modernization and refinement of CSII pumps, the use of this mode of insulin delivery increased around the turn of the century. Recent registry data demonstrated that individuals with diabetes could achieve more targeted glycemic control using pump therapy as compared to those on multiple daily injections [9–11]; yet, data from the type 1 diabetes exchange indicate that only ~60% of registry participants use CSII therapy [12].

Continuous glucose monitors: the second step in the construction of a closed-loop system

Exploration of technologies that allow for real-time sensor glucose readings through interstitial fluid measurements began nearly 30 years ago, with the first commercially available sensor being approved in 1999. Yet, in the years following its market launch, conflicting results from early clinical studies surfaced. Skepticism about the inaccuracy of the earliest generation sensors led to limited uptake in clinical practice. In 2008, the landmark Juvenile Diabetes Research Foundation (JDRF) CGM in the T1D study provided compelling results regarding the potential benefits that could be achieved with the regular use of sensor technology [13]. Importantly, while the study was conducted in three cohorts (pediatrics defined as 8–14 years old, adolescents defined as 14–25 years old, and adults defined as >25 years old), only the adult cohort showed a lowering of HbA1c levels (−0.5%) with the use of sensor therapy at the end of the 26-week trial [13]. However, secondary analyses of the data confirmed that beneficial impacts of the sensor were documentable in all study participants who, regardless of the age cohort, used the sensor at least 6 days per week [13].

In recent years, CGM use in persons with T1D has grown exponentially [12]. Several factors played a role including increased duration of wear to 1–2 weeks' time and factory calibrations in some of the systems that obviate the need for fingerstick calibration. Similarly, substantial improvements in accuracy have been achieved. Furthermore, the US Food and Drug Administration (FDA) approval of nonadjunctive CGM use, whereby treatment decisions can be made based on sensor glucose values, may have also provided the impetus for patients to embrace this technology [14,15].

The way forward: the JDRF roadmap to an artificial pancreas

With the building blocks assembled, the quest to automate insulin delivery was undertaken. JDRF created the Artificial Pancreas (AP) Project in 2006, forging a path forward to boost scientific research in creating a commercially available closed-loop system for patients with T1D. Furthermore, a "roadmap" (Fig. 16.1) was created to clearly define the steps in the process of automation, recognizing that insulin suspension based on low sensor glucose values was inherently less risky

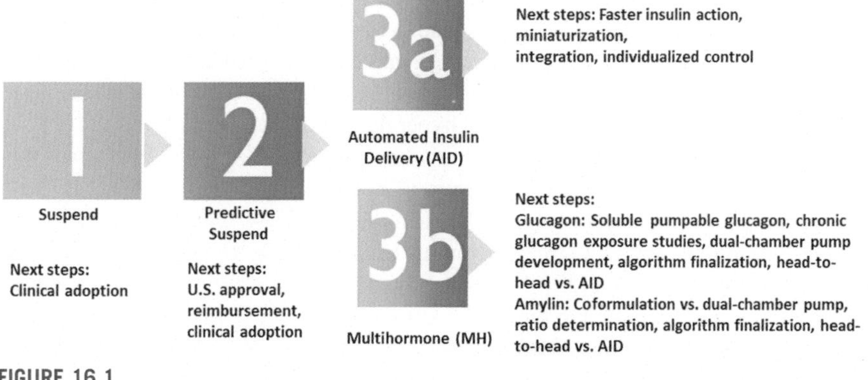

FIGURE 16.1

Revised AP road map. AP system development can be condensed from six steps to three and has bifurcated into automated insulin delivery approaches utilizing solely insulin and multihormone approaches, which may utilize insulin and glucagon, insulin and amylin, or insulin and other glucose-modulating agents.

Used with permission from Kowalski A. Pathway to artificial pancreas systems revisited: moving downstream. Diabetes Care 2015;38(6):1036–1043.

than the prospect of increasing insulin delivery if sensor glucose levels are above a target range [16,17]. This document harmonized the goals for academic researchers, industry sponsors, and regulators and provided patients' perspectives on the process.

Making the dive less deep and shorter: low glucose suspend systems

Interrupting preset basal insulin delivery when the sensor glucose reached a predefined low threshold was the first step toward closing the loop. As the fear of hypoglycemia is a well-known barrier to achieving glycemic targets, this step was critical in motivating patients, their families, and healthcare providers to attain targeted glycemia [12]. Interestingly, early data from feasibility studies of closed-loop systems provided some justification for threshold suspend systems [18,19].

Low glucose suspend (LGS) systems stop insulin delivery at a preset glucose threshold between 60 and 90 mg/dL (3.3–5 mmol/L) for 2 h unless the patient intervenes by resuming insulin delivery. Regardless of sensor glucose levels 2 h following the suspension, insulin delivery resumes. The Automation to Simulate Pancreatic Insulin Response (ASPIRE) In-Clinic study examined how the LGS feature worked if hypoglycemia was induced through exercise and demonstrated that duration and severity of hypoglycemia are reduced when the feature is activated [20].

Regulatory approval of LGS systems occurred in Europe in 2009 with the commercial availability of the Medtronic Paradigm Veo with LGS. Using real-world data extracted from insulin pump uploads, Agrawal and colleagues were able to assess nearly 50,000 patient-days of system use [21]. Although LGS events were not uncommon, occurring on 50% of days studied, the median duration of these events was ~10 min with only 11% lasting >115 min [21]. A subanalysis of 278 participants with at least 3 months of LGS use demonstrated that the feature was able to reduce hypoglycemic (<50 mg/dL) and hyperglycemic (>300 mg/dL) episodes [21].

Building on the In-Clinic ASPIRE study assessment, the ASPIRE In-Home study sought to assess the use of the LGS feature in a cohort of individuals with T1D with documented nocturnal hypoglycemia [22]. In this study, 247 participants were randomized to either sensor-augmented pump (SAP) therapy or SAP with the LGS feature for 3 months' time [22]. Nocturnal hypoglycemia was significantly reduced in the LGS group compared to the SAP control group without causing an increase in HbA1c [22]. Furthermore, a randomized controlled trial of participants with T1D and impaired hypoglycemia awareness demonstrated a reduction in severe (e.g., seizure or coma) and moderate hypoglycemic events in the LGS group. Furthermore, this group had less biochemical hypoglycemia based on sensor glucose readings, especially in the overnight period [23]. In September 2013, the first LGS system, the MiniMed 530G with Enlite, was approved by the FDA for persons with diabetes 16 years of age and older.

Similar to the European analysis of pump uploads, a retrospective review of nearly 30,000 patients accounting for 758,382 patient-days when the LGS feature was used were compared to 166,791 patient-days when the feature was not enabled. On days when the LGS feature was on, there was a 69% reduction in sensor glucose values under 50 mg/dL (2.7 mmol/L) [24]. Most suspend events occurred during the night. At the time of insulin suspension, the median sensor glucose level was 60 mg/dL (3.3 mmol/L) and rose to 87 mg/dL (4.8 mmol/L) 2 h after the suspension, and further increased to 164 mg/dL (9.1 mmol/L) 4 h after basal insulin resumed [24]. Importantly, hyperglycemia exceeding 300 mg/dL was also reduced by 5.6% when the LGS feature was used [24].

Some of the initial concern and delay in regulatory approval of LGS systems in the United States stemmed from concerns over what would occur if insulin was suspended based on inaccurate sensor glucose readings (i.e., when sensor glucose was not low). To address this, a study assessed the effect of random 2-h suspensions in the overnight period as long as blood glucose levels before bed were less than 300 mg/dL (16.7 mmol/L) and blood β-hydroxybutyrate levels were less than 0.5 mmol/L. Comparing over 100 nights of random insulin suspensions to control nights where usual basal rates were maintained, results showed fasting glucose levels were ~50 mg/dL (2.7 mmol/L) higher but there were no clinically meaningful differences in blood β-hydroxybutyrate levels [25]. This supported the contention that LGS could be safely employed in the face of inaccurate sensor glucose readings.

Stopping the plunge: suspend before low systems

Although reducing the severity and duration of hypoglycemia was an important first step, the preference would be to prevent hypoglycemia altogether. To achieve this aim, the suspension of insulin in response to a predicted low glucose value was necessary. Danne and colleagues conducted a study using in silico modeling to assess differences between responses of usual basal delivery without insulin suspensions, LGS, and a predictive low glucose suspend (PLGS) system to challenges intended to induce hypoglycemia. This demonstrated superiority of the PLGS system [26]. Subsequently, the group studied participants using exercise to induce hypoglycemia, and of the 15 evaluable episodes where the hypoglycemic threshold was met and PLGS was triggered, 80% resulted in avoided hypoglycemia [26]. In-home testing of another PLGS system demonstrated efficacy in both pediatric and adult populations by significantly reducing overnight hypoglycemia [27–29].

To date, two companies have marketed PLGS systems: Medtronic with its 640G and 670G systems (Medtronic, Northridge, California) and Tandem with its t:slim X2 with Basal IQ Technology (Tandem, San Diego, California). An in-clinic assessment of the Medtronic 640G was performed by Buckingham and colleagues in participants aged 14–75 years old. In this study, basal rates were increased in a standardized fashion to induce nocturnal hypoglycemia with the PLGS feature set at 65 mg/dL (3.6 mmol/L) [30]. The system is designed to interrupt basal insulin delivery when the sensor glucose is predicted to be below 20 mg/dL (1.1 mmol/L) above the preset threshold within 30 min. Thus, a predicted sensor glucose of 85 mg/dL (4.9 mmol/L) in the next 30 min would trigger an insulin suspension. Sixty-nine participants completed the study, and hypoglycemia was prevented in 60% of the instances [30]. Mean sensor glucose at the time of suspension was 101 mg/dL (5.6 mmol/L), and the mean duration of the suspension was 105 min. Two hours after the start of the suspension, sensor glucose levels reached 102 mg/dL (5.7 mmol/L) [30]. In a 6-month multicenter randomized controlled trial, children and adolescents with T1D were randomized to the 640G system with PLGS or SAP therapy. The results showed a significant reduction in hypoglycemia (sensor glucose < 63 mg/dL) with PLGS (1.5% vs. 2.6%) without glycemic deterioration as measured by HbA1c levels [31].

In a study by Wood and colleagues, the MiniMed 670G "suspend before low" feature was tested in children with T1D aged 7–13 in an overnight in-clinic study, using a preset limit of 65 mg/dL (3.6 mmol/L) [32]. This PLGS system was able to prevent hypoglycemia 80% of the time after exercise, without causing rebound hyperglycemia [32]. Additionally, there were no severe hypoglycemic events reported [32].

The t:slim X2 pump with Basal IQ allows for basal insulin suspension if a sensor glucose of less than 80 mg/dL (4.4 mmol/L) is predicted in the next 30 min based on CGM trends and calculated insulin on board or if the sensor glucose falls below a threshold of 70 mg/dL (3.9 mmol/L). Basal insulin resumes when the sensor readings are higher than the previous reading if the sensor glucose is no longer predicted to go below 80 mg/dL (4.4 mmol/L), if sensor data are lost for 10 min, or if insulin suspension is longer than 120 min in any 150-min period of time. Approval of this device was based on data collected in the PROLOG trial, which was a 6-week, outpatient, randomized crossover study including 103 participants and comparing the t:slim X2 pump with Basal IQ PLGS system to SAP therapy [33]. The use of the system was found to lead to a 31% reduction in time less than 70 mg/dL (3.9 mmol/L) without an increase in time spent with sensor glucose readings over 180 mg/dL (10 mmol/L) [33]. Participants in the trial found the system to be "exceptional" regardless of their previous experience with diabetes technology when queried with a usability scale [34].

Just as concerns existed regarding the risk of ketosis with an LGS system, the same held true for PLGS systems. Therefore, data were analyzed from an outpatient, home-based study of 45 adolescents and adults using a PLGS system to assess the frequency of morning ketosis after overnight insulin suspension [35]. The system suspended basal insulin delivery on 76% of the 977 study nights [35]. On the mornings following an overnight basal insulin suspension, blood ketones were found to be greater than 0.6 mmol/L 1.5% of the time [35]. When limiting the analysis to nights when insulin suspension exceeded 2 h, only 2 of the 159 nights (1.3%) had subsequent morning ketone levels greater than or equal to 0.6 mmol/L [35]. This study demonstrated the safety of predictive basal insulin suspension and concluded that routine measurements of blood or urine ketones were not necessary [35].

The algorithms: the final piece of the puzzle

With the building blocks of continuous glucose monitors and CSII pumps in place, what remained was the creation of an algorithm using sensor data to automatically adjust insulin doses. Three primary control algorithms have emerged:

(1) Proportional integral derivative (PID)
(2) Model predictive control (MPC)
(3) Fuzzy logic (FL)

The PID algorithm primarily uses a linear model, while the MPC algorithm uses a compartmental model of glucose and insulin dynamics. The FL system is designed to imitate the reasoning used to adjust insulin doses by diabetes caregivers. Fig. 16.2 [36] reviews the multitude of factors and parameters utilized in the creation of a closed-loop system.

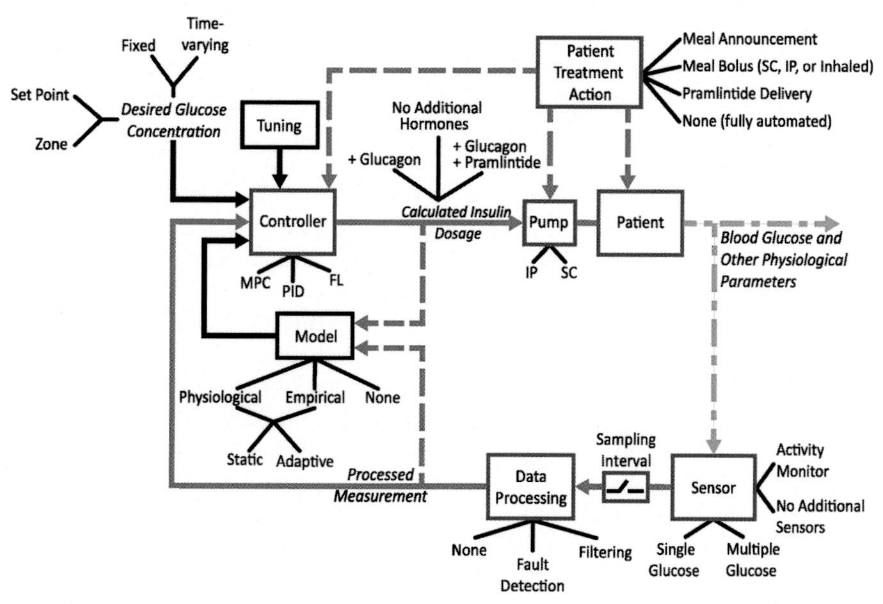

FIGURE 16.2

Taxonomy of the AP design. A specific AP configuration is created by selecting options for each of the major elements shown in the figure. *Solid lines* demonstrate connections that are always present and *dashed lines* represent connections that may only be present in some configurations. The tuning, model, and desired glucose concentration are all part of the controller, as signified by the *black arrows*. Green color distinguishes physiological states or properties from measured or digital signals. *Black lines* are used to indicate predetermined features of a block, and *blue lines* indicate signals or actions conducted during closed-loop operation.

Used with permission from Doyle FJ, 3rd, Huyett LM, Lee JB, Zisser HC, Dassau E. Closed-loop artificial pancreas systems: engineering the algorithms. Diabetes Care 2014;37(5):1191–1197.

Speeding up the process: the creation of an FDA approved simulator

Evaluation of hybrid closed-loop systems in animal models before human studies take an exceptionally long time and be very costly. Kovatchev and colleagues were able to show that in silico preclinical studies could be performed and replace animal trials through the University of Virginia-University of Padova type 1 diabetes simulator [37]. Based on these studies, the FDA approved the use of in silico studies to test closed-loop studies before clinical trials in humans. After this achievement, in silico studies projected to take years with animal trials were able to be tested in less than 6 months [38].

Early studies aimed at closing the loop

The first feasibility study of subcutaneous closed-loop automated insulin delivery was reported by Steil in 2006 and used a PID algorithm with a glucose setpoint of 120 mg/dL (6.7 mmol/L) [39]. This 30-h inpatient study compared data from open-loop (OL) therapy to what was attained with this fully closed-loop (FCL) system [39]. Participants increased time in the range of 70−180 mg/dL (3.9−10 mmol/L) from 63% in OL to 75% in FCL, with no episodes of severe hypoglycemia and similar rates of hypoglycemia in both groups [39].

Given the delay in time to peak insulin action with rapid-acting insulin analogs and the tendency for postprandial hyperglycemia noted with the FCL, Weinzimer and colleagues evaluated the feasibility of using a hybrid closed-loop approach using a PID algorithm in adolescents [40]. Seventeen adolescents with T1D underwent an initial OL assessment of glycemic control with a blinded sensor (Medtronic Datalogger CGM) with eight undergoing a 34-h FCL study. The remainder of the participants received 25%−50% of their usual mealtime bolus 15-min before eating [40]. This latter group was termed "hybrid" closed-loop, as there was a requirement to bolus for the meals [40]. Although overall mean glucose levels were similar between the two groups, postprandial hyperglycemia was more pronounced in the FCL group when compared to the hybrid closed-loop group (226 ± 51 mg/dL vs. 194 ± 47 mg/dL, respectively) [40]. Based on results from this trial, many have since adopted a hybrid closed-loop approach for other systems in development.

Control in clinic: the first closed-loop studies in rigorous research environments

To first evaluate the safety and feasibility of closed-loop insulin delivery, studies were conducted in very controlled research environments. This afforded investigators the opportunity to use bedside plasma glucose values as a means to verify sensor accuracy while also allowing close scrutiny of how plasma glucose levels were being impacted by algorithm-driven insulin delivery. Furthermore, meal preparation in a metabolic kitchen could ensure the nutritional content of meals was captured, or, in some instances, the duration and intensity of exercise. A clear picture emerged from these trials, namely that CL insulin delivery led to an increase in the percent time participants remained in the target range, with exquisite glycemic control noted in the overnight period. These trials involved a variety of patient populations, including adults [38,41–54], children [41,55], adolescents [41,43–45,55–59], and even pregnant women [60].

From transitional environments to tests at home

Thanks to the safety of these systems in inpatient research facilities, the next step involved the simulation of a home environment. For children and adolescents, the camp setting provided a perfect construct in which to assess the safety and efficacy of automated insulin delivery [61–70], while in adults hotel-based studies were utilized [70–73]. In addition, the same findings held true—the devices were safe and demonstrated an increased amount of time in the target range, paving the way for outpatient home-based studies.

The research setting moved to free-living home assessments of devices. Some studies were performed overnight only [74–83], while others opted for daytime and overnight evaluations of the systems [80,81,84–91]. Furthermore, a wide range of patient populations have been studied, including young children [90,92], adolescents [69,74,76–78,80,84,89,91,93–95], and adults [74–76,78,80–85,87,88, 94,96–98]. The picture painted by each of these trials demonstrates the benefits of closed-loop insulin delivery with increased time in range, especially in the overnight period.

From prototype to product: the MiniMed 670G system

Studies of the algorithm driving insulin delivery in the 670G system have been conducted for over 10 years' time. Early assessment of the algorithm in 10 adult participants in a research unit-based study demonstrated the feasibility of using this approach with an increase in time in target range by 12% [39]. Yet, postprandial hyperglycemia persisted, leading to the adoption of a hybrid approach that was first tested by Weinzimer and colleagues [40].

Transitioning to outpatient studies, prototype equipment was first used in an overnight camp study [64] as well as a 24-h outpatient home-based study with remote monitoring of the system [84]. Both studies demonstrated an increase in the time spent in the target range with the closed-loop controller. Subsequently, the feasibility of a pump incorporating a closed-loop algorithm was first assessed in a camp setting [65]. Although that study failed to demonstrate an improvement in glycemic control, it supported further testing of the integrated platform. This led to the assessment of the system was done over 4–5 days in a hotel-based study that included 9 adults and 15 adolescents at three clinical centers. Although time in target range did not reach statistical significance in the well-controlled adult cohort (baseline HbA1c $7.0 \pm 0.7\%$), rates of time <70 mg/dL and <60 mg/dL were reduced fourfold. Meanwhile, the adolescent cohort had higher baseline HbA1c levels ($9.0 \pm 1.1\%$), and in this group, time in the target range of 70–180 mg/dL increased by nearly 15%.

To obtain market approval, Medtronic designed a 3-month single-arm study of 124 participants to assess the safety of the system. With over 12,000 patient-days closed-loop control data in a cohort of 14–75-year-olds, there were no episodes of severe hypoglycemia or diabetic ketoacidosis [99]. Before initiating hybrid closed-loop insulin delivery, 2-week of baseline data were collected using SAP therapy [99]. Although the study was not designed with a primary efficacy endpoint, exploratory analyses demonstrated improvement in glycemic control with a mean drop in HbA1c of 0.5% [99]. Furthermore, in 94 adults aged 22–75 years old, the percent of the time in the target range rose from 68.8 ± 11.9 to $73.8 \pm 8.4\%$. Similar findings were seen in adolescents ($n = 30$, age 14–21 years) with improvements in time in target range from 60.4 ± 10.9 to $67.2 \pm 8.2\%$ [100]. Data from this study were submitted to the FDA for premarket approval in June 2016. Just over 100 days later, on September 28, 2016, the device received regulatory approval for those 14 years and over.

To assess the performance of the system in a pediatric population, a similar study design was employed in 105 youth with T1D aged 7–13 years. The median use of auto mode use was 81% of the time, and there were no episodes of severe hypoglycemia or diabetic ketoacidosis during the study period. HbA1c fell on average 0.4% from baseline with a concomitant 9% rise in time in target range over the 24 h period. By June 2018, the FDA lowered the age for which the device was approved to 7 years, and the Conformité Europénne (CE) Mark for the device mirrored this.

Most recently, data have been reported on the use of the system in very young children, between the ages of 2–6 years [101]. As had been demonstrated in the studies of both older children and adults, in this study of 42 participants, there were no episodes of severe hypoglycemia or DKA. Furthermore, time in range increased 8% primarily mediated through a reduction in time spent in the hyperglycemic range leading to an average reduction in A1c by 0.5% [101].

Exploring the equipment: components and characteristics of the 670G

Guardian sensor 3

The Medtronic 670G automates insulin delivery based on sensor glucose readings obtained with the Guardian Sensor 3. The sensor requires fingerstick calibrations at least every 12 h to display sensor data. Although it has been cleared for automated insulin delivery, it does not bear FDA approval for nonadjunctive use of sensor glucose values. Therefore, patients using this device are advised to treat based on fingerstick glucose values. Studies assessing sensor accuracy by measuring the mean absolute relative difference (MARD) have shown MARD of 8.7% [102]. In this adult study, the sensor was placed on the arm and calibrations were performed 3–4 times a day [102]. When limited to only two calibrations per day, MARD rose to 10.7% [102,103].

The Medtronic 670G insulin pump

The Medtronic 670G pump functions in two modes: manual mode or "auto mode." The manual mode includes pump therapy without the use of the Guardian 3 sensor (or conventional pump therapy) as well as sensor-augmented pump therapy with both a low glucose suspend ("suspend on low") and a predictive low glucose suspend ("suspend before low") features. "Auto mode" is the term used to indicate the system is working as a hybrid closed-loop insulin delivery method, increasing insulin delivery to curb hyperglycemia and decreasing insulin delivery for anticipated low blood sugars.

The pump can hold up to 300 units of insulin and is approved for use with rapid-acting insulin analogs. It has two bolus speed options: "standard" which delivers 1.5 units/min and "quick" providing 15 units/min. The pump is waterproof, protected to a depth of up to 12 feet for up to 24 h. At altitude, it is approved for air pressures as low as 70.33 kPa (10.2 psiA), or about 10,000 feet (3000 m) above sea level. It uses a single AA battery. Basal rates can be given as low as 0 units per hour and up to 35 units per hour, or the maximum basal rate set in the pump, whichever is lower. A maximum of 25 units may be delivered in aa single bolus.

Auto mode

The algorithm determining the basal rate modulation is housed in the insulin pump itself. To use this feature, users must receive at least 8 units of insulin per day. Only two settings may be tweaked during auto mode insulin delivery—the insulin to carbohydrate ratio and the active insulin time. Therefore, while there are preprogrammed basal rates in the pump, these do not impact the basal rates determined by the PID controller with insulin feedback. The auto basal is administered via microboluses every 5 min. The system setpoint is 120 mg/dL; however, if a correction dose is delivered, it targets glucose of 150 mg/dL. The system parameters are adjusted each evening based on the total daily dose delivered over the preceding days and fasting glucose levels. A temporary target can be set, which increases the setpoint to 150 mg/dL for activities, like exercise. It can be set in half-hour increments up to a maximum of 12 h.

To initiate auto mode, a minimum of 48 h of insulin delivery data is required. The use of the system in the manual mode for a longer period of time gives the algorithm more data upon which to derive initial algorithm parameters, leading some to suggest 1 week of manual mode use before transitioning to hybrid closed-loop insulin delivery. Meal announcements are required and, for this, the user must enter discrete carbohydrate amounts. An analysis of a subset of participants in the pivotal trial led researchers to suggest increasing the insulin to carbohydrate ratio by 10%–20% for optimal results [104].

Auto mode exits: what are they and why do they happen

Although users can choose to switch to and from auto mode, forced system exits may occur for a variety of reasons. Periods of prolonged hyperglycemia (sensor glucose >300 mg/dL for over 1 h or > 250 mg/dL for over 3 h) or maximum algorithm-derived insulin delivery for 4 h suggest that there may be a reason the user needs to reevaluate the hyperglycemia; thus, the system will exit auto mode. Similarly, minimum algorithm-derived insulin delivery for 2.5 h also leads to an auto mode exit. Finally, if there is no sensor glucose reading due to a missed calibration or the sensor is 35% off from fingerstick glucose, an exit may occur. In some instances, the system will revert to "Safe Basal," which is an algorithm-derived basal rate that can last for up to 90 min. After that time, if the issue is not resolved the system reverts back to manual mode. These exits were created for patient safety, but some have found these exits nuisances, which may cause users to discontinue automated insulin delivery.

The highs and lows of real-world use of the 670G system

Reports of clinical use of the 670G following its commercial launch have yielded somewhat conflicting results. Stone et al. published a retrospective analysis of 3141 patients who uploaded their data to Carelink during the first 10 months following the commercial launch [105]. Improvements in time spent in target range (70–180 mg/dL) were seen with auto mode when compared to manual mode (73% vs. 66%) across all age groups, though there was a reduction in time spent in both hypoglycemic (<70 mg/dL 2.1% v 2.7%) and hyperglycemic (>180 mg/dL 24.6% v 31.4%) ranges [105]. To date, the 670G system is currently used by more than 180,000 people. An analysis of more than 8-million patient-days of data showed that users on average had 71% time in target range in all age groups [106].

Yet, work from clinics implementing this therapy has painted a somewhat different picture. Indeed, data from the Barbra Davis Center following 51 youth who started the use of the system found that after 6 months, 37% of their cohort had discontinued automated insulin delivery [107]. Furthermore, youth who continued use of the system was found to have waning sensor use over time (81% vs. 71%) as well as a decrease in time in target range (61% vs. 56%) [107]. These findings have been echoed by similar observational studies conducted at both Boston Children's Hospital and Lucile Packard Children's Hospital, where discontinuation rates were 19% and 50%, respectively [108,109]. Importantly, both of these groups reported that auto mode use led to a lowering in the HbA1c level [108,109].

Similarly, an analysis of 139 adults followed at the Barbara Davis Center found that the use of auto mode led to an increase in the percent time in target range [110].

Challenges that remain

For years, closed-loop studies have provided undeniable evidence that fixed basal rates can never approximate physiologic needs due to the multitude of factors that impact glucose control. Thus, it is not surprising that with each of these systems, overnight performance is best when responses to food and activity are, often, no longer required. Yet, the greatest issue that remains is postprandial glycemic control. As current rapid-acting insulin analogs do not have the time action profile to adequately address glycemic excursions following meals, a number of strategies are being explored to overcome this issue. The use of bihormonal systems that would allow for the use of glucagon to moderate a more aggressive insulin delivery algorithm and more closely mimic normal physiology may be the key to a fully closed-loop system [70]. The creation of soluble glucagon products has been critical to this endeavor, and multiple companies have been working in this area [111]. Others have explored the use of adjunctive agents, with early feasibility studies showing that both pramlintide (amylin analog) and liraglutide (glucagon-like peptide 1 receptor agonist), have lessened postprandial hyperglycemia [44,45,50]. Indeed, the feasibility of using a fixed combination of rapid-acting insulin analog and pramlintide infusions, administered through two separate infusion pumps, has been demonstrated. Ongoing studies will assess the impact of such a fixed combination therapy [112].

Most importantly, the integration of hybrid closed-loop therapy into clinical practice requires setting realistic expectations for both providers and patients and educating providers on the differences between systems. To aide with understanding how systems differ, the CARE acronym has been proposed [113]. More recently this framework has been refined [114]. The goal is to understand five essential features of each system:

Calculate: How does the algorithm CALCULATE insulin delivery? Which components of insulin delivery are automated?
Adjust: How can the user ADJUST insulin delivery? Which parameters can be adjusted to influence insulin delivery during automation? Which parameters are fixed?
Revert: When should the user choose to REVERT to open-loop/no automation? When will the system default to open-loop/no automation?
Educate: What are the key EDUCATION points for the advanced diabetes device? How does the user optimize time using the automated features? Where can users and clinicians find additional education resources?
Sensor/Share: What are the relevant sensor characteristics for each device? What are the system capabilities for remote monitoring and cloud-based data sharing?

As more systems become commercially available, a standardized approach to compare them will help both clinicians and patients understand the benefits that each algorithm and the system components afford.

Other challenges that remain are not unique to closed-loop insulin delivery. Whenever insulin is infused in the subcutaneous tissue there is a risk of infusion set failures, which can lead to DKA if they go unrecognized. Algorithms that can detect and alert users of infusion set failures would benefit all patients on insulin pump therapy. Small studies investigating methods to tackle these issues have been conducted [115–117].

For hybrid closed-loop systems to be adopted in toddlers and children with T1D on very low insulin doses, there may be a need for diluted insulin that will allow for fine-tuned insulin dose titrations in this extremely insulin sensitive population [118]. Previous studies using diluted insulin in hybrid closed-loop systems show promising results, with reduced rates of hypoglycemia, a trend toward reduced glycemic variability, and reduced interindividual variability in time to peak insulin action with diluted insulin [119,120]. Yet, a more recent study could not corroborate this in an outpatient trial where participants were randomized diluted or standard concentration insulin [90]. Importantly, dosing frequency in that trial was in 15 min intervals, not the 5 min intervals that some systems use.

The duration that sensors and infusion sets can be worn can also pose additional hassle to those with T1D. Extended wear of both components would decrease the frequency of diabetes-related interruptions that occur. Finally, the person with diabetes must not be forgotten in these technological advancements. Indeed, to have a device integrate seamlessly into one's life would require systems to allow users to push as few buttons as possible, use their existing technology like their cell phones to administer bolus insulin, and the physical footprint of these devices should be minimized.

A bright future

Although tremendous progress has been made in the creation of closed-loop systems and commercial availability of the first-generation hybrid closed-loop was a major step, much remains to be done. Numerous academics, industry partners, and even online communities seek to create more sophisticated automation of insulin delivery (Table 16.1). Indeed, recent data from a randomized control trial of 14–71-year-old participants using the Tandem Control IQ system showed that time in the range between 70 and 180 mg/dL increased by 2.6 h (70% vs. 59%). Users were extremely satisfied with the system based on technology acceptance questionnaires, had minimal fingerstick (<0.5 per/day) throughout the study as the system uses the Dexcom G6, and used automated insulin delivery 92% of the time. FDA approval of the device is being sought.

Although commercial approval of new hybrid closed-loop systems is sought, a movement by patients has created a Do it Yourself (DIY) diabetes community. Starting with the creation of The Nightscout Project, when persons with diabetes and their loved ones created methods to allow remote monitoring of sensor glucose data, a strong online group has emerged. Moving from the viewing of sensor data to creating insulin delivery algorithms and providing instructions to others on how to create such DIY closed-loop systems has exponentially increased the community

Table 16.1 Closed-loop studies listed on *ClinicalTrials.gov* that were listed as recruiting between May and July 2019.

Pediatric studies									
Clinical trial ID	**Title**	**Study design**	**Duration (per arm)**	**Age (years)**	**Environment**	**Intervention versus control**	**Infusion device**	**Hormone**	**Sensor**
NCT03739099	Efficacy of closed-loop insulin therapy in prepubertal child in free-life (FREELIFE-KID)	Randomized Open label Parallel trial followed by a nonrandomized study extension	18 wk trial, 18 wk study extension	6–12	Outpatient, free living	"24-hr" versus "dinner and overnight" HCL use		Insulin	
NCT03671915	Diabeloop for kids (DBL4K)	Randomized Open label Crossover	6+ wk	6–12	4 d inpatient, 6 wk outpatient	HCL versus SAP	Diabeloop versus usual care SAP	Insulin	Dexcom G6
NCT03844789	A study of t:slim X2 with control-IQ technology (DCLP5)	Randomized Open label Parallel	4 mo + optional 3 mo extension	6–13	Outpatient	HCL versus usual care (pump or injections)	t:slim X2 with control IQ versus usual care	Insulin	Dexcom G6
NCT02925299	Day and night closed-loop in young people with T1D (DAN05)	Randomized Open label Parallel	6 mo	6–18	Outpatient	HCL versus CSII	640G (FlorenceM, Cambridge MPC)	Insulin	Enlite 3
NCT02871089	Closed loop from onset in T1D	Randomized Open label Parallel	24 mo	10–16	Outpatient free living	HCL versus MDI	ForenceM, Cambridge MPC	Insulin	
NCT03428932	Glycemic control and the brain in children with T1D	Randomized Parallel group	6 mo	14–18	Outpatient	HCL versus standard care (CSII or MDI)	Medtronic 670G	Insulin	Enlite 3 with GST3C transmitter
Adult studies									
NCT03234491	Improving postprandial blood glucose control with Afrezza during closed-loop therapy	Randomized Open label Crossover	3 d	18–29 phase 1, 18–50 phase 2	Inpatient	Insulin versus low dose Afrezza versus high dose Afrezza	DiAS with insulin only versus DiAS Afrezza low dose versus DiAs Afrezza high dose	Insulin (subcutaneous), Insulin (inhaled)	
NCT03774186	Pregnancy intervention with a closed-loop system (PICLS) study (PICLS)	Randomized Single-blind Parallel group	10+ mo	18–45	Outpatient	HCL versus SAP		Insulin	

(*continued*)

Table 16.1 Closed-loop studies listed on *ClinicalTrials.gov* that were listed as recruiting between May and July 2019.—*cont'd*

Pediatric studies									
Clinical trial ID	**Title**	**Study design**	**Duration (per arm)**	**Age (years)**	**Environment**	**Intervention versus control**	**Infusion device**	**Hormone**	**Sensor**
NCT03577158	SAFE-AP: automatic control of blood glucose under announced and unannounced exercise (SAFE-AP3)	Randomized Single-blind Prospective Crossover	6 wk	18–65	Outpatient, inpatient for exercise tests	HCL with or without exercise mitigation module versus CSII ± sensor		Insulin	
NCT03494010	Hybrid closed-loop insulin delivery system in T1D candidates for living donor kidney transplant	Prospective Observational cohort	Up to 12 mo	18–65	Outpatient	HCL	Medtronic 670G	Insulin	
NCT03849612	A study of an automated insulin delivery system in adult participants with T1D	Single arm Feasibility	5 d	18–65	Inpatient		AID system	Insulin (Lispro)	
NCT03858062	Algorithm to control postprandial, postexercise, and night glucose excursions in portable closed-loop format (APPEL5)	Randomized Crossover	14 d	18–75	Outpatient	Bihormonal CL (insulin + glucagon) (no meal announcements) versus CSII (+/− sensor)	Inreda diabetic	Insulin, glucagon	
NCT03816761	Research study to look at fast-acting insulin aspart with the insulin pump system "iLet" in adults with T1D	Randomized Double blind Crossover Clinical trial	14 d	18–75		Default tmax = 65 min versus tmax = 40 or 50 min	iLet	Insulin (FiAsp)	
NCT02814123	Effect of basal-bolus closed-loop coadministration of insulin and pramlintide on improving the glycemic control in T1D	Randomized 3-way crossover Clinical trial	24 h	18+	Inpatient	RAI versus RAI + pramlintide versus regular insulin + pramlintide		Rapid-acting insulin, Regular insulin, Pramlintide	

(*continued*)

Table 16.1 Closed-loop studies listed on *ClinicalTrials.gov* that were listed as recruiting between May and July 2019.—*cont'd*

Pediatric studies									
Clinical trial ID	**Title**	**Study design**	**Duration (per arm)**	**Age (years)**	**Environment**	**Intervention versus control**	**Infusion device**	**Hormone**	**Sensor**
NCT03800875	Triple-hormone (insulin-pramlintide-glucagon) closed-loop strategy to regulate glucose levels without carbohydrate counting	Randomized Crossover Clinical trial	27 h	18+	Inpatient	Triple hormone CL (insulin, glucagon, pramlintide) without meal announcements versus insulin only CL		Insulin (FiAsp), Glucagon, Pramlintide	
NCT03554486	Evaluation of Fiasp (fast acting insulin aspart) in 670G hybrid closed-loop therapy	Randomized Double-blinded Crossover Clinical trial	4 wk	18+	Outpatient	FiAsp versus aspart	670G	Insulin (FiAsp)	Guardian 3
NCT03738852	Mechanisms for the restoration of hypoglycemia awareness	Randomized Single-blinded Parallel group Clinical trial	3 mo	18+	Outpatient	HCL 3 mos versus HCL 4 wks versus standard of care	670G	Insulin	Guardian 3
NCT03767790	Model predictive control (MPC) artificial pancreas versus sensor-augmented pump (SAP)/predictive low glucose suspend (PLGS) with different food choices in the outpatient setting	Randomized crossover	4 wk	18+	Outpatient	HCL versus SAP/ PLGS	iAPS	Insulin	Dexcom G6
NCT03258853	Feasibility of outpatient closed-loop control with the bionic pancreas in cystic fibrosis-related diabetes	Randomized Open label Crossover	7 d	18+	Outpatient	Bihormonal CL versus HCL versus usual care (pump or injections)	Bihormonal bionic pancreas versus insulin only bionic pancreas versus usual care	Insulin, Glucagon	
NCT03977727	Flasp versus Novolog in type 1 diabetics using 670G Medtronic pump	Randomized Open label Crossover	7 wk	18+	Outpatient	FiAsp versus Novolog	670G HCL with Fiasp versus 670G HCL with Novolog	Insulin	Guardian 3

(*continued*)

Table 16.1 Closed-loop studies listed on *ClinicalTrials.gov* that were listed as recruiting between May and July 2019.—*cont'd*

Pediatric studies									
Clinical trial ID	**Title**	**Study design**	**Duration (per arm)**	**Age (years)**	**Environment**	**Intervention versus control**	**Infusion device**	**Hormone**	**Sensor**
NCT02846831	Closed-loop control of glucose levels (artificial pancreas) for 12 days in adults with T1D	Randomized Open label 2-way crossover Clinical trial	12 d	18+	Outpatient Free living	HCL versus SAP with PLGS	Tandem insulin pump versus MiniMed Paradigm Veo	Insulin	Dexcom G5 versus Enlite
NCT03424044	Three-way crossover closed-loop study with xeris glucagon	Randomized 3-way crossover clinical trial	9 h inpatient, 67 h outpatient	21–50	Inpatient and outpatient	Insulin only HCL versus insulin + glucagon HCL versus CSII with PLGS	Omnipod (APC controller with exercise detection)	Insulin, Glucagon	Dexcom G5
NCT03215914	Hybrid closed-loop insulin delivery system in hypoglycemia (Aim2)	Single arm Open-label Clinical trial	22 mo	25–70	Outpatient, inpatient for clamp		670G	Insulin	Guardian 3
NCT03353792	Using an artificial pancreas system in older adult T1D patients (T1DM AP)	Randomized Parallel group Clinical trial	8–10 wk	50–75	Outpatient	HCL versus CSII ± sensor		Insulin	iPRO (Medtronic)
Pediatric and adult studies									
NCT03216460	Insulet artificial pancreas free-living IDE3	Single arm Open label Observational Free living Feasibility	2–5 d	2–85	Outpatient	HCL versus Standard therapy	Omnipod (MPC)	Insulin	Dexcom G4Share
NCT02748018	Multicenter trial in adult and pediatric patients with T1D using hybrid closed loop system at home	Randomized Parallel group Adaptive study	6 mo	2–80	Outpatient	HCL versus SAP or MDI or CSII	670G	Insulin	Guardian 3
NCT03674281	The VRIF trial: hypoglycemia reduction with automated insulin delivery system	Nonrandomized crossover Clinical trial	8–10 wks	6-10, 65+	Outpatient	HCL versus SAP	Tandem t:slim X2 with control IQ	Insulin	Dexcom G6
NCT03959423	Safety evaluation of the advanced hybrid closed loop (AHCL) system	Single arm Open label Clinical trial	3 mos	7–75	Outpatient	Advanced HCL	Medtronic 670G 4.0 HCL (AHCL) with Guardian 3	Insulin	Guardian 3
NCT02776696	Comparison of two closed-loop strategies for glucose control in T1D The DREMED Trial-2	Segment 1, part 1—single arm, pilot Segment 1 part 2—crossover	36 hr	10–40	Clinic (part 1) Camp (part 2)	AHCL versus HCL		Insulin	

(*continued*)

Table 16.1 Closed-loop studies listed on *ClinicalTrials.gov* that were listed as recruiting between May and July 2019.—*cont'd*

Pediatric studies									
Clinical trial ID	**Title**	**Study design**	**Duration (per arm)**	**Age (years)**	**Environment**	**Intervention versus control**	**Infusion device**	**Hormone**	**Sensor**
		Segment 2—Randomized Crossover	2 d	10–40	Camp	AHCL versus HCL		Insulin	
		Segment 3 – Parallel group	12 d	10–40	Camp	AHCL versus HCL		Insulin	
		Segment 4 -single arm	5 d camp, 21 d home	14–40	Camp	AHCL		Insulin	
NCT03993366	Pramlintide and Fiasp closed loop with a simple meal announcement	Randomized Open label 2-way Crossover Clinical trial	24 hr	12+	Inpatient	HCL	FiAsp + pramlintide versus FiAsp	Insulin + pramlintide, Insulin	
NCT02985866	The international diabetes closed loop (iDCL) trial: protocol 1	Randomized Open label Parallel group Clinical trial	3 mo	14+	Outpatient	HCL (InControl diabetes management platform) versus SAP		Insulin	
NCT03040414	Fuzzy logic-automated insulin regulation (FLAIR)	Randomized Open-label Crossover	3 mo	14–30	Outpatient	AHCL versus HCL	Medtronic 670G 4.0 AHCL versus Medtronic 670G 3.0 HCL	Insulin	Guardian

AHCL, *advanced hybrid closed loop;* CSII, *continuous subcutaneous insulin infusion;* d, *days;* HCL, *hybrid closed-loop;* hr, *hours;* mo, *month;* RAI, *rapid-acting insulin analog;* SAP, *sensor-augmented pump;* T1D, *type 1 diabetes;* wk, *week.*

size. Currently, an open artificial pancreas system (open APS) and loop have garnered thousands of users. The JAEB center for health research is conducting a study entitled, "An Observational Study of Individuals with Type 1 Diabetes Using the Loop System for Automated Insulin Delivery." Tidepool, a nonprofit organization, which seeks to make diabetes data more accessible through downloads and data reports, is working to obtain regulatory approval of the loop system.

Patient considerations

It is important to understand how hybrid closed-loop technology psychologically affects those with T1D and if patient expectations are being met with the new technology. Although hybrid closed-loop therapy allows for the potential to decrease diabetes burden, improve glycemic control, and as a result, reduce complications, early studies have shown a lack of trust in the system [121]. One of the challenges in early trials was perceived CGM inaccuracy [122,123], with person-specific factors, including time since diagnosis and self-perception of T1D management playing a role in this viewpoint. Prior negative experience with system components, such as early generation sensors, made individuals in these studies less apt to want to use a newer device from the same company. Conversely, those who were comfortable with their current devices were less likely to be open to trying a new device.

The degree that a user wants to engage with a hybrid closed-loop system has been found to be a wide spectrum. Some users want more control over the hybrid closed-loop system, with the ability to override the system frequently, find ways to work around the system by using it incorrectly to get the result they want, or users wanted to input more detailed information. Others may enjoy having to make fewer decisions and think about diabetes less by allowing the system to function but may also become deskilled and less vigilant over time [122]. Thus, creating different types of systems that require more or less user feedback will be beneficial to meet the needs of the individual with T1D.

To increase access and usability, hybrid closed-loop technology needs to be easy to understand and adaptable for individuals of all education levels and available in multiple languages. Currently, only older versions of the Medtronic pumps (530G, paradigm revel, paradigm 522/722, paradigm 515/715) allow for languages other than English. This poses a limitation for families who may not be comfortable reading English on a medical device.

Conclusion

The past 40 years have seen technology integrate into how we manage type 1 diabetes, with a technological revolution currently underway. Just as the technology we use in our personal lives has rapidly evolved, the same holds true for diabetes devices. To date, closed-loop systems have been successful in increasing time in the target range, albeit with the caveat of requiring a premeal bolus. Although only one system is currently approved by regulatory bodies, it is anticipated a number of systems will soon come to market and DIY systems will become commercially available. In the end, the technology that will most help a person with diabetes is the one he/she chooses to use; therefore, continued innovation and choice will be critical to allow uptake and successful use of these devices. There is great hope that future closed-loop therapies are likely to continue to improve psychological and physiological outcomes for those with T1D.

Short biography

Laura Nally (short biography): Dr. Nally is an Instructor at Yale University School of Medicine. She completed medical school at UC Davis School of Medicine, pediatrics residency training at Children's Hospital of Orange County, and pediatric endocrinology fellowship at Stanford University. Jennifer Sherr (short biography): Dr. Sherr is an Associate Professor at Yale University School of Medicine. Following her diagnosis of type 1 diabetes in 1987, Dr. Sherr decided to pursue a career in pediatric endocrinology. She completed a joint BA/MD program between Rutgers University and the University of Medicine and Dentistry of New Jersey—Robert Wood Johnson Medical School. Her pediatric residency and pediatric endocrinology fellowship were completed at Yale University, where she also obtained a Ph.D. in Investigative Medicine through the Yale Graduate School of Arts & Sciences.

Disclosures

LMN has received free supplies from Dexcom for investigator initiated studies. JLS reports receiving consulting fees from Medtronic Diabetes and serving on advisory boards for Bigfoot Biomedical, Cecelia Health, Eli Lilly (Nasal Glucagon), Insulet, and the T1D Fund.

References

[1] The Diabetes Control and Complications Trial Research Group. The effect of intensive treatment of diabetes on the development and progression of long-term complications in insulin-dependent diabetes mellitus. New England Journal of Medicine 1993; 329(14):977—86.

[2] Nathan DM, Zinman B, Cleary PA, Backlund JY, Genuth S, Miller R, et al. Modern-day clinical course of type 1 diabetes mellitus after 30 years' duration: the diabetes control and complications trial/epidemiology of diabetes interventions and complications and Pittsburgh epidemiology of diabetes complications experience (1983—2005). Archives of Internal Medicine 2009;169(14):1307—16.

[3] Hessler DM, Fisher L, Polonsky WH, Masharani U, Strycker LA, Peters AL, et al. Diabetes distress is linked with worsening diabetes management over time in adults with

type 1 diabetes. Diabetic Medicine: A Journal of the British Diabetic Association 2017;34(9):1228–34.

[4] Hagger V, Hendrieckx C, Sturt J, Skinner TC, Speight J. Diabetes distress among adolescents with type 1 diabetes: a systematic review. Current Diabetes Reports 2016;16(1):9.

[5] Kadish AH. Automation control of blood sugar. I. A servomechanism for glucose monitoring and control. The American Journal of Medical Electronics 1964;3:82–6.

[6] Clemens AH, Chang PH, Myers RW. The development of biostator, a glucose controlled insulin infusion system (GCIIS). Hormone and Metabolic Research 1977;(Suppl. 7):23–33.

[7] Pickup JC, Keen H, Parsons JA, Alberti KG. Continuous subcutaneous insulin infusion: an approach to achieving normoglycaemia. British Medical Journal 1978;1(6107):204–7.

[8] Tamborlane WV, Sherwin RS, Genel M, Felig P. Reduction to normal of plasma glucose in juvenile diabetes by subcutaneous administration of insulin with a portable infusion pump. New England Journal of Medicine 1979;300(11):573–8.

[9] Sherr JL, Hermann JM, Campbell F, Foster NC, Hofer SE, Allgrove J, et al. Use of insulin pump therapy in children and adolescents with type 1 diabetes and its impact on metabolic control: comparison of results from three large, transatlantic paediatric registries. Diabetologia 2016;59(1):87–91.

[10] Szypowska A, Schwandt A, Svensson J, Shalitin S, Cardona-Hernandez R, Forsander G, et al. Insulin pump therapy in children with type 1 diabetes: analysis of data from the SWEET registry. Pediatric Diabetes 2016;17(Suppl. 23):38–45.

[11] Karges B, Schwandt A, Heidtmann B, Kordonouri O, Binder E, Schierloh U, et al. Association of insulin pump therapy vs insulin injection therapy with severe hypoglycemia, ketoacidosis, and glycemic control among children, adolescents, and young adults with type 1 diabetes. Journal of the American Medical Association 2017;318(14):1358–66.

[12] Foster NC, Beck RW, Miller KM, Clements MA, Rickels MR, DiMeglio LA, et al. State of type 1 diabetes management and outcomes from the T1D exchange in 2016–2018. Diabetes Technology and Therapeutics 2019;21(2):66–72.

[13] Tamborlane WV, Beck RW, Bode BW, Buckingham B, Chase HP, Clemons R, et al. Continuous glucose monitoring and intensive treatment of type 1 diabetes. New England Journal of Medicine 2008;359(14):1464–76.

[14] Aleppo G, Ruedy KJ, Riddlesworth TD, Kruger DF, Peters AL, Hirsch I, et al. REPLACE-BG: a randomized trial comparing continuous glucose monitoring with and without routine blood glucose monitoring in adults with well-controlled type 1 diabetes. Diabetes Care 2017;40(4):538–45.

[15] FDA advisory panel votes to recommend non-adjunctive use of of Dexcom G5 mobile CGM. Diabetes Technology and Therapeutics 2016;18(8):512–6.

[16] Kowalski AJ. Can we really close the loop and how soon? Accelerating the availability of an artificial pancreas: a roadmap to better diabetes outcomes. Diabetes Technology and Therapeutics 2009;11(Suppl. 1):S113–9.

[17] Kowalski A. Pathway to artificial pancreas systems revisited: moving downstream. Diabetes Care 2015;38(6):1036–43.

[18] Elleri D, Allen JM, Nodale M, Wilinska ME, Acerini CL, Dunger DB, et al. Suspended insulin infusion during overnight closed-loop glucose control in children and adolescents with type 1 diabetes. Diabetic Medicine 2010;27(4):480–4.

[19] Cengiz E, Sherr JL, Weinzimer SA, Tamborlane WV. Clinical equipoise: an argument for expedited approval of the first small step toward an autonomous artificial pancreas. Expert Review of Medical Devices 2012;9(4):315–7.

[20] Garg S, Brazg RL, Bailey TS, Buckingham BA, Slover RH, Klonoff DC, et al. Reduction in duration of hypoglycemia by automatic suspension of insulin delivery: the in-clinic ASPIRE study. Diabetes Technology and Therapeutics 2012;14(3):205–9.

[21] Agrawal P, Welsh JB, Kannard B, Askari S, Yang Q, Kaufman FR. Usage and effectiveness of the low glucose suspend feature of the Medtronic Paradigm Veo insulin pump. Journal of Diabetes Science and Technology 2011;5(5):1137–41.

[22] Bergenstal RM, Klonoff DC, Garg SK, Bode BW, Meredith M, Slover RH, et al. Threshold-based insulin-pump interruption for reduction of hypoglycemia. New England Journal of Medicine 2013;369(3):224–32.

[23] Ly TT, Nicholas JA, Retterath A, Lim EM, Davis EA, Jones TW. Effect of sensor-augmented insulin pump therapy and automated insulin suspension vs standard insulin pump therapy on hypoglycemia in patients with type 1 diabetes: a randomized clinical trial. Journal of the American Medical Association 2013;310(12):1240–7.

[24] Agrawal P, Zhong A, Welsh JB, Shah R, Kaufman FR. Retrospective analysis of the real-world use of the threshold suspend feature of sensor-augmented insulin pumps. Diabetes Technology and Therapeutics 2015;17(5):316–9.

[25] Sherr JL, Palau Collazo M, Cengiz E, Michaud C, Carria L, Steffen AT, et al. Safety of nighttime 2-hour suspension of Basal insulin in pump-treated type 1 diabetes even in the absence of low glucose. Diabetes Care 2014;37(3):773–9.

[26] Danne T, Tsioli C, Kordonouri O, Blaesig S, Remus K, Roy A, et al. The PILGRIM study: in silico modeling of a predictive low glucose management system and feasibility in youth with type 1 diabetes during exercise. Diabetes Technology and Therapeutics 2014;16(6):338–47.

[27] Maahs DM, Calhoun P, Buckingham BA, Chase HP, Hramiak I, Lum J, et al. A randomized trial of a home system to reduce nocturnal hypoglycemia in type 1 diabetes. Diabetes Care 2014;37(7):1885–91.

[28] Buckingham BA, Raghinaru D, Cameron F, Bequette BW, Chase HP, Maahs DM, et al. Predictive low-glucose insulin suspension reduces duration of nocturnal hypoglycemia in children without increasing ketosis. Diabetes Care 2015;38(7):1197–204.

[29] Calhoun PM, Buckingham BA, Maahs DM, Hramiak I, Wilson DM, Aye T, et al. Efficacy of an overnight predictive low-glucose suspend system in relation to hypoglycemia risk factors in youth and adults with type 1 diabetes. Journal of Diabetes Science and Technology 2016;10(6):1216–21.

[30] Buckingham BA, Bailey TS, Christiansen M, Garg S, Weinzimer S, Bode B, et al. Evaluation of a predictive low-glucose management system in-clinic. Diabetes Technology and Therapeutics 2017;19(5):288–92.

[31] Abraham MB, Nicholas JA, Smith GJ, Fairchild JM, King BR, Ambler GR, et al. Reduction in hypoglycemia with the predictive low-glucose management system: a long-term randomized controlled trial in adolescents with type 1 diabetes. Diabetes Care 2018;41(2):303–10.

[32] Wood MA, Shulman DI, Forlenza GP, Bode BW, Pinhas-Hamiel O, Buckingham BA, et al. In-clinic evaluation of the MiniMed 670G system "suspend before low" feature in children with type 1 diabetes. Diabetes Technology and Therapeutics 2018;20(11):731–7.

[33] Forlenza GP, Li Z, Buckingham BA, Pinsker JE, Cengiz E, Wadwa RP, et al. Predictive low-glucose suspend reduces hypoglycemia in adults, adolescents, and children with type 1 diabetes in an at-home randomized crossover study: results of the PROLOG trial. Diabetes Care 2018;41(10):2155–61.

[34] Pinsker JELZ, Buckingham BA, Forlenza GP, Cengiz E, Church MM, et al. Exceptional usability of Tandem t:slim X2 with Basal-IQ predictive low-glucose suspend (PLGS)—the PROLOG study. Orlando, Florida: American Diabetes Association Scientific Session; 2018.

[35] Beck RW, Raghinaru D, Wadwa RP, Chase HP, Maahs DM, Buckingham BA, et al. Frequency of morning ketosis after overnight insulin suspension using an automated nocturnal predictive low glucose suspend system. Diabetes Care 2014;37(5):1224–9.

[36] Doyle 3rd FJ, Huyett LM, Lee JB, Zisser HC, Dassau E. Closed-loop artificial pancreas systems: engineering the algorithms. Diabetes Care 2014;37(5):1191–7.

[37] Kovatchev BP, Breton M, Man CD, Cobelli C. In silico preclinical trials: a proof of concept in closed-loop control of type 1 diabetes. Journal of Diabetes Science and Technology 2009;3(1):44–55.

[38] Kovatchev B, Cobelli C, Renard E, Anderson S, Breton M, Patek S, et al. Multinational study of subcutaneous model-predictive closed-loop control in type 1 diabetes mellitus: summary of the results. Journal of Diabetes Science and Technology 2010;4(6): 1374–81.

[39] Steil GM, Rebrin K, Darwin C, Hariri F, Saad MF. Feasibility of automating insulin delivery for the treatment of type 1 diabetes. Diabetes 2006;55(12):3344–50.

[40] Weinzimer SA, Steil GM, Swan KL, Dziura J, Kurtz N, Tamborlane WV. Fully automated closed-loop insulin delivery versus semiautomated hybrid control in pediatric patients with type 1 diabetes using an artificial pancreas. Diabetes Care 2008; 31(5):934–9.

[41] Nimri R, Danne T, Kordonouri O, Atlas E, Bratina N, Biester T, et al. The "Glucositter" overnight automated closed loop system for type 1 diabetes: a randomized crossover trial. Pediatric Diabetes 2013;14(3):159–67.

[42] El-Khatib FH, Russell SJ, Nathan DM, Sutherlin RG, Damiano ER. A bihormonal closed-loop artificial pancreas for type 1 diabetes. Science Translational Medicine 2010;2(27):27ra.

[43] Nimri R, Atlas E, Ajzensztejn M, Miller S, Oron T, Phillip M. Feasibility study of automated overnight closed-loop glucose control under MD-logic artificial pancreas in patients with type 1 diabetes: the DREAM Project. Diabetes Technology and Therapeutics 2012;14(8):728–35.

[44] Sherr JL, Patel NS, Michaud CI, Palau-Collazo MM, Van Name MA, Tamborlane WV, et al. Mitigating meal-related glycemic excursions in an insulin-sparing manner during closed-loop insulin delivery: the beneficial effects of adjunctive pramlintide and liraglutide. Diabetes Care 2016;39(7):1127–34.

[45] Weinzimer SA, Sherr JL, Cengiz E, Kim G, Ruiz JL, Carria L, et al. Effect of pramlintide on prandial glycemic excursions during closed-loop control in adolescents and young adults with type 1 diabetes. Diabetes Care 2012;35(10):1994–9.

[46] Haidar A, Legault L, Dallaire M, Alkhateeb A, Coriati A, Messier V, et al. Glucose-responsive insulin and glucagon delivery (dual-hormone artificial pancreas) in adults with type 1 diabetes: a randomized crossover controlled trial. Canadian Medical Association Journal 2013;185(4):297–305.

[47] Hovorka R, Kumareswaran K, Harris J, Allen JM, Elleri D, Xing D, et al. Overnight closed loop insulin delivery (artificial pancreas) in adults with type 1 diabetes: crossover randomised controlled studies. BMJ British Medical Journal 2011:342.

[48] Russell SJ, El-Khatib FH, Nathan DM, Magyar KL, Jiang J, Damiano ER. Blood glucose control in type 1 diabetes with a bihormonal bionic endocrine pancreas. Diabetes Care 2012;35(11):2148–55.

[49] Van Bon AC, Jonker LD, Koebrugge R, Koops R, Hoekstra JB, DeVries JH. Feasibility of a bihormonal closed-loop system to control postexercise and postprandial glucose excursions. Journal of Diabetes Science and Technology 2012;6(5):1114–22.

[50] Renukuntla VS, Ramchandani N, Trast J, Cantwell M, Heptulla RA. Role of glucagon-like peptide-1 analogue versus amylin as an adjuvant therapy in type 1 diabetes in a closed loop setting with ePID algorithm. Journal of Diabetes Science and Technology 2014;8(5):1011–7.

[51] Zisser H, Dassau E, Lee JJ, Harvey RA, Bevier W, Doyle III FJ. Clinical results of an automated artificial pancreas using technosphere inhaled insulin to mimic first-phase insulin secretion. Journal of Diabetes Science and Technology 2015;9(3):564–72.

[52] Atlas E, Nimri R, Miller S, Grunberg EA, Phillip M. MD-logic artificial pancreas system: a pilot study in adults with type 1 diabetes. Diabetes Care 2010;33(5):1072–6.

[53] Pinsker JE, Lee JB, Dassau E, Seborg DE, Bradley PK, Gondhalekar R, et al. Randomized crossover comparison of personalized MPC and PID control algorithms for the artificial pancreas. Diabetes Care 2016;39(7):1135–42.

[54] Breton M, Farret A, Bruttomesso D, Anderson S, Magni L, Patek S, et al. Fully integrated artificial pancreas in type 1 diabetes: modular closed-loop glucose control maintains near normoglycemia. Diabetes 2012;61(9):2230–7.

[55] Hovorka R, Allen JM, Elleri D, Chassin LJ, Harris J, Xing D, et al. Manual closed-loop insulin delivery in children and adolescents with type 1 diabetes: a phase 2 randomised crossover trial. Lancet 2010;375(9716):743–51.

[56] Castle JR, Engle JM, El Youssef J, Massoud RG, Yuen KC, Kagan R, et al. Novel use of glucagon in a closed-loop system for prevention of hypoglycemia in type 1 diabetes. Diabetes Care 2010;33(6):1282–7.

[57] El-Khatib FH, Russell SJ, Magyar KL, Sinha M, McKeon K, Nathan DM, et al. Autonomous and continuous adaptation of a bihormonal bionic pancreas in adults and adolescents with type 1 diabetes. The Journal of Cinical Endocrinology and Metabolism 2014;99(5):1701–11.

[58] O'Grady MJ, Retterath AJ, Keenan DB, Kurtz N, Cantwell M, Spital G, et al. The use of an automated, portable glucose control system for overnight glucose control in adolescents and young adults with type 1 diabetes. Diabetes Care 2012;35(11):2182–7.

[59] Ruiz JL, Sherr JL, Cengiz E, Carria L, Roy A, Voskanyan G, et al. Effect of insulin feedback on closed-loop glucose control: a crossover study. Journal of Diabetes Science and Technology 2012;6(5):1123–30.

[60] Murphy HR, Kumareswaran K, Elleri D, Allen JM, Caldwell K, Biagioni M, et al. Safety and efficacy of 24-h closed-loop insulin delivery in well-controlled pregnant women with type 1 diabetes A randomized crossover case series. Diabetes Care 2011;34(12):2527–9.

[61] Del Favero S, Boscari F, Messori M, Rabbone I, Bonfanti R, Sabbion A, et al. Randomized summer camp crossover trial in 5- to 9-year-old children: outpatient wearable artificial pancreas is feasible and safe. Diabetes Care 2016;39(7):1180–5.

[62] Ly TT, Breton MD, Keith-Hynes P, De Salvo D, Clinton P, Benassi K, et al. Overnight glucose control with an automated, unified safety system in children and adolescents with type 1 diabetes at diabetes camp. Diabetes Care 2014;37(8):2310–6.
[63] Ly TT, Buckingham BA, DeSalvo DJ, Shanmugham S, Satin-Smith M, DeBoer MD, et al. Day-and-Night closed-loop control using the unified safety system in adolescents with type 1 diabetes at camp. Diabetes Care 2016;39(8):e106–7.
[64] Ly TT, Keenan DB, Roy A, Han J, Grosman B, Cantwell M, et al. Automated overnight closed-loop control using a proportional-integral-derivative algorithm with insulin feedback in children and adolescents with type 1 diabetes at diabetes camp. Diabetes Technology and Therapeutics 2016;18(6):377–84.
[65] Ly TT, Roy A, Grosman B, Shin J, Campbell A, Monirabbasi S, et al. Day and night closed-loop control using the integrated Medtronic hybrid closed-loop system in type 1 diabetes at diabetes camp. Diabetes Care 2015;38(7):1205–11.
[66] Phillip M, Battelino T, Atlas E, Kordonouri O, Bratina N, Miller S, et al. Nocturnal glucose control with an artificial pancreas at a diabetes camp. New England Journal of Medicine 2013;368(9):824–33.
[67] Russell SJ, Hillard MA, Balliro C, Magyar KL, Selagamsetty R, Sinha M, et al. Day and night glycaemic control with a bionic pancreas versus conventional insulin pump therapy in preadolescent children with type 1 diabetes: a randomised crossover trial. The Lancet Diabetes and Endocrinology 2016;4(3):233–43.
[68] Breton MD, Chernavvsky DR, Forlenza GP, DeBoer MD, Robic J, Wadwa RP, et al. Closed-loop control during intense prolonged outdoor exercise in adolescents with type 1 diabetes: the artificial pancreas Ski study. Diabetes Care 2017;40(12):1644–50.
[69] Haidar A, Legault L, Matteau-Pelletier L, Messier V, Dallaire M, Ladouceur M, et al. Outpatient overnight glucose control with dual-hormone artificial pancreas, single-hormone artificial pancreas, or conventional insulin pump therapy in children and adolescents with type 1 diabetes: an open-label, randomised controlled trial. The lancet Diabetes and endocrinology 2015;3(8):595–604.
[70] Russell SJ, El-Khatib FH, Sinha M, Magyar KL, McKeon K, Goergen LG, et al. Outpatient glycemic control with a bionic pancreas in type 1 diabetes. New England Journal of Medicine 2014;371(4):313–25.
[71] Brown SA, Kovatchev BP, Breton MD, Anderson SM, Keith-Hynes P, Patek SD, et al. Multinight "bedside" closed-loop control for patients with type 1 diabetes. Diabetes Technology and Therapeutics 2015;17(3):203–9.
[72] Kovatchev BP, Renard E, Cobelli C, Zisser HC, Keith-Hynes P, Anderson SM, et al. Safety of outpatient closed-loop control: first randomized crossover trials of a wearable artificial pancreas. Diabetes Care 2014;37(7):1789–96.
[73] Kovatchev BP, Renard E, Cobelli C, Zisser HC, Keith-Hynes P, Anderson SM, et al. Feasibility of outpatient fully integrated closed-loop control first studies of wearable artificial pancreas. Diabetes Care 2013;36(7):1851–8.
[74] Sharifi A, De Bock MI, Jayawardene D, Loh MM, Horsburgh JC, Berthold CL, et al. Glycemia, treatment satisfaction, cognition, and sleep quality in adults and adolescents with type 1 diabetes when using a closed-loop system overnight versus sensor-augmented pump with low-glucose suspend function: a randomized crossover study. Diabetes Technology and Therapeutics 2016;18(12):772–83.
[75] Haidar A, Rabasa-Lhoret R, Legault L, Lovblom LE, Rakheja R, Messier V, et al. Single-and dual-hormone artificial pancreas for overnight glucose control in type 1 diabetes. The Journal of Clinical Endocrinology 2016;101(1):214–23.

[76] Hovorka R, Elleri D, Thabit H, Allen JM, Leelarathna L, El-Khairi R, et al. Overnight closed-loop insulin delivery in young people with type 1 diabetes: a free-living, randomized clinical trial. Diabetes Care 2014;37(5):1204−11.

[77] Nimri R, Muller I, Atlas E, Miller S, Kordonouri O, Bratina N, et al. Night glucose control with MD-logic artificial pancreas in home setting: a single blind, randomized crossover trial—interim analysis. Pediatric Diabetes 2014;15(2):91−9.

[78] Nimri R, Muller I, Atlas E, Miller S, Fogel A, Bratina N, et al. MD-Logic overnight control for 6 weeks of home use in patients with type 1 diabetes: randomized crossover trial. Diabetes Care 2014;37(11):3025−32.

[79] Thabit H, Elleri D, Leelarathna L, Allen JM, Lubina-Solomon A, Stadler M, et al. Unsupervised home use of overnight closed-loop system over 3 to 4 weeks - pooled analysis of randomized controlled studies in adults and adolescents with type 1 diabetes. Diabetes, Obesity and Metabolism 2014.

[80] Thabit H, Tauschmann M, Allen JM, Leelarathna L, Hartnell S, Wilinska ME, et al. Home use of an artificial beta cell in type 1 diabetes. New England Journal of Medicine 2015;373(22):2129−40.

[81] Anderson SM, Raghinaru D, Pinsker JE, Boscari F, Renard E, Buckingham BA, et al. Multinational home use of closed-loop control is safe and effective. Diabetes Care 2016;39(7):1143−50.

[82] Kropff J, Del Favero S, Place J, Toffanin C, Visentin R, Monaro M, et al. 2 month evening and night closed-loop glucose control in patients with type 1 diabetes under free-living conditions: a randomised crossover trial. The Lancet Diabetes and Endocrinology 2015;3(12):939−47.

[83] Thabit H, Lubina-Solomon A, Stadler M, Leelarathna L, Walkinshaw E, Pernet A, et al. Home use of closed-loop insulin delivery for overnight glucose control in adults with type 1 diabetes: a 4-week, multicentre, randomised crossover study. The Lancet Diabetes and Endocrinology 2014;2(9):701−9.

[84] de Bock MI, Roy A, Cooper MN, Dart JA, Berthold CL, Retterath AJ, et al. Feasibility of outpatient 24-hour closed-loop insulin delivery. Diabetes Care 2015;38(11): e186−7.

[85] Forlenza GP, Deshpande S, Ly TT, Howsmon DP, Cameron F, Baysal N, et al. Application of zone model predictive control artificial pancreas during extended use of infusion set and sensor: a randomized crossover-controlled home-use trial. Diabetes Care 2017;40(8):1096−102.

[86] Tauschmann M, Allen JM, Wilinska ME, Thabit H, Acerini CL, Dunger DB, et al. Home use of day-and-night hybrid closed-loop insulin delivery in suboptimally controlled adolescents with type 1 diabetes: a 3-week, free-living, randomized crossover trial. Diabetes Care 2016;39(11):2019−25.

[87] El-Khatib FH, Balliro C, Hillard MA, Magyar KL, Ekhlaspour L, Sinha M, et al. Home use of a bihormonal bionic pancreas versus insulin pump therapy in adults with type 1 diabetes: a multicentre randomised crossover trial. The Lancet 2017;389(10067): 369−80.

[88] Leelarathna L, Dellweg S, Mader JK, Allen JM, Benesch C, Doll W, et al. Day and night home closed-loop insulin delivery in adults with type 1 diabetes: three-center randomized crossover study. Diabetes Care 2014;37(7):1931−7.

[89] Tauschmann M, Thabit H, Bally L, Allen JM, Hartnell S, Wilinska ME, et al. Closed-loop insulin delivery in suboptimally controlled type 1 diabetes: a multicentre, 12-week randomised trial. Lancet 2018;392(10155):1321−9.

[90] Tauschmann M, Allen JM, Nagl K, Fritsch M, Yong J, Metcalfe E, et al. Home use of day-and-night hybrid closed-loop insulin delivery in very young children: a multicenter, 3-week, randomized trial. Diabetes Care 2019;42(4):594–600.

[91] de Bock M, Dart J, Hancock M, Smith G, Davis EA, Jones TW. Performance of medtronic hybrid closed-loop iterations: results from a randomized trial in adolescents with type 1 diabetes. Diabetes Technology and Therapeutics 2018;20(10):693–7.

[92] DeBoer MD, Breton MD, Wakeman C, Schertz EM, Emory EG, Robic JL, et al. Performance of an artificial pancreas system for young children with type 1 diabetes. Diabetes Technology and Therapeutics 2017;19(5):293–8.

[93] Tauschmann M, Allen JM, Wilinska ME, Thabit H, Stewart Z, Cheng P, et al. Day-and-night hybrid closed-loop insulin delivery in adolescents with type 1 diabetes: a free-living, randomized clinical trial. Diabetes Care 2016;39(7):1168–74.

[94] Spaic T, Driscoll M, Raghinaru D, Buckingham BA, Wilson DM, Clinton P, et al. Predictive hyperglycemia and hypoglycemia minimization: in-home evaluation of safety, feasibility, and efficacy in overnight glucose control in type 1 diabetes. Diabetes Care 2017;40(3):359–66.

[95] Thabit H, Elleri D, Leelarathna L, Allen J, Lubina-Solomon A, Stadler M, et al. Unsupervised overnight closed loop insulin delivery during free living: analysis of randomised cross-over home studies in adults and adolescents with type 1 diabetes. Lancet 2015;385(Suppl. 1):S96.

[96] Thabit H, Elleri D, Leelarathna L, Allen JM, Lubina-Solomon A, Stadler M, et al. Unsupervised home use of an overnight closed-loop system over 3–4 weeks: a pooled analysis of randomized controlled studies in adults and adolescents with type 1 diabetes. Diabetes, Obesity and Metabolism 2015;17(5):452–8.

[97] Blauw H, van Bon A, Koops R, DeVries J. Performance and safety of an integrated bihormonal artificial pancreas for fully automated glucose control at home. Diabetes, Obesity and Metabolism 2016;18(7):671–7.

[98] Del Favero S, Bruttomesso D, Di Palma F, Lanzola G, Visentin R, Filippi A, et al. First use of model predictive control in outpatient wearable artificial pancreas. Diabetes Care 2014;37(5):1212–5.

[99] Bergenstal RM, Garg S, Weinzimer SA, Buckingham BA, Bode BW, Tamborlane WV, et al. Safety of a hybrid closed-loop insulin delivery system in patients with type 1 diabetes. Journal of the American Medical Association 2016;316(13):1407–8.

[100] Garg SK, Weinzimer SA, Tamborlane WV, Buckingham BA, Bode BW, Bailey TS, et al. Glucose outcomes with the in-home use of a hybrid closed-loop insulin delivery system in adolescents and adults with type 1 diabetes. Diabetes Technology and Therapeutics 2017;19(3):155–63.

[101] Lee SWSJ, Coredero T, Kaufman F. Glycemic outcomes during MiniMed 670G system use in children aged 2–6 years with type 1 diabetes. Berlin, Germany: Advanced Technologies and Treatments forr Diabetes Conference; 2019.

[102] Christiansen MP, Garg SK, Brazg R, Bode BW, Bailey TS, Slover RH, et al. Accuracy of a fourth-generation subcutaneous continuous glucose sensor. Diabetes Technology and Therapeutics 2017;19(8):446–56.

[103] Slover RH, Tryggestad JB, DiMeglio LA, Fox LA, Bode BW, Bailey TS, et al. Accuracy of a fourth-generation continuous glucose monitoring system in children and adolescents with type 1 diabetes. Diabetes Technology and Therapeutics 2018; 20(9):576–84.

[104] Messer LH, Forlenza GP, Sherr JL, Wadwa RP, Buckingham BA, Weinzimer SA, et al. Optimizing hybrid closed-loop therapy in adolescents and emerging adults using the MiniMed 670G system. Diabetes Care 2018;41(4):789−96.

[105] Stone MP, Agrawal P, Chen X, Liu M, Shin J, Cordero TL, et al. Retrospective analysis of 3-month real-world glucose data after the MiniMed 670G system commercial launch. Diabetes Technology and Therapeutics 2018;20(10):689−92.

[106] MiniMed(TM) 670G system real-world data on 8 million patient days shows 71 percent time in range across all age groups. Medtronic; 2019 [press release]. Online.

[107] Berget CML, Vigers T, Wadwa RP, Slover RH, Pyle L, Driscoll KA, Forlenza GP. Hybrid closed loop therapy in the real world: 6 month clinical observation of youth with type 1 diabetes. Berlin, Germany: Advanced Technologies and Treatments of Diabetes; 2019.

[108] Goodwin WG, Lyons J, Oladunjoye A, Steil G. Challenges in implementing hybrid closed loop insulin pump therapy (Medtronic 670G) in a "real world" clinical setting. The Journal of Clinical Endocrinology and Metabolism 2019;3(1). OR14-O15.

[109] Lal RBM, Maahs DM, Buckingham BA, Jeannine L, Chmielewski A, Peterson K, Wilson DM. 670G clinical experience. San Francisco, CA: American Diabetes Association Annual Scientific Meeting; 2019.

[110] Akturk HKGD, Joseph H, Garg SK, Snell-Bergeon JK. Improvement in time-in-range (TIR) with real-life use of hybrid closed-loop system in ptients with type 1 diabetes. San Francisco, California: American Diabetes Association Annual Scientific Meeting; 2019.

[111] Newswanger B, Ammons S, Phadnis N, Ward WK, Castle J, Campbell RW, et al. Development of a highly stable, nonaqueous glucagon formulation for delivery via infusion pump systems. Journal of Diabetes Science and Technology 2015;9(1): 24−33.

[112] Riddle MC, Nahra R, Han J, Castle J, Hanavan K, Hompesch M, et al. Control of postprandial hyperglycemia in type 1 diabetes by 24-hour fixed-dose coadministration of pramlintide and regular human insulin: a randomized, two-way crossover study. Diabetes Care 2018;41(11):2346−52.

[113] Messer LH, Forlenza GP, Wadwa RP, Weinzimer SA, Sherr JL, Hood KK, et al. The dawn of automated insulin delivery: a new clinical framework to conceptualize insulin administration. Pediatric Diabetes 2017;19(1):14−7.

[114] Messer LH, Berget C, Forlenza GP. A clinical guide to advanced diabetes devices and closed-loop systems using the CARES paradigm. Diabetes Technology and Therapeutics 2019;21(8):462−9.

[115] Cescon M, DeSalvo DJ, Ly TT, Maahs DM, Messer LH, Buckingham BA, et al. Early detection of infusion set failure during insulin pump therapy in type 1 diabetes. Journal of Diabetes Science and Technology 2016;10(6):1268−76.

[116] Facchinetti A, Del Favero S, Sparacino G, Cobelli C. An online failure detection method of the glucose sensor-insulin pump system: improved overnight safety of type-1 diabetic subjects. IEEE Transactions on Biomedical Engineering 2013;60(2): 406−16.

[117] Facchinetti A, Del Favero S, Sparacino G, Cobelli C. Detecting failures of the glucose sensor-insulin pump system: improved overnight safety monitoring for type-1 diabetes. Conference Proceedings: Annual International Conference of the IEEE Engineering in Medicine and Biology Society IEEE Engineering in Medicine and Biology Society Annual Conference 2011;2011:4947−50.

[118] Sherr JL. Closing the loop on managing youth with type 1 diabetes: children are not just small adults. Diabetes Care 2018;41(8):1572–8.

[119] Elleri D, Allen JM, Tauschmann M, El-Khairi R, Benitez-Aguirre P, Acerini CL, et al. Feasibility of overnight closed-loop therapy in young children with type 1 diabetes aged 3–6 years: comparison between diluted and standard insulin strength. BMJ Open Diabetes Research and Care 2014;2(1):e000040.

[120] Ruan Y, Elleri D, Allen JM, Tauschmann M, Wilinska ME, Dunger DB, et al. Pharmacokinetics of diluted (U20) insulin aspart compared with standard (U100) in children aged 3–6 years with type 1 diabetes during closed-loop insulin delivery: a randomised clinical trial. Diabetologia 2015;58(4):687–90.

[121] Tanenbaum ML, Iturralde E, Hanes SJ, Suttiratana SC, Ambrosino JM, Ly TT, et al. Trust in hybrid closed loop among people with diabetes: perspectives of experienced system users. Journal of Health Psychology 2017:1–10. https://doi.org/10.1177/1359105317718615.

[122] Tanenbaum ML, Hanes SJ, Miller KM, Naranjo D, Bensen R, Hood KK. Diabetes device use in adults with type 1 diabetes: barriers to uptake and potential intervention targets. Diabetes Care 2017;40(2):181–7.

[123] Ramchandani N, Arya S, Ten S, Bhandari S. Real-life utilization of real-time continuous glucose monitoring: the complete picture. Journal of Diabetes Science and Technology 2011;5(4):860–70.

Index

'*Note:* Page numbers followed by "t" indicate tables and "f" indicate figures.'

D

E

N

P

R

T

U

V

W

Z